DATE DUE

FUNCTIONAL ANATOMY OF THE NEUROENDOCRINE HYPOTHALAMUS

The Ciba Foundation is an international scientific and educational charity. It was established in 1947 by the Swiss chemical and pharmaceutical company of CIBA Limited—now CIBA-GEIGY Limited. The Foundation operates independently in London under English trust law.

The Ciba Foundation exists to promote international cooperation in biological, medical and chemical research. It organizes about eight international multidisciplinary symposia each year on topics that seem ready for discussion by a small group of research workers. The papers and discussions are published in the Ciba Foundation symposium series. The Foundation also holds many shorter meetings (not published), organized by the Foundation itself or by outside scientific organizations. The staff always welcome suggestions for future meetings.

The Foundation's house at 41 Portland Place, London W1N 4BN, provides facilities for meetings of all kinds. Its Media Resource Service supplies information to journalists on all scientific and technological topics. The library, open five days a week to any graduate in science or medicine, also provides information on scientific meetings throughout the world and answers general enquiries on biomedical and chemical subjects. Scientists from any part of the world may stay in the house during working visits to London.

Ciba Foundation Symposium 168

FUNCTIONAL ANATOMY OF THE NEURO-ENDOCRINE HYPOTHALAMUS

A Wiley-Interscience Publication

1992

JOHN WILEY & SONS

Chichester · New York · Brisbane · Toronto · Singapore

Published in 1992 by John Wiley & Sons Ltd
Baffins Lane, Chichester
West Sussex PO19 1UD, England

Other Wiley Editorial Offices

John Wiley & Sons, Inc., 605 Third Avenue,
New York, NY 10158-0012, USA

Jacaranda Wiley Ltd, G.P.O. Box 859, Brisbane,
Queensland 4001, Australia

John Wiley & Sons (Canada) Ltd, 22 Worcester Road,
Rexdale, Ontario M9W 1L1, Canada

John Wiley & Sons (SEA) Pte Ltd, 37 Jalan Pemimpin #05-04,
Block B, Union Industrial Building, Singapore 2057

Suggested series entry for library catalogues:
Ciba Foundation Symposia

Ciba Foundation Symposium 168
x + 300 pages, 53 figures, 4 tables

Library of Congress Cataloging-in-Publication Data

Functional anatomy of the neuroendocrine hypothalamus.
p. cm.—(Ciba Foundation symposium; 168)
Edited by Derek J. Chadwick and Joan Marsh.
"A Wiley–Interscience publication."
Includes bibliographical references and index.
ISBN 0-471-93440-2
1. Hypothalamus—Physiology. 2. Hypothalamus—Anatomy.
3. Neuroendocrinology. 4. Neuroanatomy. I. Chadwick, Derek.
II. Marsh, Joan. III. Series.
[DNLM: 1. Hypothalamus—anatomy & histology—congresses.
2. Hypothalamus—physiology—congresses. W3 C161F v. 168]
QP383.7F86 1992
612.8′262—dc20
DNLM/DLC
for Library of Congress 92-5731
CIP

British Library Cataloguing in Publication Data

A catalogue record for this book is
available from the British Library

ISBN 0 471 93440 2

Phototypeset by Dobbie Typesetting Limited, Tavistock, Devon.
Printed and bound in Great Britain by Biddles Ltd, Guildford.

Contents

Symposium on Functional anatomy of the neuroendocrine hypothalamus, held in collaboration with the Hungarian Academy of Sciences at The Hotel Gellért, Budapest, Hungary 8–10 October 1991

Editors: Derek J. Chadwick (Organizer) and Joan Marsh

Participants

E. Babarczy Department of Pathophysiology, Albert Szen-Györgyi Medical University, Semmelweis utca 1, Postafick 531, H-6701 Szeged, Hungary

C. A. Barraclough Department of Physiology, Center for Studies in Reproduction, School of Medicine, University of Maryland, 655 West Baltimore Street, Baltimore, MD 21201, USA

H. M. Charlton Department of Human Anatomy, University of Oxford, South Parks Road, Oxford OX1 3QX, UK

I. J. Clarke Prince Henry's Institute of Medical Research, PO Box 152, Clayton, Victoria 3168, Australia

J. J. Dreifuss Department of Physiology, University Medical Centre, 9 Avenue de Champel, CH-1211 Geneva 4, Switzerland

J. P. Epelbaum Unite de dynamique des Systèmes Neuroendocriniens, INSERM U159, Centre Paul Broca, 2ter rue d'Alésia, F-75014 Paris, France

H. Gainer Laboratory of Neurochemistry, National Institute of Neurological Disorders & Strokes, Building 36, 9000 Rockville Pike, Bethesda, MD 20892, USA

C. E. Grosvenor Department of Molecular & Cell Biology, The Pennsylvania State University, University Park, PA 16802, USA

B. Halász 2nd Department of Anatomy, Semmelweis University Medical School, Tüzoltó utca 58, H-1094 Budapest IX, Hungary

G. Hatton Department of Neuroscience, University of California, Riverside, CA 92521, USA

R. M. Lechan Division of Endocrinology, Department of Medicine, Box No 268, New England Medical Center, 750 Washington Street, Boston, MA 02111, USA

S. L. Lightman Neuroendocrinology Unit, Charing Cross & Westminster Medical School, Charing Cross Hospital, Fulham Palace Road, London W6 8RF, UK

Zs. Liposits Department of Anatomy, University Medical School, Szigeti utca 12, H-7643 Pécs, Hungary

G. B. Makara Institute of Experimental Medicine, Hungarian Academy of Sciences, PO Box 67, H-1450 Budapest, Hungary

P. L. Mellon* The Salk Institute, PO Box 85800, San Diego, CA 92186-5800, USA

K. E. Moore Department of Pharmacology & Toxicology, Michigan State University, Life Sciences Building, East Lansing, MI 48824-1317, USA

G. Nagy 2nd Department of Anatomy, Semmelweis University Medical School, Tüzoltó utca 58, H-1094 Budapest IX, Hungary

C. M. Oliver Laboratoire de Neuroendocrinologie Expérimentale, INSERM U297, UER de Médecine Nord, Boulevard Pierre Dramard, F-13326 Marseille Cedex 15, France

M. Palkovits Laboratory of Neuromorphology, Department of Anatomy 1, Semmelweis University Medical School, Tüzoltó utca 58, H-1094 Budapest IX, Hungary

D. W. Pfaff Laboratory of Neurobiology and Behavior, The Rockefeller University, 1230 York Avenue, New York, NY 10021-6399, USA

T. M. Plant Department of Physiology, University of Pittsburgh School of Medicine, Pittsburgh, PA 15261, USA

D. A. Poulain Laboratoire de Neuroendocrinologie Morphofonctionnelle, INSERM CJF 91.10, Université de Bordeaux II, 146 rue Léo Saignat, F-330676, Bordeaux Cedex, France

---

**Present address:* Departments of Reproductive Medicine & Neurosciences 0674, Center for Molecular Genetics, UCSD School of Medicine, 9500 Gilman Drive, La Jolla, CA 92093-0674, USA

I. C. A. F. Robinson Department of Neurophysiology & Neuropharmacology, National Institute of Medical Research, The Ridgeway, Mill Hill, London NW7 1AA, UK

P. E. Sawchenko Laboratory of Neuronal Structure & Function, The Salk Institute, PO Box 85800, San Diego, CA 92186-5800, USA

A. I. Smith Baker Medical Research Institute, Commercial Road, PO Box 348, Prahran, Melbourne, Victoria 3181, Australia

D. T. Theodosis Laboratoire de Neuroendocrinologie Morphofonctionnelle, INSERM CJF 91.10, Université de Bordeaux II, 146 rue Léo Saignat, F-33076 Bordeaux Cedex, France

W. S. Young III Laboratory of Cell Biology, National Institute of Mental Health, Building 36, Bethesda, MD 20892, USA

R. T. Zoeller Department of Anatomy & Neurobiology, University of Missouri-Columbia, Medical Science Building M304, Columbia, MO 65212, USA

Preface

In the autumn of 1991, the Ciba Foundation, in collaboration with the Hungarian Academy of Sciences, organized a symposium and open meeting in Budapest on The functional anatomy of the neuroendocrine hypothalamus. For the Foundation, this was a most important occasion because it was the first time that one of its meetings had been held in a formerly Eastern Bloc country. Our choice of location was, of course, a recognition of the profound changes that have taken place (and continue to take place) in Central and Eastern Europe. It gives me great pleasure to record here the help, hospitality and support given by the Hungarian Academy of Sciences and, in particular, by Professor István Láng (its Secretary General), Dr János Pusztai and Ms Georgine Pechlof, together with Professor Béla Halász, our scientific adviser in Hungary.

The topic selected for the meetings (originally proposed by Professor Halaśz and Professor Miklós Palkovits) seemed particularly apposite, given the important contributions which Hungarian scientists have made to neuroanatomy generally and to neuroendocrinology in particular for over 40 years. The venue chosen for the symposium, the Hotel Gellért, also seemed especially fitting. As Professor Láng pointed out in his welcoming address to the participants, Bishop Gellért came from Italy 950 years ago to help spread Christianity in Hungary, which helped to unite the developed parts of Europe. Much more recently, it was at a meeting of ministers of COMECON countries held at the Gellért that the disbanding of this organization was announced. Thus, in this historic setting, 28 participants (including six from Hungary) assembled in October 1991 for the Ciba Foundation's 299th symposium.

I very much hope that this record of the wide-ranging papers presented at the symposium and the discussion they aroused will contribute in some measure to the traffic in ideas and the collaboration among scientists throughout the world that is fundamental to the progress of science and is a key element of the Ciba Foundation's mission. I also hope that it will serve as a fitting testimony to an occasion memorable both for the warmth and hospitality of our Hungarian hosts and for their deep commitment to embracing the new opportunities and challenges.

Derek J. Chadwick
Director, The Ciba Foundation

Introduction

Professor S. L. Lightman

Neuroendocrinology Unit, Charing Cross & Westminster Medical School, Charing Cross Hospital, Fulham Palace Road, London W6 8RF, UK

Those aspects of brain function which are controlled by the hypothalamus have been of great interest to scientists for an extremely long time. Aristotle (Parts of Animals II vii) tells us that the brain is present in order to preserve the animal organism as a whole. He describes how it produces sleep by accumulating blood after a meal, and controls temperature by acting as a cooling unit for circulating blood. These were, of course, startling and revelatory ideas at the time. What is particularly remarkable is that it is only recently that we have had the scientific ability to make significant advances on these hypotheses.

Once we began to gain a scientific understanding of hypothalamic function, it was not long before this knowledge was put to clinical use. Although the present symposium covers the more fundamental aspects of hypothalamic function, I thought it would be worthwhile pointing out the great help this knowledge has already provided for the treatment of human disorders.

The discovery that gonadotropin-releasing hormone (GnRH) is secreted in pulses resulting in the release of luteinizing hormone (LH) and follicle-stimulating hormone (FSH) has resulted in two major types of clinical application. First, since we discovered that steady infusion of GnRH (rather than pulses) results in down-regulation of the pituitary and loss of responsiveness to GnRH, we have been able to use long-acting superactive analogues of GnRH to down-regulate the pituitary and effectively turn off LH and FSH secretion, leading to a reversible chemical castration. This is now used in the treatment of sex-hormone dependent cancers such as carcinoma of the prostate and more recently breast and ovarian cancer. It can also be used to stop temporarily the progression of premature puberty, which can be a distressing disorder in very young children. Conversely, we can use the knowledge about the pulsatile release of GnRH to treat patients who fail to reach puberty or to ovulate properly because of poor hypothalamic function. GnRH given in regular pulses by a pump can result in normal gonadal function and male and female fertility.

There are many other peptides in the hypothalamus and one of particular clinical relevance is somatostatin. We know that growth hormone-releasing hormone (GHRH) releases growth hormone while somatostatin inhibits the release of growth hormone. This knowledge can be applied to humans in several ways: in patients who have high levels of growth hormone, for example those with acromegaly, we can give somatostatin (or an analogue with a longer plasma half-life) and inhibit growth hormone secretion. In addition to its useful effects in patients with growth hormone excess, there is now evidence that this agent may have important growth-inhibiting effects on a variety of human tumours.

The converse action to somatostatin is produced by the GHRH. In addition to releasing growth hormone from the pituitary, GHRH is trophic. Patients who do not make GHRH have very few growth hormone-secreting cells. When they are treated chronically with GHRH, the number of these cells increases. In patients with very low levels of GHRH and consequently of growth hormone, there are only a few small pulses of growth hormone during the normal period of growth hormone secretion at night, and the patient does not grow. If they are treated with GHRH for six to 12 months, the growth hormone-secreting cells increase in number and the patient starts getting normal endogenous pulses of growth hormone. These are just examples of how knowledge of hypothalamic function can be readily transcribed into the clinical situation.

The present symposium will not deal with aspects of clinical importance but will concentrate on the more basic details of hypothalamic function and we shall address many fascinating questions about the regulation of the hypothalamus. At the single cell level we shall be asking whether individual hypothalamic cells have their own inherent pattern of hormone release, and how we can investigate responses to neurally active agents. How are hypothalamic cells aware of their environment? Do they organize their own afferent connections? How do they remember signals received from earlier exposure to various agents? How do hypothalamic cells know when events such as puberty are due to be initiated? What is the role of coexpressed peptides, can they be differentially expressed and what is the physiological value of this co-localization? Similarly we will question the meaning of detected changes in mRNA in hypothalamic cells.

Other areas of current interest include the mechanism by which trans-synaptic activation of hypothalamic neurons interacts with hormonal feedback and the levels at which this hormonal feedback occurs. Finally, there is increasing evidence that hypothalamic anatomy can actively adjust to changing neuroendocrine conditions. Many other questions will undoubtedly be raised but it is already clear that we shall need to discuss issues ranging from basic anatomy down to molecular biology. I for one am looking forward to a very exciting few days of science.

Peptidergic neurotransmitters in the endocrine hypothalamus

M. Palkovits

Laboratory of Neuromorphology, Department of Anatomy 1, Semmelweis University Medical School, Tüzoltó utca 58, H-1094 Budapest IX, Hungary

Abstract. More than 20 neuropeptides have been localized in the endocrine hypothalamus. They may exert a neurohormonal effect on the pituitary or innervate other neurons (intranuclear, intrahypothalamic or extrahypothalamic) and act as neurotransmitters. Many of the hypothalamic neuropeptides are synthesized as inactive precursors that are activated by proteolysis during axonal transport from the cell body to the synapse. Studies in which the paraventricular nuclei were bilaterally destroyed have shown that the neuroendocrine cells in the hypothalamus show functional plasticity and cells that do not usually make detectable quantities of a particular neuropetide may be activated to do so. Within the hypothalamic nuclei are dense networks of synaptic connections among neurons synthesizing the same or different neuropeptides. These local circuits may coordinate the activities of peptidergic neurons in a hypothalamic nucleus. Hypothalamic neurons project axons to the median eminence–pituitary stalk and the posterior pituitary, also to nuclei within the hypothalamus and to extrahypothalamic areas such as the lower brainstem. Peptidergic neurons in the hypothalamus can have combined neurohormonal and neurotransmitter activities mediated by axon terminals on portal capillaries and other hypothalamic nuclei. Double labelling immunohistochemistry has been used to demonstrate reciprocal connections between peptidergic neurons in the hypothalamus, such as those synthesizing growth hormone-releasing hormone and somatostatin.

1992 Functional anatomy of the neuroendocrine hypothalamus. Wiley, Chichester (Ciba Foundation Symposium 168) p 3–15

Hormone-producing neurons, generally termed neuroendocrine cells, are present in the supraoptic, paraventricular, medial preoptic, periventricular and arcuate nuclei. Cells in these nuclei synthesize releasing and release-inhibiting hormones, which are axonally transported down through the median eminence and released into the blood stream. Several other neuropeptides are also synthesized in these nuclei, where they are found co-localized in the same cells with other neuropeptides or neurotransmitters.

More than 20 neuropeptides have been localized in the endocrine hypothalamus (Palkovits 1986). Besides neurohormonal activity, they may have neurotransmitter (neuromodulator) activity influencing hypothalamo–hypophysial regulatory mechanisms, and/or they may constitute a neuronal link between the endocrine hypothalamus and extrahypothalamic brain regions, especially the limbic system and autonomic regulatory centres in the lower brainstem.

The following neuropeptides are known to be present in the endocrine hypothalamus (Palkovits 1988):

(1) Hypophysiotropic hormones: luteinizing hormone-releasing hormone (LHRH), thyrotropin-releasing hormone (TRH), corticotropin-releasing factor (CRF), growth hormone-releasing hormone (GHRH), growth hormone release-inhibiting hormone (somatostatin), vasopressin and oxytocin;

(2) Brain-born anterior pituitary hormones;

(3) Opioid peptides—endorphins, enkephalins and dynorphins;

(4) Brain-born gastrointestinal peptides—neuropeptide Y, vasoactive intestinal polypeptide, cholecystokinin, tachykinins and galanin;

(5) Brain-born vasoactive peptides—neurotensin, bradykinin, atrial and brain natriuretic peptides and angiotensin II.

One group of neuropeptidergic neurons projects to the posterior lobe. These neurons arise mainly from the supraoptic, paraventricular and accessory magnocellular nuclei and from cells scattered in the perifornical and lateral hypothalamic areas and the bed nucleus of the stria terminalis. The majority of these fibres contain vasopressin and oxytocin, but other neuropeptides such as enkephalins, dynorphins, galanin, cholecystokinin and angiotensin II have also been demonstrated in the fibres of the supraopticohypophysial tract and in nerve terminals in the posterior pituitary.

Another group of peptidergic neurons is located mainly in the arcuate nucleus, but some cells are also found in the neighbouring ventral premamillary, ventromedial and dorsomedial nuclei and in the caudal part of the lateral hypothalamus. After entering the median eminence, axons of these neurons (also called tuberoinfundibular neurons) cross the hypothalamo–hypophysial path in the internal layer and reach the lateral part of the external layer of the median eminence, where they make contacts with the capillaries of the primary portal plexus. Many of these neurons are peptidergic: they contain GHRH, pro-opiomelanocortin (POMC) (containing adrenocorticotropic hormone, β-endorphin and α-melanocyte-stimulating hormone), neuropeptide Y, galanin, substance P and dynorphins (Hökfelt et al 1987).

The third group of peptidergic neurons of the endocrine hypothalamus constitutes parvicellular cells in the paraventricular nucleus as well as cells in the periventricular and medial preoptic nuclei. Fibres of these neurons enter the median eminence from an anterolateral direction and pass through the lateral retrochiasmatic area. They contain LHRH, TRH, CRF and somatostatin, as well as galanin, cholecystokinin, dynorphins, enkephalins, atrial natriuretic peptide and angiotensin II.

Some special features of peptidergic neurons in the endocrine hypothalamus, their neurohormonal and neurotransmitter activities are summarized briefly in this paper.

Activation of peptide hormones and enzymes during axonal transport

It is well known that numerous hypothalamic neuropeptides are synthesized as prohormone precursors that must undergo limited proteolysis to yield the smaller, physiologically active peptides. The prohormone processing occurs during axonal transport of maturing secretory vesicles from the cell body to the synapse (Brownstein et al 1980) and requires several enzymes acting in a proteolytic cascade. Carboxypeptidase H is one of the enzymes necessary for processing prohormones such as provasopressin, prooxytocin, proenkephalin and others in the hypothalamo–hypophysial system. This enzyme, like the prohormone precursors, is synthesized as a proenzyme, processed and activated during axonal transport. In the supraoptic nucleus, where the neuronal cell bodies are located, the predominant species of carboxypeptidase H is an inactive 65 kDa protein. In the median eminence and pituitary stalk, which contain axons running from the supraoptic neurons to the posterior pituitary, both the inactive 65 kDa and the active 55 kDa forms of carboxypeptidase H are present. Nerve terminals in the posterior pituitary contain primarily the active enzyme. Thus, the biosynthesis of active peptide hormone requires the simultaneous processing of proenzyme and prohormone (Hook et al 1990).

Plasticity of neuroendocrine cells in the hypothalamus

Studies of various types have all confirmed that CRF-containing neurons in the paraventricular nucleus are responsible for the release of ACTH and β-endorphin from the anterior pituitary. A bilateral paraventricular lesion destroys the majority of input from CRF neurons to the median eminence. In a few weeks, cells in the hypothalamus may take over the function of paraventricular neurons and control of the pituitary–adrenal axis is largely restored (Makara et al 1981, 1986). We hypothesized that the bilateral paraventricular lesion activates other cells in the hypothalamus to synthesize CRF. The existence of CRF-synthesizing cells was investigated three days after hypothalamic transections of rats with six-week-old bilateral paraventricular lesions. In these animals, accumulations of CRF immunostaining were observed in the proximal portions of the transected axons and in neuronal perikarya. By this technique, CRF-containing neurons have been demonstrated in the supraoptic and perifornical nuclei that project to the median eminence (Palkovits et al 1991). These cells contain CRF mRNA (Palkovits et al 1992) and become activated by paraventricular lesions which eliminate the major CRF pool in the hypothalamo–hypophysial system.

Intranuclear organization of peptidergic neurons

A high degree of intrinsic organization of hypothalamic nuclei has been reported. A fairly large percentage (50–60%) of synaptic connections in the supraoptic and paraventricular nuclei is of local origin (Léránth et al 1975, van den Pol 1982, Kiss et al 1983). These connections may be established by recurrent (initial) collaterals of projection axons or by axon terminals of neighbouring interneurons. The existence of either of these types of connections does not exclude the other.

A local network of fibres seems to be a common feature of neuroendocrine hypothalamic nuclei. Local GHRH–GHRH (Horváth & Palkovits 1988), ACTH–ACTH (Kiss & Williams 1983), substance P–substance P (Tsuruo et al 1984) and POMC–POMC (Chen & Pelletier 1983) synapses have been demonstrated in the arcuate nucleus; LHRH–LHRH synapes are found in the medial preoptic nucleus (Léránth et al 1985); somatostatin–somatostatin synapses occur in the periventricular nucleus (Jew et al 1984, Epelbaum et al 1986), CRF–CRF synapses in the paraventricular nucleus (Liposits et al 1985) and oxytocin–oxytocin synapses in the supraoptic nucleus (Theodosis 1985).

Local circuit neurons may synchronize the activities of peptidergic neurons in a hypothalamic nucleus to integrate or coordinate them as a functional unit. Recurrent axon collaterals of hormone-producing peptidergic neurons may provide the morphological basis for an ultrashort feedback mechanism.

Intranuclear connections exist between neurons that synthesize different peptides. Synaptic specializations between magnocellular (oxytocin or vasopressin) and parvicellular (CRF) neuronal elements have been visualized in the paraventricular nucleus of rats (Léránth et al 1983). Neuropeptide Y immunoreactivity was localized in axon terminals forming synaptic contacts with both perikarya and dendrites that contain ACTH immunoreactivity in the arcuate nucleus (Csiffáry et al 1990).

Hypothalamic peptidergic neurons with intra- and extrahypothalamic projections

As well as projecting to the median eminence–pituitary stalk or posterior pituitary, peptidergic neurons in the endocrine hypothalamus project to several intra- and extrahypothalamic brain nuclei.

Intrahypothalamic projections. Peptidergic projections from the arcuate nucleus may innervate other hypothalamic neurons (Sawchenko et al 1982, Kiss et al 1984, Bai et al 1985, Liposits et al 1988); this nucleus also provides POMC-neuronal inputs to almost every other brain region. Intra- and extrahypothalamic oxytocin immunoreactive fibres arise exclusively in the paraventricular and accessory magnocellular nuclei. Synaptic relations exist between CRF- and somatostatin-containing hypothalamic neurons (Hisano & Daikoku 1991).

Extrahypothalamic projections. Hypothalamic peptidergic neurons send descending fibres to autonomic centres in the lower brainstem and to the intermediolateral cell column in the spinal cord. Fibres from the parvicellular paraventricular nucleus may contain CRF, somatostatin, enkephalin and vasopressin, while fibres containing oxytocin and vasopressin arise in the caudal, dorsal magnocellular portion of the nucleus (Sawchenko & Swanson 1982). POMC-immunoreactive fibres in the lower brainstem are partly of arcuate nucleus origin (Palkovits et al 1987).

Combined neurohormonal and neurotransmitter activity of peptidergic neurons in the endocrine hypothalamus

A fairly dense network of fibres can be seen in the medial-basal hypothalamus by using the Golgi impregnation technique (Palkovits 1980). Axons passing through that area can have collaterals which terminate in the arcuate nucleus or along the external layer of the median eminence. Such collaterals may represent the morphological basis of the possible combined (neurohormonal and neurotransmitter) activity of petidergic neurons. If this is the case, the axon or axonal branches may terminate in the pericapillary space along portal capillaries in the external layer of the median eminence (or in the posterior pituitary) while their axon collaterals innervate neurons in the arcuate or other nuclei of the hypothalamus. After microinjection of an anterograde tracer, *Phaseolus vulgaris* leukoagglutinin, into the parvicelluar subdivisions of the paraventricular nucleus, fine, labelled, varicose fibres were observed not only in the external layer of the median eminence but also throughout the length of the arcuate nucleus. Recent studies provide evidence for the existence of oxytocin–β-endorphin neuronal connections in the arcuate nucleus (Csiffáry et al 1992). Besides the dense bundle of oxytocin immunoreactive fibres in the internal layer of the median eminence, fine, varicose oxytocin-containing fibres were observed in the external layer and also in the arcuate nucleus. By double immunostaining, oxytocin immunopositive fibres were shown to make synaptic contacts with β-endorphin immunoreactive arcuate neurons. Similarly to oxytocin immunoreactive neurons, CRF-containing paraventricular neurons may also project both to the median eminence and to other neurons in the central nervous system (Rho & Swanson 1987).

Reciprocal connections between peptidergic neurons in the hypothalamus

GHRH-containing neuronal perikarya have been shown by light microscopic immunohistochemistry to be distributed in the arcuate nucleus and to project short axons to the external layer of the median eminence. Somatostatin-containing neurons with median eminence–pituitary stalk projections are present in the anterior hypothalamus, along the third ventricle in the periventricular nucleus.

Using a double labelling immunohistochemical technique with both light and electron microscopy, Horváth et al (1989) demonstrated direct and reciprocal synaptic connections between these two peptidergic systems. Somatostatin immunoreactive axons making synaptic contacts with the perikarya and dendrites of GHRH neurons were often found in the arcuate nucleus. Conversely, GHRH immunoreactive neurons in the arcuate nucleus project to the periventricular hypothalamic nucleus where they innervate somatostatin immunopositive neuronal cells. These morphological findings strongly suggest that GHRH and somatostatin may regulate growth hormone synthesis and release not only in the anterior pituitary but also at the level of the hypothalamus.

References

Bai FL, Yamano M, Shiotani Y et al 1985 An arcuato–paraventricular and dorsomedial hypothalamic neuropeptide Y-containing system which lacks noradrenaline in the rat. Brain Res 331:172–175

Brownstein MJ, Russell JT, Gainer H 1980 Synthesis, transport and release of posterior pituitary hormones. Science (Wash DC) 207:373–378

Chen YY, Pelletier G 1983 Demonstration of contacts between proopiomelanocortin neurons in the rat hypothalamus. Neurosci Lett 43:271–276

Csiffáry A, Görcs TJ, Palkovits M 1990 Neuropetide Y innervation of ACTH-immunoreactive neurons in the arcuate nucleus of rats: a correlated light and electron microscopic double immunolabeling study. Brain Res 506:215–222

Csiffáry A, Ruttner Z, Tóth Z, Palkovits M 1992 Oxytocin nerve fibers innervate β-endorphin neurons in the arcuate nucleus of the rat hypothalamus. Neuroendocrinology, in press

Epelbaum J, Tapia-Arancibia L, Alonso G, Astier H, Kordon C 1986 The anterior periventricular hypothalamus is the site of somatostatin inhibition of its own release: an in vitro and immunocytochemical study. Neuroendocrinology 44:255–259

Hisano S, Daikoku S 1991 Existence of mutual synaptic relations between corticotropin-releasing factor-containing and somatostatin-containing neurons in the rat hypothalamus. Brain Res 545:265–275

Hökfelt T, Meister B, Everitt B et al 1987 Chemical neuroanatomy of the hypothalamo–pituitary axis: focus on multimessenger systems. In: McCann SM, Weiner RI (eds) Integrative neuroendocrinology: molecular, cellular and clinical aspects. Karger, Basel p 1–34

Hook VYH, Affolter H-U, Palkovits M 1990 Carboxypeptidase H in the hypothalamo–neurohypophyseal system: evidence for processing and activation of a prohormone-processing enzyme during axonal transport. J Neurosci 10:3219–3226

Horváth S, Palkovits M 1988 Synaptic interconnections among growth hormone-releasing hormone (GHRH)-containing neurons in the arcuate nucleus of the rat hypothalamus. Neuroendocrinology 48:471–476

Horváth S, Palkovits M, Görcs TJ, Arimura A 1989 Electron microscopic immunocytochemical evidence for the existence of bidirectional synaptic connections between growth hormone-releasing hormone- and somatostatin-containing neurons in the hypothalamus of the rat. Brain Res 481:8–15

Jew JY, Léránth C, Arimura A, Palkovits M 1984 Preoptic LH-RH and somatostatin in the rat median eminence. An experimental light and electron microscopic immunocytochemical study. Neuroendocrinology 38:169–175

Kiss JZ, Williams TH 1983 ACTH-immunoreactive boutons form synaptic contacts in the hypothalamic arcuate nucleus of rat: evidence for local opiocortin connections. Brain Res 263:142–146

Kiss JZ, Palkovits M, Záborszky L, Tribollet E, Szabó D, Makara GB 1983 Quantitative histological studies on the hypothalamic paraventricular nucleus in rats. II. Number of local and certain afferent nerve terminals. Brain Res 265:11–20

Kiss JZ, Cassell MD, Palkovits M 1984 Analysis of the ACTH/β-End/α-MSH-immunoreactive afferent input to the hypothalamic paraventricular nucleus of rat. Brain Res 324:91–99

Léránth C, Záborszky L, Marton J, Palkovits M 1975 Quantitative studies on the supraoptic nucleus in the rat. I. Synaptic organization. Exp Brain Res 22:509–523

Léránth C, Antoni FA, Palkovits M 1983 Ultrastructural demonstration of ovine CRF-like immunoreactivity (oCRF-LI) in the rat hypothalamus: processes of magnocellular neurons establish membrane specializations with parvocellular neurons containing oCRF-LI. Regul Pept 6:179–188

Léránth C, Seguta LMG, Palkovits M, MacLusky NJ, Shanabrough M, Naftolin F 1985 The LH-RH-containing neuronal network in the preoptic area of the rat: demonstration of LH-RH-containing nerve terminals in synaptic contact with LH-RH neurons. Brain Res 345:332–336

Liposits Z, Paull WK, Sétáló G, Vigh S 1985 Evidence for local corticotropin releasing factor (CRF)-immunoreactive neuronal circuits in the paraventricular nucleus of the rat hypothalamus. An electron microscopic immunohistochemical analysis. Histochemistry 83:5–16

Liposits Z, Sievers L, Paull WK 1988 Neuropeptide Y- and ACTH-immunoreactive innervation of corticotropin-releasing factor (CRF)-synthesizing neurons in the hypothalamus of the rat. An immunocytochemical analysis at the light and electron microscopic levels. Histochemistry 88:227–234

Makara GB, Stark E, Kárteszi M, Palkovits M, Rappay GY 1981 Effects of paraventricular lesions on stimulated ACTH release and CRF in stalk-median eminence of the rat. Am J Physiol 240:E441–E446

Makara GB, Stark E, Kapocs G, Antoni FA 1986 Long-term effects of hypothalamic paraventricular lesion on CRF content and stimulated ACTH secretion. Am J Physiol 250:E319–E324

Palkovits M 1980 Functional anatomy of the 'endocrine' brain. In: Motta M (ed) The endocrine functions of the brain. Raven Press, New York, p 1–19

Palkovits M 1986 Neuropeptides in the median eminence. Neurochem Int 9:131–137

Palkovits M 1988 Neuropeptides in the brain. In: Martini L, Ganong WF (eds) Frontiers in neuroendocrinology. Raven Press, New York vol 10:1–44

Palkovits M, Mezey E, Eskay RL 1987 Pro-opiomelanocortin-derived peptides (ACTH/β-endorphin/α-MSH) in brainstem baroreceptor areas of the rat. Brain Res 436:323–338

Palkovits M, Kovács K, Makara GB 1991 Corticotropin-releasing hormone-containing neurons in the hypothalamo–hypophyseal system in rats six weeks after bilateral lesions of the paraventricular nucleus. Neuroscience 42:841–851

Rho J-H, Swanson LW 1987 Neuroendocrine CRF motoneurons: intrahypothalamic axon terminals shown with a new retrograde-Lucifer-immuno method. Brain Res 436:143–147

Sawchenko PE, Swanson LW 1982 Immunohistochemical identification of neurons in the paraventricular nucleus of the hypothalamus that project to the medulla or to the spinal cord in the rat. J Comp Neurol 205:260–272

Sawchenko PE, Swanson LW, Joseph SA 1982 The distribution and cells of origin of ACTH(1-39)-stained varicosities in the paraventricular and supraoptic nuclei. Brain Res 232:365–374

Theodosis DT 1985 Oxytocin-immunoreactive terminals synapse on oxytocin neurons in the supraoptic neurons. Nature (Lond) 313:682–684
Tsuruo Y, Hisano S, Daikoku S 1984 Morphological evidence for synaptic junctions between substance P-containing neurons in the arcuate nucleus of the rat. Neurosci Lett 46:65–69
van den Pol AN 1982 The magnocellular and parvocellular paraventricular nucleus of rat: intrinsic organization. J Comp Neurol 206:317–345

DISCUSSION

Lightman: When you made the lesions in the paraventricular nucleus and saw the increase in CRF in the cells in the supraoptic nucleus, it is important to know how specific this response was. Was there a specific increase in the CRF derived from the supraoptic nucleus cells that project to the median eminence and could mediate the stress response? Or was the response common to all CRF-containing cells, including those which project to the neural lobe? Was there, for instance, also a large increase in CRF peptide in the neural lobe itself?

Palkovits: The increase in CRF is specific because these CRF immunostained neurons definitely project to the median eminence (high level accumulation of CRF appears only after transection of the axons). If you look at the immunostaining with anti-CRF in the median eminence, it is in the external not the internal layer. Secondly, the *in situ* hybridization technique indicates a very nice increase in the amount of CRF mRNA in the supraoptic and perifornical nuclei. Thirdly, we saw CRF immunostaining in only two places, the supraoptic nucleus and the perifornical nucleus. In these nuclei, cells contain oxytocin which co-localizes with CRF. These cells are activated to synthesize CRF by transection of their fibres in the paraventricular nucleus-lesioned rats. We didn't see any changes in the amounts of CRF in other parts of the brain, including the limbic system.

Hatton: Most of the earlier work with [^{3}H]amino acids has shown that there is a direct input from the subfornical organ to the supraoptic nucleus; it also projects to the AV3V region. You have suggested that the subfornical organ neurons do not project to the supraoptic directly. Could you comment on this?

Palkovits: It doesn't change the general picture. There is some debate about whether those projections from the subfornical region down to the supraoptic nucleus are relayed by neurons in the so-called AV3V region. I wish to give an example of the existence of a humoral–neuronal–humoral circuit in the brain. The neuronal part might comprise not just a single neuron, but a multiple neuronal chain. So the humoral part of the circuit is followed by a neuronal part, then another humoral component.

Hatton: How important do you think the large percentage of 'intrinsic', as you called them, connections really is? I no longer think there is such a large number and in fact the evidence has never really shown intrinsic connections in the supraoptic nucleus; these connections are apparently intrinsic to the supraoptic *plus* the perinuclear zone.

From studies of knife cuts that sever the subfornical organ efferent, there is an increase in the number of synapses in the supraoptic nucleus within 24 hours. That would lead one to believe that there were more intrinsic connections when, in fact, there should have been fewer.

Palkovits: We have made several control measurements. You have to take into account synaptic splitting, retrograde degeneration and other kinds of changes in the neurons. We checked post-operatively at Day 1, Day 3, Day 7 and so on. I am not giving absolute values, just a range: 50–70% or 40–60% of the synaptic boutons remain intact, which is still a high number.

There are many other good examples of a high number of intrinsic nerve terminals in the central nervous system. After total transection of neuronal inputs to the hippocampus by fornical transections, we still found that at least 70% of the synaptic contacts remained normal.

Theodosis: Miklós, your 'isolated' supraoptic nuclei often included perinuclear tissue. The hypothalamic tissue immediately adjacent to the supraoptic may contain relay neurons that directly innervate the supraoptic neurons. GABAergic and cholinergic neurons have been visualized in this area. The 60% synapses you see remaining after lesion may not therefore be intranuclear but could derive from the interneurons just outside the nucleus.

Palkovits: I agree. But inside the nucleus we can find 7000 cells and around the nucleus we find about 70 cells.

Theodosis: Yes, but 70 cells can make many synapses!

Gainer: How great is the distance between the edge of this perinuclear region and the other border of the supraoptic nucleus? I might consider a neuron spanning such a distance to be an intrinsic interneuron.

Hatton: We are talking of hundreds of μm. Secondly, you can't immunostain the supraoptic nucleus and find many vasopressinergic or oxytocinergic terminals. So what are these intrinsic terminals? They can't be from the vasopressin or oxytocin cells. Some other cells have to be involved, as Dionysia (Theodosis) was saying, such as the GABAergic neurons. I just want to clarify what's intrinsic to the nucleus versus what's intrinsic to some region of the hypothalamus. I really don't think we have a good handle on that, as yet. If you sever the subfornical organ efferent with a knife cut up at the level of the lamina terminalis, you get an increase in the number of multiple synapses in the supraoptic nucleus. We think that's a reaction to the knife cut—that the supraoptic nucleus is getting some information about the absence of inputs.

Palkovits: I am not sure about that. You get a number of neural synaptic boutons because you can see them after transection. They are present in intact animals but they are not stained or active. Look at the supraoptic nucleus in animals after transection of the pituitary stalk. At the proper time, you will see a great number of terminals immunostained for oxytocin or vasopressin. We saw galanin staining after unilateral transection, not only in the supraoptic cell bodies—there was also a huge immunostained neuronal network in the nucleus, indicating the existence of local collaterals and/or terminals filled retrogradely. In normal conditions you cannot see these because the major axonal flow down to the pituitary is intact.

Hatton: The traditional degeneration studies have tried to use that technique to show that there was a large number of inputs from some structure. So if you cut the so-called input from the preoptic to the supraoptic or from the subfornical organ to the supraoptic, as Renaud et al (1983) did and as we have done (Weiss et al 1988), you should see a diminution in the number of terminals, because we know there's massive input. But you don't see any, you see maybe 0.01% decrease in the number of synapses. I would say that may be due to a compensatory increase in the number of these multiple synapses, which could lead us astray.

Poulain: There is another question in that we do not know the origin of the synaptic input. There may be one GABAergic neuron at the border of the nucleus sending dozens of synapses into the nucleus. We don't see GABAergic neurons within the nucleus, so they must come from outside, making a network very close to the supraoptic nucleus.

Palkovits: Probably. One thing I tried to express is that if you cannot see or cannot stain a substance in neuronal cells in normal conditions, it doesn't mean that these cells are unable to produce that substance under any conditions.

Gainer: I seem to have a lot more confidence in anatomy than the anatomists do! If these GABAergic cells are only 100 μm away, could they not be considered as part of the intrinsic structure of that nucleus?

Theodosis: No. The structure of the hypothalamus, even only 100 μm away from the supraoptic nucleus, is not at all the same as that of the nucleus!

Gainer: Maybe we should change our view of the structure of the nucleus. Maybe we have to consider that 500 μm away is still part of the intrinsic circuitry of the supraoptic nucleus. If it is true that 60% of the axons and terminals are GABAergic and they originate from the population of cells out there, then functionally that region is part of the circuitry of the nucleus.

Sawchenko: It is by no means clear that the perinuclear zone is the solitary source of the GABAergic innervation. In fact, we have some evidence to the contrary, which suggests that its sources may be more widely distributed within the hypothalamus (Roland et al 1991). It is also not clear that the supraoptic nucleus is the only projection field of those perinuclear GABAergic cells. Thus, any redefinition of what constitutes the supraoptic nucleus would seem premature at present.

Gainer: But I am still stuck with the 60%. Do you accept the value of 60% for the fraction of terminals remaining after deafferentation? Are these exclusively GABAergic neurons?

Sawchenko: We have not examined this directly. Two factors are worth considering in this regard. First, it is likely, in these deafferentation experiments, that a significant fraction of these perinuclear afferents is spared. Secondly, some plastic reorganization, some compensatory synaptogenesis, may be occurring after the lesion. I am not aware of any experiments that have attempted to tease apart the relative contributions of these two factors.

Hatton: Another thing to be considered with the supraoptic nucleus—I am not sure if this applies elsewhere—is that there is a relatively neat separation of inputs that come into the cell body region (all the earlier work was done on the cell body region) from those that come in exclusively to the dendrites, particularly the ventral dendritic zone. There are also inputs to what I call the α dendrites which are more dorsal and medial. Some dendritic afferents may come from places like the olfactory bulbs. We have recently shown direct monosynaptic input from the olfactory bulb just to the dendritic zone, not to the cell body zone at all. Paul Sawchenko has lots of peptidergic separations where the inputs come into one part of the nucleus and not to another.

Theodosis: I think these nuclei need to be studied again as they were here in Hungary in the 1970s. We lack rigorous synaptic density analyses of these nuclei under different conditions, including lesions. Whenever I look at the literature, I always come back to the beautiful, painstaking studies done in the 1970s by Miklos (Palkovits) and others here (Záborszky et al 1975). Although we are trying to do such studies, they are painstaking and difficult, especially because these neurons change constantly in form and size. Until such analyses have been done, we should be careful when comparing numbers of synapses.

Dreifuss: I wish to raise two questions relating to axon collaterals arising from neurons in the supraoptic and paraventricular nuclei that synthesize oxytocin or vasopressin. First, did you see recurrent collaterals which feed back onto cells within these nuclei?

Palkovits: Not in the supraoptic nucleus, but in other places.

Dreifuss: The second question is, are there axons arising from cells in the magnocellular endocrine nuclei of the hypothalamus and giving off two or more branches travelling to widely separate areas within the brain, for example, towards the pituitary gland on the one hand and towards the brainstem on the other?

Palkovits: One example is the arcuate nucleus. There are POMC neurons there which are able to innervate thousands of other CNS neurons. They do not have one-to-one synaptic connections with other cells; one POMC cell in the arcuate nucleus may project to amygdala and other parts of the brain. I am not saying that these cells are catecholamine-containing neurons, where a single cell can provide 50 000–60 000 terminals or terminal-like varicosities.

Sawchenko: It seems clear from both the electrophysiological and the anatomical points of view, that in the paraventricular nucleus the number of cells that gives rise to neuroendocrine and intracerebral projections is very small.

I would question whether so small a contingent could represent a major integrative mechanism in this nucleus.

Dreifuss: As far as the magnocellular neurons are concerned, there seems to be little evidence for either recurrent collaterals or collaterals which go to different places in the brain.

Palkovits: I agree. Plenty of reports based on the multiple retrograde labelling technique show an accumulation of more than one tracer substance in the same cell body. So, there is evidence for axon collaterals that terminate in at least 3–4 different regions. We have to remember that a single cell in the hypothalamus has at least 600 terminals on the dendrites or on the cell bodies, whereas extrahypothalamic ones have up to 3000–4000 terminals per cell. In the cerebellum, this number goes up to 50 000 terminals per cell.

Liposits: I have two questions about the hypophysiotropic neurons. How should we consider the parvocellular neurosecretory systems that are also known to terminate in the posterior pituitary? Can we consider those hormones as hypophysiotropic in nature? What kind of communication exists between the posterior pituitary and the anterior pituitary?

The other thing is what is the current status of the brainstem? Can we consider the monoaminergic cells as hypophysiotropic neurons? Using immunocytochemistry we can see noradrenergic and adrenergic fibres in the median eminence. You also published degenerating profiles in the median eminence after lesioning noradrenergic/adrenergic cells (Palkovits et al 1980).

Palkovits: First of all I don't like the term hypophysiotropic, it is too broad in concept. There are reciprocal connections between the posterior and the anterior lobes. I am not saying how important they are, but substances may go from the anterior lobe to the posterior.

There are several aminergic inputs to the hypothalamus. They all participate in innervation of the endocrine hypothalamus. There are also many peptidergic inputs from the lower brainstem and from the limbic system, directly or through relay centres.

Makara: The neural lobe may have an effect on the anterior lobe. There are two pieces of evidence for this. One is from studies of neural lobectomy and ACTH secretion, the other from measurements we made of plasma concentrations of corticosterone as an index of ACTH secretion after neural lobe stimulation (Makara et al 1980). We found that tiny, localized neural lobe stimulation activated ACTH release. That means either CRF or vasopressin released in the neural lobe has access to ACTH cells in the anterior lobe or CRF neurons and vasopressin neurons in the neural lobe have collaterals in the median eminence and release neuropeptides there to reach the anterior lobe through the normal way, i.e. the portal circulation. I don't know which is the correct explanation, but certainly the neural lobe is connected to the hypophysiotropic system. This is a point to which we will come back experimentally to look again with newer techniques at the role of the neural lobe in ACTH and prolactin secretion. The ablation techniques have many problems of regeneration in the median eminence, so maybe stimulation should be applied again.

References

Makara GB, Stark E, Kárteszi M, Fellinger E, Rappay G, Szabó D 1980 Effect of electrical stimulation of the neurohypophysis on ACTH release in rats with hypothalamic lesions. Neuroendocrinology 31:237–243

Palkovits M, Záborszky L, Feminger A et al 1980 Noradrenergic innervation of the rat hypothalamus: experimental, biochemical and electron microscopic studies. Brain Res 191:161–171

Renaud LP, Rogers J, Sgro S 1983 Terminal degeneration in supraoptic nucleus following subfornical organ lesions: ultrastructural observations in the rat. Brain Res 275:365–368

Roland BL, Brown MR, Sawchenko PE 1991 Distribution and origins of GABAergic projections to the paraventricular nucleus. Soc Neurosci Abstr 17:1188

Weiss ML, Tweedle GD, Marzban F, Modney BK, Hatton GI 1988 Rapid formation of new double synapses in the rat supraoptic nucleus in response to interruption of subfornical organ efferent projections. Soc Neurosci Abstr 14:1133

Záborszky L, Léránth Cs, Makara GB, Palkovits M 1975 Quantitation studies on the supraoptic nucleus in the rat. II. Afferent fiber connections. Exp Brain Res 22:525–540

Co-localization of neuroactive substances in the endocrine hypothalamus

P. E. Sawchenko, T. Imaki and W. Vale

Laboratory of Neuronal Structure & Function, The Salk Institute, and The Clayton Foundation for Research—California Division, San Diego CA 92186-5800, USA

Abstract. In addition to their characterizing secretory products, both magnocellular and parvocellular neurosecretory neurons are now known to express other neuroactive substances. Parvocellular neurons that make corticotropin-releasing factor (CRF) for example are capable of synthesizing at least seven neuropeptides. Some of these, like arginine vasopressin (AVP), interact with CRF at the level of the anterior pituitary to promote corticotropin secretion, and, like CRF, are regulated negatively by glucocorticoids and positively by at least some stressors. Others are inert in these two contexts but are responsive to various challenges. Magnocellular neurosecretory oxytocin- and AVP-containing neurons are capable of producing similarly broad and distinctive complements of neuroactive principles. These are typically expressed at levels far lower than those of the nonapeptides, suggesting local modulatory effects on oxytocin and/or AVP secretion at the level of the posterior lobe. Differential regulation of coexisting molecules within magnocellular neurons by systemic challenges and steroid hormones has also been described. Secretory products of magnocellular neurons may gain access to the anterior pituitary via exocytotic release at the level of the median eminence or through vascular links between the posterior and anterior lobes, suggesting another form of 'co-localization' by which the two neurosecretory cell types may interact in the control of stress and perhaps other pituitary-mediated responses.

1992 Functional anatomy of the neuroendocrine hypothalamus. Wiley, Chichester (Ciba Foundation Symposium 168) p 16–42

The traditional view of hypothalamo-pituitary relationships as comprising a series of linearly organized, closed-loop systems functioning in parallel, with each bearing a single neurochemical signature, has been rendered untenable by work carried out over the past two decades. One factor that has contributed to our enhanced appreciation of the complexity of the central representation of neuroendocrine systems, and of the interactions among them, is the recognition that many individual neurosecretory neurons possess the ability to express multiple biologically active molecules. For nowhere else in the brain are reports of co-localization more pervasive than for certain components of the neuroendocrine hypothalamus. While in some instances this has not advanced beyond phenomenology, other manifestations have provided glimpses into an intricate, and quite malleable, manner of organization which is characterized by situation-specific alterations in the expression of neuroactive substances and interactions among cell types and molecules to achieve common, and physiologically meaningful, ends.

Here we summarize recent evidence on the nature and possible functional significance of co-localization in the magnocellular and parvocellular neurosecretory systems. For the latter, we shall focus on the central limb of the hypothalamo-pituitary-adrenal (HPA) axis, or the corticotropin-releasing factor (CRF) neuron, since the bulk of the available data pertain to this cell type, which provides the major stimulatory drive to pituitary corticotropin (ACTH) secretion. Moreover, because the principal secretory products of the magnocellular system also participate in the control of the HPA axis, this provides a convenient venue in which to appreciate the interplay between the two classes of neurosecretory neurons in effecting appropriately integrated hypothalamic responses.

Cellular organization: the paraventricular nucleus

The paraventricular nucleus (PVN) is unique among hypothalamic cell groups in housing substantial populations of cells that participate directly in the control of anterior and posterior pituitary secretions. Converging evidence from a variety of disciplines has identified this nucleus as the predominant source of CRF in hypophysial-portal plasma (Antoni 1986); parvocellular neurosecretory neurons expressing other hypophysiotropic factors, including somatostatin, thyrotropin-releasing hormone, dopamine and growth hormone-releasing hormone are also represented in the nucleus (Swanson et al 1986). In addition, the PVN is acknowledged, along with the supraoptic nucleus, as a principal seat of magnocellular neurosecretory neurons, which synthesize the nonapeptide hormones oxytocin and arginine vasopressin (AVP) for release into the general circulation from terminals in the posterior lobe. A third major cell type comprises neurons that give rise to long descending projections to the brainstem and spinal cord, which include sensory and motor structures associated with the autonomic nervous system (see Swanson et al 1986). These three visceromotor populations are essentially separate and exhibit a high degree of topographic organization. Upon this high degree of anatomical organization is imposed a somewhat imprecise manner of chemical coding.

While CRF, oxytocin and AVP are best known for their unique and closely regulated hormonal roles, it is now clear that their expression within the PVN is not limited to a single cell type (Table 1). Each peptide, for example, is expressed in a subset of autonomic-related projection neurons of this nucleus

TABLE 1 Co-localization of neuroactive substances in neuroendocrine neurons of the paraventricular nucleus

Magnocellular		*Parvocellular*
Vasopressin	*Oxytocin*	*Corticotropin-releasing factor*
Vasopressin	Oxytocin	Corticotropin-releasing factor
Dynorphin	Corticotropin-releasing factor	Vasopressin
Angiotensin II	Dynorphin	Angiotensin II
Enkephalin	Cholecystokinin	Cholecystokinin
Tyrosine hydroxylase (dopamine?)	Enkephalin	Enkephalin
Galanin	Thyrotropin-releasing hormone	Neurotensin
PHI/VIP		Galanin
		PHI/VIP
		GABA

Citations documenting the co-localization patterns summarized above include: Ceccatelli et al 1989, Meister et al 1988, 1990, Sawchenko et al 1984a,b, Tsuruo et al 1988. GABA, γ-aminobutyric acid; PHI/VIP, peptide histidine isoleucine/vasoactive intestinal polypeptide.

(Sawchenko & Swanson 1982, Sawchenko 1987a). Moreover, each of the neuroendocrine cell types in which oxytocin, AVP and CRF are expressed has at least the ability to synthesize multiple additional neuroactive substances. This ability does not appear to be a promiscuous attribute of all neurosecretory neurons; we are unaware of clear demonstrations of co-localized molecules in hypophysiotropic somatostatin- or thyrotropin-releasing hormone cells.

It is important to point out that many of the observations on co-localization are based on immunohistochemical staining of animals pretreated with colchicine. Owing to its ability to impede axonal transport, colchicine has been assumed to allow assessments of the *capability* of cells to express a given antigen because it causes low-abundance antigens to accumulate in the cell body, thus rendering them detectable by immunohistochemical techniques. In addition to the uncertain sensitivity of any immunohistochemical method, recent studies by Hökfelt and colleagues (e.g. Cortes et al 1990) have emphasized that colchicine treatment is itself aptly considered to cause stress, which is an important regulator of neuropeptide gene expression in the cell types considered here.

Parvocellular neurosecretory system

Coexisting neuroactive substances are invariably described in immunohistochemical preparations as occupying subsets of cells positively immunostained for CRF in the parvocellular division of the PVN (e.g. Hökfelt et al 1983, Sawchenko et al 1984a, Whitnall 1988, Ceccatelli et al 1989). Among the more likely functional roles they may play are (1) to modulate the actions of CRF via effects exerted directly on corticotrophs; (2) to target independently a distinct, or multiple, anterior pituitary cell type(s); or (3) to exert synaptic or extrajunctional influences on the release of other hypophysiotropic hormones at the level of the median eminence. Any of these would broaden substantially the integrative capabilities of the system, but the physiological roles and/or sites of action of most of the coexisting peptides in the neuroendocrine system have remained elusive. Here, however, we will focus on one intensively studied example of co-localization in the CRF neuron whose potential relevance to function is quite well established.

Although CRF has been identified as the factor which sets the stimulatory tone of basal and stress-elicited corticotropin (ACTH) secretion, it is clear that additional secretagogues, notably oxytocin and AVP, are present in portal plasma and play significant roles in imparting the situation-specific drive to the axis (Plotsky 1991). The manner in which the 'posterior pituitary' nonapeptides may gain access to the portal vasculature remains controversial. Potential insight into this problem was provided when it was shown that many hypophysiotropic CRF neurons can express AVP under conditions of corticosteroid withdrawal following adrenalectomy (Fig. 1, Kiss et al 1984, Sawchenko et al 1984b, Young et al 1986). This suggested a novel form of plasticity that is relevant to function, in that it identified a manner of organization that could explain the documented synergism between the two peptides in stimulating ACTH secretion (Plotsky 1991) under conditions (adrenalectomy) that must be interpreted by the brain as demanding an adrenal response.

Further studies have shown that the steroid regulation of CRF and AVP expression in parvocellular neurosecretory neurons appears to be mediated predominantly by glucocorticoids and to exhibit a marked specificity for particular molecules and a particular cell type. Staining for other peptides, such as neurotensin and enkephalin, which can be expressed in the CRF neuron but which appear to possess no corticotropin-releasing activity is not discernibly enhanced in adrenalectomized rats (Fig. 2, Sawchenko 1987b). Neither does adrenalectomy exert pronounced effects on CRF and AVP expression in magnocellular neurosecretory or autonomic-related projection neurons in the PVN (Sawchenko et al 1984b, Swanson & Simmons 1989, Young et al 1988). By contrast, angiotensin II (Lind et al 1985) and cholecystokinin (Mezey et al 1986), two additional, albeit weaker, secretagogues for corticotropin, can be detected in parvocellular CRF-immunoreactive neurons in response to various combined manipulations that necessarily include adrenalectomy. Thus far, it appears to be a general rule that regulatory influences are exerted differentially, and on the basis of connectivity and/or functional associations of cell groups, rather than uniformly on each cell group that may share a given biochemical phenotype.

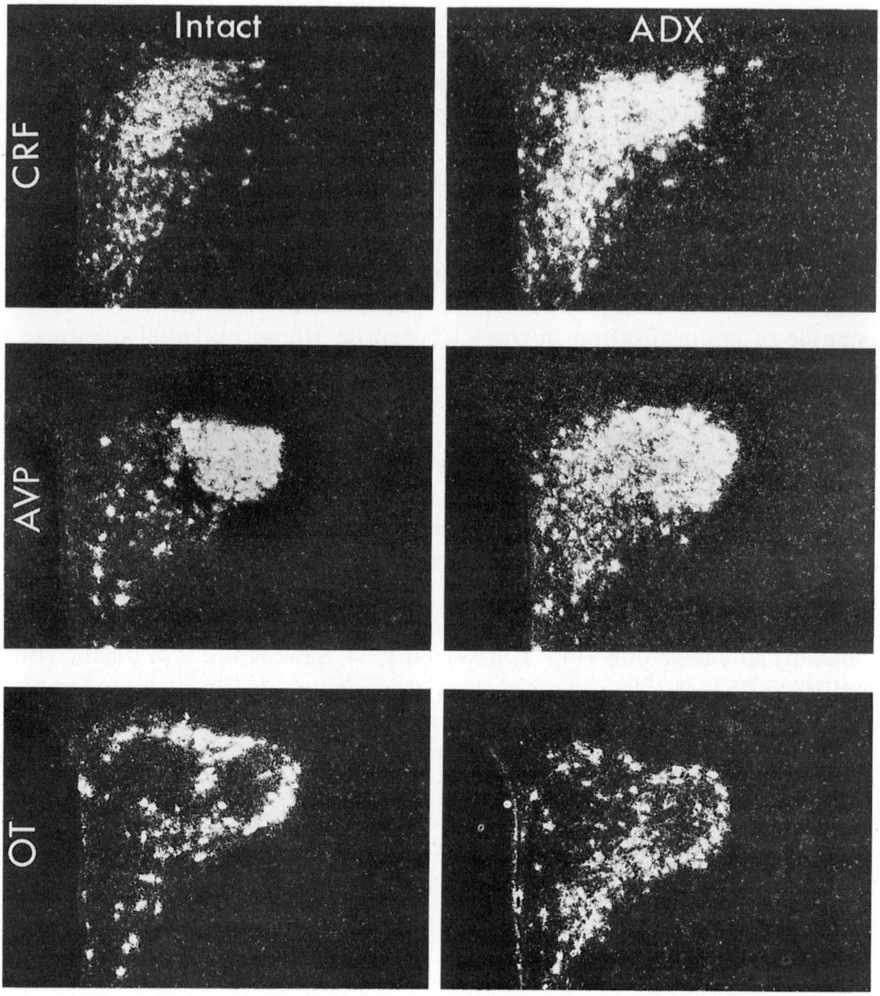

FIG. 1. Dark field photomicrographs of hybridization histochemical material to show the cellular localization of mRNAs encoding corticotropin-releasing factor (CRF), arginine vasopressin (AVP) and oxytocin (OT) in the paraventricular nucleus in intact and adrenalectomized (ADX) rats. In response to steroid withdrawal, CRF mRNA is up-regulated and AVP mRNA ostensibly induced in the parvocellular division of the nucleus. Oxytocin mRNA is not significantly affected by this treatment.

Regulatory influences on CRF and AVP in the parvocellular system are not limited to those exerted by steroids. So-called 'short-loop' feedback effects of circulating ACTH on hypothalamic corticotropin-releasing activity were

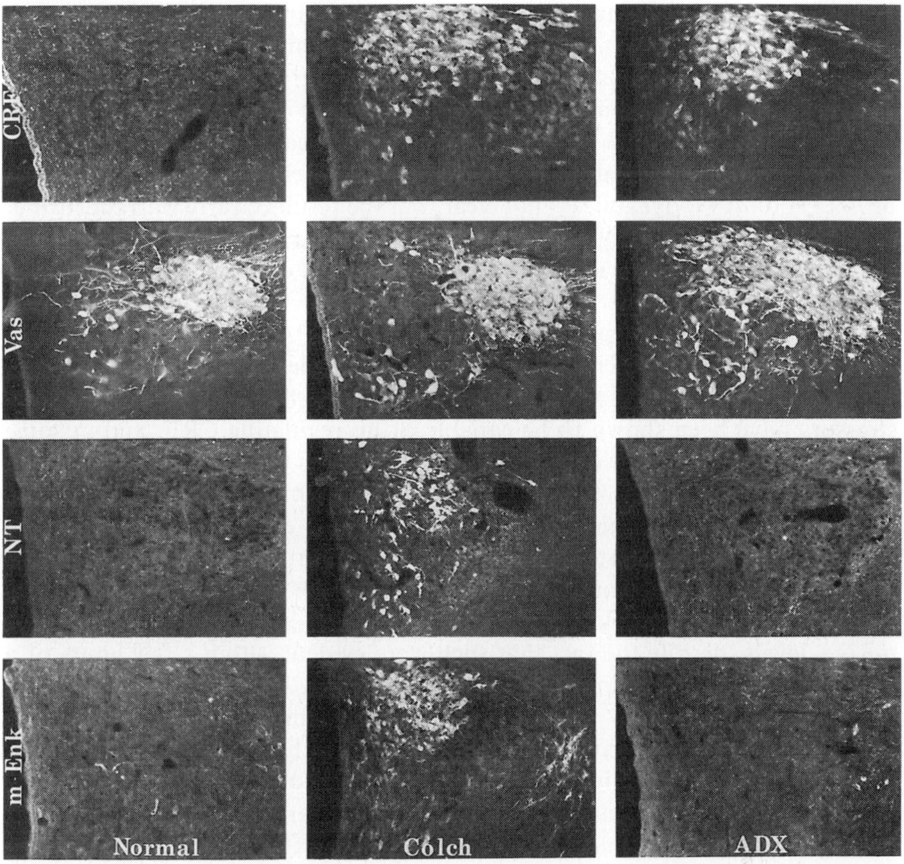

FIG. 2. Specificity of the effect of steroid withdrawal on secretagogues for corticotropin. Fluorescence photomicrographs to compare the distribution of corticotropin-releasing factor (CRF), vasopressin (Vas), neurotensin (NT) and met-enkephalin (m-Enk) immunoreactivity at a similar level of the paraventricular nucleus in untreated (Normal), colchicine-treated (Colch) and adrenalectomized (ADX) rats. In the untreated animals, only magnocellular vasopressin cells are prominent. Colchicine treatment reveals the ability of parvocellular neurons to display CRF, m-Enk and NT immunostaining. Adrenalectomy enhances CRF and vasopressin expression in parvocellular neurons selectively. The effects of adrenalectomy appear to be specific to CRF and vasopressin, which alone among these peptides are stimulators of corticotropin secretion.

postulated some time ago (Yates & Maran 1974). Prompted by the observation that hypophysectomy tended to be more effective than adrenalectomy in enhancing CRF and AVP immunostaining, we attempted to create short-loop effects by administering ACTH (or vehicle) to rats that were both hypophysectomized and adrenalectomized. Replacement doses of ACTH that mimicked moderate stress levels of the hormone were surprisingly effective in mitigating the enhancement of AVP (70% reduction of positively stained cells) and to a lesser extent (32% reduction) CRF immunoreactivity (Sawchenko 1989). Whether this might be indicative of a preferential negative feedback influence of ACTH on AVP dynamics is not clear, but the results clearly warrant further evaluation of the potency of ACTH, relative to that of glucocorticoids, as a negative regulator of the synthesis and/or secretion of corticotropin-releasing peptides.

The principal counterpoint to the brake on HPA axis function imposed by corticosteroids is generally thought to be provided via the array of neural inputs that convey stress-related information to the PVN. Several diverse stressors have been shown to be capable of up-regulating CRF, and in some instances AVP, mRNA levels in parvocellular neurons (Lightman & Young 1988, Harbuz & Lightman 1989, Herman et al 1989, Imaki et al 1991, see also de Goeij et al 1991). This provides a basis for an explanation of continued axis output in chronic stress in the face of heightened negative feedback. In our experiments using an intermittent foot-shock paradigm (Imaki et al 1991), we failed to detect any acute (within 24 hours) stress-induced alteration in CRF mRNA levels, despite the fact that such were evident within a few hours of a single foot-shock session in some extrahypothalamic CRF-expressing cell groups. This might be taken to indicate that any stress-induced drive to the axis is blunted by raised corticosteroid levels. Others have, however, described significant stress-induced increases in CRF mRNA expression in the PVN within several hours (e.g. Lightman & Young 1988, Harbuz & Lightman 1989). It is not clear whether such discrepancies may be attributable to differences in the sensory pathways that might be invoked by different stressors or to differences in the intensity of various challenges. Also remaining to be explored is the important issue of whether the positive (neurogenic) and negative (steroid feedback) influences are simply competitive, or whether the former is somehow capable of diminishing the sensitivity of the CRF neuron to steroid effects.

In summary, the capability for CRF and AVP to be jointly expressed in parvocellular neurosecretory neurons would seem to represent a physiologically meaningful mechanism by which the stress response may be modulated. It is doubtful, however, that this ability of parvocellular neurons is sufficient to account for the high AVP:CRF ratios commonly reported in portal plasma (e.g. Plotsky & Sawchenko 1987). Moreover, oxytocin is also widely recognized as a secretagogue for ACTH, and there is as yet no indication that oxytocin can be expressed by any substantial complement of parvocellular neurosecretory neurons. Magnocellular neurons would seem at present to constitute the only viable candidates for such an influence (Fig. 3). Potential mechanisms by which

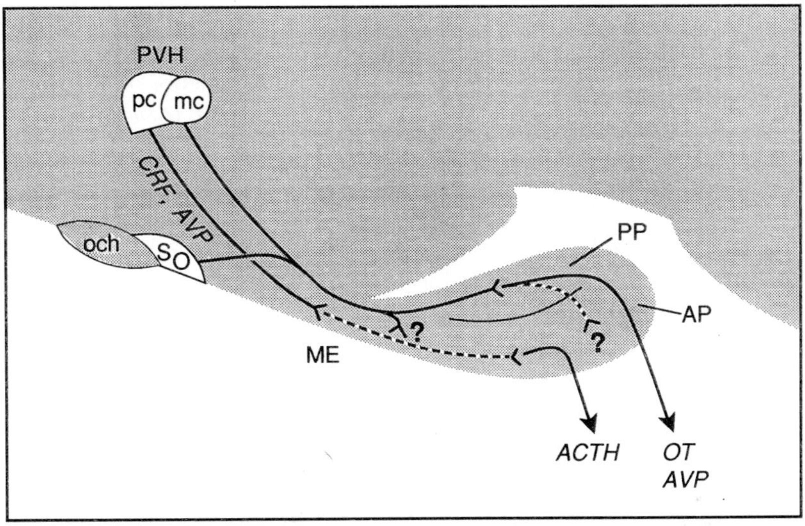

FIG. 3. Schematic sagittal view of the hypothalamus and pituitary to show the organization of magnocellular (mc) and parvocellular (pc) neurosecretory systems, highlighting potential sites at which interactions relevant to the control of pituitary corticotropin (ACTH) secretion may occur. Corticotropin-releasing factor (CRF) is produced by cells in the parvocellular division of the paraventricular nucleus (PVH) which project to the external zone of the median eminence (ME), ultimately reaching the anterior pituitary (AP) via the portal vasculature (dashed line). Magnocellular neurosecretory projections from the PVN and the supraoptic nucleus (SO) pass through the internal lamina of the median eminence en route to the posterior lobe (PP), where oxytocin (OT) and vasopressin (AVP) are released into the systemic circulation. OT and AVP of magnocellular origin may gain access to the portal vasculature by way of exocytotic release at the median eminence or via vascular links between the posterior and anterior lobes (question marks). och, optic chiasma.

oxytocin and/or AVP of magnocellular origin could gain access to the anterior lobe include the release of secretory granules in proximity to the portal plexus via exocytosis from pre-terminal portions of magnocellular axons coursing through the internal lamina of the median eminence (Buma & Nieuwenhuys 1987, see also Holmes et al 1986), and proposed direct vascular links permitting blood flow from the neurohypophysis to the anterior lobe (Page 1986). The manner in which either mechanism might be regulated and integrated with the factors that control the release of oxytocin and AVP into the general circulation is completely obscure.

Magnocellular neurosecretory system

Reports of co-localization in magnocellular neurosecretory oxytocin and AVP neurons are no less pervasive than for their parvocellular counterparts (Table 1).

Some magnocellular cells have even been reported to be capable of expressing markers for both nonapeptide hormones (Kiyama & Emson 1990, see Young, this volume). Under basal conditions, at least, assessments of the degree of oxytocin–AVP co-localization range from non-existent to vanishingly low, raising questions as to the physiological significance of this particular manifestation of the phenomenon.

More typically, variously sized subsets of magnocellular oxytocin and AVP neurons have been shown to be able to express a number of neuropeptides and even markers for some non-peptidergic neurotransmitters. Oxytocin- and AVP-expressing neurons display distinctive complements of co-localized principles, which tend to be preferentially associated with one or the other magnocellular cell type (e.g. Meister et al 1990, Fig. 4). One significant exception to this is prodynorphin-derived peptides, which in our experiments have proven to be capable of being expressed extensively in both (Sawchenko & Levin 1987).

In the magnocellular system, indices of basal levels of expression of co-localized peptides in the perikarya and axon terminals are generally low. The posterior pituitary content of co-localized principles typically ranges between two and four orders of magnitude lower than that of either nonapeptide hormone (Brownstein & Mezey 1986), making it unlikely that these function as secreted hormones of consequence. Instead, autocrine or paracrine effects on oxytocin or AVP release are indicated, and such have been described for many coexisting peptides; the best characterized example is the inhibition of oxytocin secretion by opiate peptides (Bicknell 1985). Also warranting consideration are indirect effects exerted, for example, on pituicytes, which appear capable of dynamically modulating access of magnocellular axon terminals to the basal laminae surrounding perivascular space in the neural lobe (Hatton 1988).

As in the parvocellular system, regulation of coexisting peptides within magnocellular neurons may be influenced by the steroid environment. Evidence from a variety of sources, including steroid receptor localization in identified magnocellular neurons (see e.g. Warembourg & Poulain 1991), would suggest that gonadal steroids, specifically oestrogens, are principally involved. We (Sawchenko & Levin 1987) have found that treatment of ovariectomized rats with oestradiol enhances immunohistochemical staining for three peptides (dynorphin(1–8), angiotensin II and galanin) in magnocellular vasopressinergic cells, while other peptides (dynorphin, cholecystokinin, CRF) co-localized with oxytocin were not affected. That these findings may hold some relevance to normal physiology is indicated by the fact that similar results were seen in normally cycling female rats killed at oestrus but not at other phases of the oestrous cycle.

As a coarse parallel to the situation in the parvocellular system, several systemic challenges known to target oxytocin and/or AVP synthesis and secretion have also been shown to modify coexisting peptide or mRNA levels (Sherman et al 1986, Lightman & Young 1987, Meister et al 1990). The best

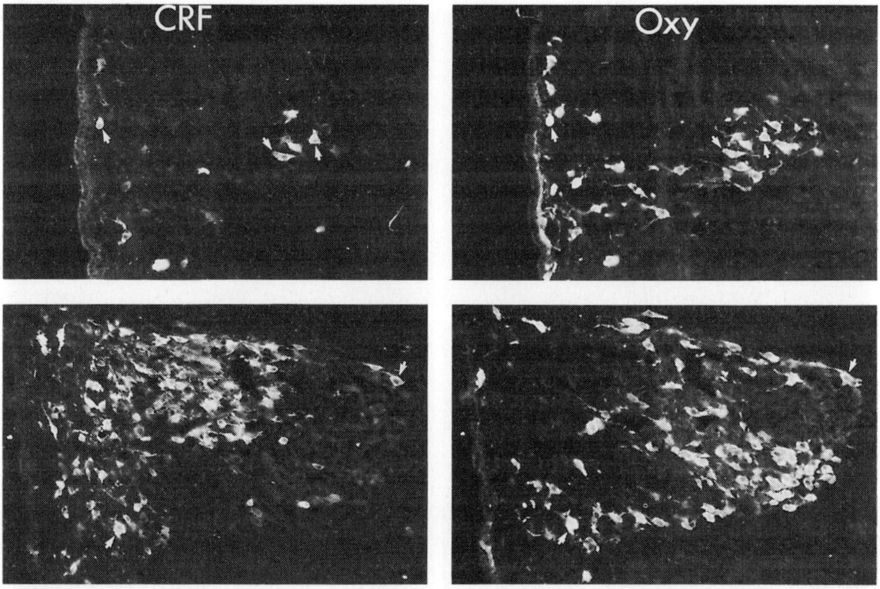

FIG. 4. Co-localization of oxytocin (Oxy) and corticotropin-releasing factor (CRF) immunoreactivity in the paraventricular nucleus. Fluorescence photomicrographs of single sections through the rostral (anterior magnocellular; top) and mid-caudal (bottom) aspects of the nucleus from a colchicine-treated rat to demonstrate co-localization of CRF-immunoreactivity in a subpopulation of oxytocin-containing cells. Arrows show examples of doubly labelled neurons.

studied challenge is the hyperosmolarity associated with salt-loading, which increases synthesis and secretion of both oxytocin and AVP. In addition, salt-loading up-regulates several mRNAs co-localized in one or both cell types, including dynorphin (Sherman et al 1988), CRF (Young 1986), cholecystokinin (Sherman et al 1988), tyrosine hydroxylase (Young et al 1987) and galanin (Meister et al 1989). The net impact of such responses on posterior lobe output of oxytocin and AVP is not yet clear.

The effects of salt-loading on CRF gene and peptide expression in the magnocellular system are interesting (Fig. 5), in that the massive increases in expression in the magnocellular system are accompanied by a marked down-regulation in the parvocellular neurosecretory system (Young 1986, Imaki et al 1992). This yields a pattern of CRF expression that is virtually indistinguishable from that of oxytocin. The effects on both cell types may represent adaptive responses to hyperosmotic challenge. CRF has been reported to increase AVP secretion from isolated neuro-intermediate lobe (Alzein et al 1984) and several lines of evidence suggest

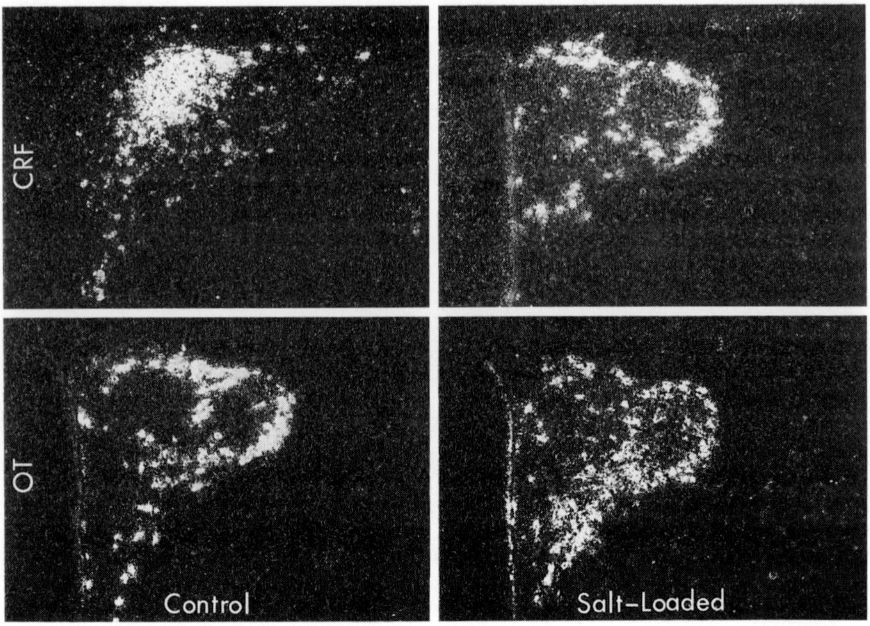

FIG. 5. Effects of salt-loading on corticotropin-releasing factor (CRF) mRNA in the paraventricular nucleus. Dark field photomicrographs of comparable levels through the nucleus from control (left) and 5-day salt-loaded (right) rats hybridized with cRNA probes for CRF (top) and oxytocin (OT) (bottom) mRNA. Salt-loading results in a marked increase in CRF mRNA levels in the magnocellular division of the nucleus (ring of cells near the right-centre of each field), and decreased levels in the parvocellular division (adjoining the ventricle at the left-centre of each field). In the salt-loaded animal the CRF and oxytocin distributions are similar.

that AVP secretion is under a tonic inhibitory control by glucocorticoids (e.g. Raff 1987).

Discussion

In a review summarizing the status of co-localized principles in the magnocellular neurosecretory system, Brownstein & Mezey (1986) described the array of neuroactive substances that had at that time been localized to oxytocin and AVP neurons. They concluded that such cataloguing was only a first step towards understanding the functional significance of the phenomenon. The field has since been dominated by studies of the manner in which co-localized principles are regulated. The literature is now replete with examples of impressive, and often differential, control of coexisting molecules in response to ever more physiologically relevant stimuli, in combinations that can sometimes be construed as conforming to current perceptions of 'physiological sense'.

The ability of the parvocellular neurosecretory CRF neuron to express co-secretagogues for corticotropin represents an appealing link to function, and the fact that both CRF and AVP can be up-regulated in response to stress places the phenomenon in the realm of physiology. But how important a source of AVP in hypophysial portal plasma are the parvocellular neurons? And what of the other resident peptides in this cell type? Do those which display no substantial direct effect on anterior pituitary secretions exert modulatory influences at the level of the median eminence? If so, which hypophysiotropic fibre types are targeted by them?

Magnocellular neurosecretory neurons display a similarly impressive ability to express multiple neuroactive substances. All indications are that these act locally, at the level of the neurohypophysis, to modify oxytocin and/or AVP secretion. But what is the significance of so many such modulators? How can we begin to investigate the ways in which they might interact in anything approaching a physiological setting?

Clearly, the challenge to associate the phenomenology directly with function remains.

Acknowledgements

The work from our laboratory that was summarized here was supported by USPHS grants NS-21182 and DK-26741, and was conducted in part by the Clayton Foundation for Research—California Division. P. E. S. and W. V. are Clayton Foundation Investigators.

References

Alzein M, Jeandel L, Lutz-Bucher B, Koch B 1984 Evidence that CRF stimulates vasopressin secretion from isolated neurointermediate pituitary. Neuroendocrinol Lett 6:151

Antoni FA 1986 Hypothalamic control of adrenocorticotropin secretion: advances since the discovery of 41-residue corticotropin-releasing factor. Endocr Rev 7:351–378

Bicknell RJ 1985 Endogenous opioid peptides and hypothalamic neuroendocrine neurons. J Endocrinol 107:437–446

Brownstein MJ, Mezey E 1986 Multiple chemical messengers in hypothalamic magnocellular neurons. In: Hökfelt T, Fuxe K, Pernow B (eds) Coexistence of neuronal messengers: a new principle in chemical transmission. Elsevier, Amsterdam (Prog Brain Res 68) p 161–189

Buma P, Nieuwenhuys R 1987 Ultrastructural demonstration of oxytocin and vasopressin release sites in the neural lobe and median eminence of the rat by tannic acid and immunogold methods. Neurosci Lett 74:151–157

Ceccatelli S, Eriksson M, Hökfelt T 1989 Distribution and coexistence of corticotropin-releasing factor-, neurotensin-, enkephalin-, cholecystokinin-, galanin- and vasoactive intestinal polypeptide/peptide histidine isoleucine-like peptides in the parvocellular part of the paraventricular nucleus. Neuroendocrinology 49:309–323

Cortés R, Ceccatelli S, Schalling M, Hökfelt T 1990 Differential effects of intracerebroventricular colchicine administration on the expression of mRNAs for neuropeptides and neurotransmitter enzymes, with special emphasis on galanin: an in situ hybridization study. Synapse (NY) 6:369–391

de Goeij DC, Kvetnansky R, Whitnall MH et al 1991 Repeated stress-induced activation of corticotropin-releasing factor neurons enhances vasopressin stores and colocalization with corticotropin-releasing factor in the median eminence of rats. Neuroendocrinology 53:150–159

Harbuz MS, Lightman SL 1989 Responses of hypothalamic and pituitary mRNA to physical and psychological stress in the rat. J Endocrinol 122:705–711

Herman JP, Schafer MK-H, Sladek CD, Day R, Young EA, Akil H 1989 Chronic electroconvulsive shock treatment elicits up-regulation of CRF and AVP in select populations of neuroendocrine cells. Brain Res 501:235–246

Hatton GI 1988 Pituicytes, glia, and the control of terminal secretion. J Exp Biol 139:67–79

Holmes MC, Antoni FA, Aguilera G, Catt KJ 1986 Magnocellular axons in passage through the median eminence release vasopressin. Nature (Lond) 319: 326–329

Hökfelt T, Fahrenkrug J, Tatemoto K et al 1983 The PHI (PHI-27) corticotropin-releasing factor/enkephalin immunoreactive hypothalamic neuron: possible morphological basis for integrated control of prolactin, corticotropin and growth hormone secretion. Proc Natl Acad Sci USA 80:895–898

Imaki T, Nahon J-L, Rivier C, Sawchenko PE, Vale W 1991 Differential regulation of corticotropin-releasing factor mRNA in rat brain cell types by glucocorticoids and stress. J Neurosci 11:585–599

Imaki T, Vale W, Sawchenko PE 1992 Regulation of corticotropin-releasing factor mRNA in neuroendocrine and autonomic neurons by osmotic stimulation and volume expansion. Neuroendocrinology, in press

Kiss JZ, Mezey E, Skirboll L 1984 Corticotropin-releasing factor-immunoreactive neurons of the paraventricular nucleus become vasopressin positive after adrenalectomy. Proc Natl Acad Sci USA 81:1854–1858

Kiyama H, Emson PC 1990 Evidence for the co-expression of oxytocin and vasopressin messenger ribonucleic acids in magnocellular neurosecretory cells: simultaneous demonstration of two neurophysin messenger ribonucleic acids by hybridization histochemistry. J Neuroendocrinol 2:257–259

Lightman SL, Young WS III 1987 Vasopressin, oxytocin, dynorphin, enkephalin and corticotropin-releasing factor mRNA stimulation in the rat. J Physiol (Camb) 394:23–29

Lightman SL, Young WS III 1988 Corticotropin-releasing factor, vasopressin and pro-opiomelancortin mRNA responses to stress and opiates in the rat. J Physiol (Camb) 403:511–523

Lind RW, Swanson LW, Sawchenko PE 1985 Anatomical evidence that neural circuits related to the subfornical organ contain angiotensin II. Brain Res Bull 15:79–82

Meister B, Cortes R, Villar MJ, Hökfelt T 1989 Peptide and transmitter enzymes in hypothalamic magnocellular neurons after administration of hyperosmotic stimuli: comparison between messenger RNA and peptide/protein levels. Cell Tissue Res 260:279–297

Meister B, Hökfelt T, Geffard M, Oertel W 1988 Glutamic acid decarboxylase- and γ-aminobutyric acid-like immunoreactivities in corticotropin-releasing factor-containing parvocellular neurons of the hypothalamic paraventricular nucleus. Neuroendocrinology 48:516–526

Meister B, Villar MJ, Ceccatelli S, Hökfelt T 1990 Localization of chemical messengers in magnocellular neurons of the hypothalamic supraoptic and paraventricular nuclei: an immunohistochemical study using experimental manipulations. Neuroscience 37:603–633

Mezey E, Reseine TD, Skirboll L, Beinfeld M, Kiss JZ 1986 Role of cholecystokinin in corticotropin release: coexistence with vasopressin and corticotropin-releasing factor in cells of the rat hypothalamic paraventricular nucleus. Proc Natl Acad Sci USA 83:3510–3513

Page RB 1986 The pituitary portal system. In: Ganten D, Pfaff D (eds) Morphology of the hypothalamus and its connections. Springer-Verlag, Berlin (Curr Top Neuroendocrinol 7) p 1–47

Plotsky PM 1991 Pathways to the secretion of adrenocorticotropin: a view from the portal. J Neuroendocrinol 3:1–9

Plotsky PM, Sawchenko PE 1987 Hypophysial-portal plasma levels, median eminence content and immunohistochemical staining of corticotropin releasing factor, arginine vasopressin and oxytocin following pharmacological adrenalectomy. Endocrinology 120:1361–1369

Raff H 1987 Glucocorticoid inhibition of neurohypophyseal vasopressin secretion. Am J Physiol 252:R635–R642

Sawchenko PE 1987a Evidence for differential regulation of corticotropin-releasing factor and vasopressin immunoreactivities in parvocellular neurosecretory and autonomic-related projections of the paraventricular nucleus. Brain Res 437: 253–263

Sawchenko PE 1987b Adrenalectomy-induced enhancement of CRF and vasopressin immunoreactivity in parvocellular neurosecretory neurons: anatomic, peptide and steroid specificity. J Neurosci 7:1093–1106

Sawchenko PE 1989 Short-loop feedback effects of adrenocorticotropic hormone on corticotropin-releasing factor and vasopressin-immunoreactivity in the paraventricular nucleus. Soc Neurosci Abstr 15:134

Sawchenko PE, Levin MC 1987 Immunohistochemical evidence for modulation by estrogen of neuropeptide coexpression in the magnocellular neurosecretory system of the female rat. Soc Neurosci Abstr 13:1371

Sawchenko PE, Swanson LW 1982 Immunohistochemical identification of neurons in the paraventricular nucleus of the hypothalamus that project to the medulla or the spinal cord in the rat. J Comp Neurol 205:260–272

Sawchenko PE, Swanson LW, Vale WW 1984a Corticotropin-releasing factor: co-expression within distinct subsets of oxytocin-, vasopressin-, and neurotensin-immunoreactive neurons in the hypothalamus of the male rat. J Neurosci 4:1118–1129

Sawchenko PE, Swanson LW, Vale WW 1984b Co-expression of corticotropin-releasing factor and vasopressin immunoreactivity in parvocellular neurosecretory neurons of the adrenalectomized rat. Proc Natl Acad Sci USA 81:1883–1887

Sherman TG, Day R, Civelli O et al 1988 Regulation of hypothalamic magnocellular neuropeptides and their mRNAs in the Brattleboro rat: coordinate responses to further osmotic challenge. J Neurosci 8:3785–3796

Swanson LW, Sawchenko PE, Lind RW 1986 Regulation of multiple peptides in CRF parvocellular neurosecretory neurons: implications for the stress response. In: Hökfelt T, Fuxe K, Pernow B (eds) Coexistence of neuronal messengers: a new principle in chemical transmission. Elsevier, Amsterdam, (Prog Brain Res 68) p 161–189

Swanson LW, Simmons DM 1989 Differential steroid hormone and neural influences on peptide mRNA levels in CRH cells of the paraventricular nucleus: a hybridization histochemical study in the rat. J Comp Neurol 285:413–435

Tsuruo Y, Ceccatelli S, Villar MJ et al 1988 Coexistence of TRH with other neuroactive substances in the rat central nervous system. J Chem Neuroanat 1:235–253

Warembourg M, Poulain D 1991 Presence of estrogen receptor immunoreactivity in the oxytocin-containing magnocellular neurons projecting to the neurohypophysis in the guinea pig. Neuroscience 40:41–53

Whitnall MH 1988 Distributions of pro-vasopressin expressing and pro-vasopressin deficient CRH neurons in the paraventricular hypothalamic nucleus of colchicine-treated normal and adrenalectomized rats. J Comp Neurol 275:13–28

Yates FE, Maran JW 1974 Stimulation and inhibition of ACTH release. In: Knobil E, Sawyer WH (eds) Handbook of Physiology. Oxford University Press, New York (Am Physiol Soc) Section 7, vol 4:367–404

Young WS III 1986 Corticotropin-releasing factor mRNA in the hypothalamus is affected differently by drinking saline and by dehydration. FEBS (Fed Eur Biochem Soc) Lett 208:158–162

Young WS III 1992 Regulation of gene expression in the hypothalamus:hybridization histochemical studies. In: Functional anatomy of the neuroendocrine hypothalamus. Wiley, Chichester (Ciba Found Symp 168) p 127–143

Young WS III, Mezey E, Siegel RE 1986 Quantitative *in situ* hybridization histochemistry reveals increased levels of corticotropin-releasing factor mRNA after adrenalectomy in rats. Neurosci Lett 70:198–203

Young WS III, Warden M, Mezey E 1987 Tyrosine hydroxylase mRNA is increased by hyperosmotic stimuli in the paraventricular and supraoptic nuclei. Neuroendocrinology 46:439–444

Young WS III, Mezey E, Siegel RE 1988 Vasopressin and oxytocin mRNAs in adrenalectomized and Brattleboro rats: analysis by quantitative *in situ* hybridization histochemistry. Mol Brain Res 1:231–241

DISCUSSION

Lightman: Have you ever noticed any differential regulation of CRF and vasopressin in the parvocellular cells in the paraventricular nucleus (PVN)? In rats that have a chronic stimulus due to an immunologically mediated inflammation from adjuvant-induced arthritis, there are increased levels of vasopressin in the portal blood but decreased levels of CRF, which are mirrored by decreased levels of CRF mRNA in the PVN. This could be due either to an increase in the number of cells making vasopressin or to the amount of vasopressin synthesized and released by individual cells. Another possibility is the recruitment of magnocellular cells, which might release vasopressin from terminals in the median eminence. Do you have any evidence for a differential increase or decrease of vasopressin or CRF in those parvocellular cells at any time?

Sawchenko: The only relevant situation is the salt-loading paradigm and there we have not extensively examined what is happening to parvocellular vasopressin expression. Whether we could see a commensurate decrease is doubtful because we are starting from such low levels. In response to perturbation of neural inputs, or any perturbation we care to make of the steroid environment, markers for both peptides tend to respond in tandem, but those for vasopressin always exhibit the greater sensitivity. I continue to be perplexed by the fact that two peptides in the same cell that serve complementary functions appear to be regulated under all circumstances in the same way.

Clarke: Paul, you showed us that after adrenalectomy there was increased expression of CRF and AVP in the PVN (Fig. 1). You suggested this might be due to the removal of negative feedback. The *in situ* data showed that stress caused an increase in CRF and AVP mRNAs and you suggested that might be due to a corticosteroid clamp. It seems to me you can't have it both ways—either the negative feedback effect of corticosteroids is to inhibit the expression of CRF and AVP or it's not. An alternative explanation for your results might be that some of the effects are transcriptional and some are post-transcriptional. Do you think that's possible? For example, the immunocytochemical data might be showing an effect that is post-transcriptional, whereas the stress-related increases might be due to transcriptional regulation.

Sawchenko: In both the steroid manipulations and the stress paradigms, we saw the effects on CRF and AVP expression by both immunocytochemistry and hybridization histochemistry. In the paradigm we are using, my assumption would be that with time the exposure to chronic stress or ongoing stress eventually has the power to overcome the steroid-mediated brake on the synthesis of these peptides in parvocellular neurons. We don't see effects of stress on CRF and AVP expression acutely. I suspect that is because the system has not been driven long enough or hard enough for us to see up-regulation of these two genes.

Stafford Lightman and Scott Young showed some time ago that the effects of stress and steroids are largely exerted by independent mechanisms. I suspect we are looking at steroid effects and trans-synaptic effects that both have access to transcriptional control mechanisms and that prolonged stress can actually overcome the normal brake on synthesis that's provided by feedback. Both effects are clearly manifest at the mRNA level: we can make no strong statements about the extent to which these involve influences on transcription or message stability because we are only assessing relative levels of mRNA.

Smith: Several groups have demonstrated co-localization of AVP and CRF in neurons projecting to the median eminence both before and after adrenalectomy. However, the reported percentage of CRF-positive neurons also staining for AVP and the percentage change following adrenalectomy vary among these studies. Could these differences be due to antibody specificity? For example, in some studies the vasopressin antibodies recognize only fully processed vasopressin, while in others I believe the antibodies would recognize intermediate precursor forms of vasopressin as well as the processed product. Have you any evidence that precursor processing may be regulated? Have you looked chromatographically at the immunoreactive products of vasopressin in the median eminence?

Sawchenko: Neither I nor anyone to my knowledge has done that kind of chromatographic analysis. Vasopressin expression in the parvocellular system has been demonstrated using antisera that recognize processed vasopressin as well as its associated neurophysin and glycopeptide.

Gainer: Those studies were done by Mark Whitnall in our laboratory. There is always a good correlation between the mRNA data and the data on expression of the peptide, as studied using anti-neurophysin antibodies which are specific for the prohormone, or using anti-vasopressin antibodies specific for cleaved but either non-amidated or amputated vasopressin. So there is no reason to believe a post-translational processing event plays a role in these particular cells in these circumstances.

I would like to know what you think of our idea (see references below) that normally there are two populations of cells, one of which contains CRF and vasopressin all the time, the other of which contains only CRF. Those results were originally established for the median eminence where it's really clear: it doesn't require colchicine treatment or anything to see that about 50% of the nerve terminals that have CRF also contain vasopressin (Whitnall et al 1987a,b). Recent studies have indicated that certain stressors selectively stimulate release from that population (Whitnall 1989).

Sawchenko: I find the data convincing. I am perplexed by the fact that Mark is able to detect AVP in the perikarya of parvocellular neurons in animals treated with colchicine. Using hybridization methods, which are somewhat more sensitive in phenotyping these neurons, we see very little evidence of this in the unperturbed state. The caveat that needs to be placed on a lot of the results that have been generated using the colchicine model is derived from the recent work from Hökfelt's lab showing that colchicine treatment is itself a stressor. A failure to observe co-localization in a colchicine-treated animal may not necessarily be evidence that the cells of interest are not able to produce a particular peptide; it may indicate that its expression may be down-regulated in that situation. Likewise, it is unclear whether any enhancement may be attributed to the transport-blocking or the stress-inducing properties of colchicine.

Gainer: But no colchicine was necessary for the median eminence studies, so there we are free of any colchicine problem. Mark was able to see AVP immunoreactivity in the normal cell bodies using electron microscopic immunocytochemistry (Whitnall & Gainer 1988). By light microscopy, this is not possible. Using electron microscopy, one has the sensitivity to detect a single vesicle immunocytochemically. So there are two answers: colchicine treatment is not necessary to demonstrate this for the median eminence and he uses electron microscopy for sensitivity of detection and cellular resolution.

Palkovits: Paul, you showed dopamine co-localization with vasopressin. In the 1970s, Larry Swanson showed that there are hundreds of dopaminergic cells in the PVN. Is that dopamine acting as a neurotransmitter there? What's happening to the tyrosine hydroxylase activity in the vasopressin cells? It is definitely not transported down to the posterior lobe.

Sawchenko: The tyrosine hydroxylase immunoreactive cells in the PVN that Larry Swanson and I described some years ago were overwhelmingly localized to the parvocellular division of the nucleus. The initial demonstration of the ability of magnocellular neurons to express the enzyme under certain circumstances was by Kiss & Mezey (1986). At present, I think it is prudent to refer to these neurons as tyrosine hydroxylase positive rather than dopaminergic because I am not aware that dopamine itself has been identified in magnocellular neurons: they do have the ability to express the biosynthetic enzyme. I do believe, however, that the dopamine content of the posterior lobe is relatively high.

Palkovits: Yes, but definitely nothing from the PVN.

Moore: When rats are permitted to drink only saline or are deprived of drinking water for 1–3 days, there is an increase in the activity of tyrosine hydroxylase in the posterior pituitary (Alper et al 1980) and in the content of dopamine in the neural lobe of the pituitary (Manzanares et al 1990). I don't know if anyone has been able to localize the source of the increased tyrosine hydroxylase or dopamine, but salt-loading and dehydration probably induce changes in dopaminergic neurons projecting to the posterior pituitary.

Palkovits: What is the origin of the dopaminergic neurons, the PVN or the arcuate nucleus?

Moore: That's not been determined. Dopaminergic neurons projecting to the posterior pituitary were initially believed to originate in the rostral arcuate nucleus (Björklund et al 1973), but we have recently found that they originate, at least in part, in the more rostral periventricular nucleus (Goudreau et al 1990). On the other hand, Young et al (1987) reported that salt-loading increased the amount of mRNA for tyrosine hydroxylase in the paraventricular and supraoptic nuclei.

Palkovits: But not in an animal treated with monosodium glutamate. It indicates that the posterior pituitary dopamine is more likely to originate from the arcuate nucleus than the magnocellular cells.

Paul, do any of your manipulations, such as salt-loading or adrenalectomy, that affect concentrations of vasopressin in the PVN cause any changes in the vasopressinergic cells in the suprachiasmatic nucleus?

Sawchenko: We didn't analyse that systematically. My general impression is that if vasopressin expression is altered in the suprachiasmatic nucleus, the change is much less dramatic than it is in the PVN.

Palkovits: I agree, I couldn't find any procedure that caused a significant change in the vasopressin staining in the suprachiasmatic nucleus.

Gainer: Paul, you said that the concentrations of vasopressin and oxytocin in normal portal blood are orders of magnitude higher than the amounts that could come from the cells in the external zone of the median eminence.

Therefore, these peptides were most likely secreted from the magnocellular neurons. If this is the norm, that there is so much resident oxytocin and vasopressin in portal blood, then what is the point of an up-regulation in response to stress?

Sawchenko: It seems to be the norm and at least in terms of oxytocin I am not aware of any candidates other than magnocellular neurons as sources for the peptide in the portal vasculature. Ferenc Antoni and George Fink have looked at portal levels of vasopressin in paraventricular-lesioned animals and seen relatively minor reductions. There is potentially the capability for bloodflow from the posterior pituitary to the anterior lobe but that is somewhat mitigated by evidence that neural lobectomy doesn't markedly affect portal levels of vasopressin in primates.

My sense is that the data are just incomplete. Studies of the unperturbed animal don't necessarily tell us what these cell populations might be doing in a stress situation. Until we know that, we can't address the relative importance of the several potential sources of the magnocellular peptides in portal plasma.

Lightman: A major problem in the investigation of portal blood responses to stress in the rat model is the fact that these animals are anaesthetized and already under a very severe degree of surgical stress. Thus, to show any changes in vasopressin or CRF in this model, where the levels are already very high, would be difficult. Any data obtained would not necessarily bear a close relationship to the physiological changes that occur when a conscious healthy rat responds to stress.

Barraclough: In the rat model used to collect portal blood, the entire pituitary gland was removed to gain access to stalk portal blood. As a consequence, retrograde flow of pituitary hormones to the brain could not occur in this preparation.

Clarke: There is an even more important issue: when Paul Plotsky did those studies, he sectioned the stalk. So he was taking blood from a stump of the median eminence that had cut ends which normally project to the posterior lobe. It is possible that what he was measuring was vasopressin and oxytocin which were simply leaking out of the cut neurons. Certainly, if you look downstream in the portal blood system, in the sheep for example, you can't detect oxytocin in portal blood. This leads me to believe that in the rat model the AVP and oxytocin results are an artifact.

Makara: When a cell is firing, does it always express the same set of neuropeptides? Is the expression of these neuropeptides cell specific or do different stimuli cause the expression of different neuropeptides in a given cell type?

Sawchenko: That's a general question and I don't know that we have enough data to give a general answer. It addresses the issue of whether any given challenge that may be conveyed neuronally is necessarily going to affect the transcription of genes for effector molecules which that neuron is expressing.

Paul Plotsky and I infused hypertonic saline into the jugular vein and monitored the changes in plasma osmolality, c-*fos* expression, vasopressin secretion and the amount of vasopressin mRNA. Over a six hour infusion we saw an immediate rise in plasma osmolality, an immediate induction of c-*fos* expression in both magnocellular cell types, and an immediate secretion of vasopressin from these neurons. However, we saw no evidence for any up-regulation of vasopressin mRNA. It may be that in this example we just didn't carry on for long enough. Such observations do raise questions as to whether stimulation of a neuronal pathway need necessarily exert transcriptional influences.

Lightman: That's a difficult example, because there is already so much vasopressin message in those cells in the basal state that one might not expect to see a change in six hours. It may be just that the techniques used cannot detect small changes in such a large pool.

Theodosis: In fact, there is enough vasopressin peptide stored in the terminals to take care of such a situation without inducing new transcription.

Zoeller: Transcription of the vasopressin and oxytocin genes may not be necessary to support elevated secretion over a six hour period, but the increased expression of c-*fos* indicates a cellular response to heightened activity that may mediate several secondary transcriptional effects, which could be important in preparing the cell for further secretion. Examples might include changing the number or type of channels or increasing the synthesis of cellular machinery supporting synthesis, such as rough endoplasmic reticulum and Golgi—the ultrastructural correlates of raised secretion.

Young: Dr Makara asked whether we can dissociate the expression of different genes within the same neurons. Dr Palkovits and I have made lesions and examined the expression of galanin and vasopressin in the magnocellular neurons. There was a decrease in vasopressin expression and at the same time there was an increase in galanin. In the hyperosmolar state, these two genes are both up-regulated.

To get at some of these control mechanisms of regulation will take various manipulations. For example, the question about whether or not dopamine synthesis is occurring in the posterior pituitary because of tyrosine hydroxylase that is transported there from the magnocellular neurons could be examined by making lesions of dopaminergic cells in the arcuate nucleus using monosodium glutamate, then salt-loading and doing the studies Ken Moore suggested.

There is evidence that, for example, angiotensin II is not made in magnocellular neurons but in glia and then it is taken up by magnocellular neurons in the paraventricular and supraoptic nuclei. Do you have any comments about that or any other neuropeptides or transmitters made in glia that are transported to neurons?

Sawchenko: Larry Swanson used a probe against angiotensinogen to look at the mRNA encoding this enzyme in normal and steroid-manipulated (i.e. adrenalectomized with and without corticosterone replacement) animals. He couldn't resolve individual neurons because the apparent hybridization over glial cells was robust. People have often invoked the argument that a given localization or staining pattern could be attributable to a cell having incorporated and sequestered peptide released locally. I have never found this to be convincing. I think the weight of the available evidence suggests that both magnocellular and parvocellular neurons have the ability to express angiotensin II. That is based solely on immunohistochemical data. Hybridization studies have not yet helped to resolve this issue. I am not aware of any other situation in this system where local non-neuronal elements may play the kind of role you suggest.

Zoeller: Superimposed on the issue of selective regulation of different cellular processes, there are dramatic shifts in selective groups of neurons within the parvocellular nucleus—down-regulation of expression in some cells, up-regulation of expression in other cells. Hence, there is a higher order of integration within the parvocellular nucleus itself. How do you think that's being integrated? Is it being integrated locally or from afferents that arise in a very different spot?

Sawchenko: Oxytocin neurons, magnocellular vasopressin neurons and these parvocellular CRF neurons receive largely distinct and non-overlapping complements of afferents from a number of well defined sites, particularly in the brainstem and to a lesser degree in the forebrain. This probably accounts for much of the specificity with which different stressors target particular cell types.

Charlton: In general terms, the stresses Paul Sawchenko is talking about are fairly acute. But consider a stress, for example experimental allergic encephalitis in a rat, where the majority of the animals get paralysis in the hind limbs and then 99–100% of animals will recover. If you adrenalectomize them, they all die. So there is some means whereby the body knows that there is inflammation somewhere in the central nervous system. In that particular case, glucocorticoid levels are high in the acute phase with a subsequent lymphopenic response, so there may be a drive for the system to produce CRF and ignore the corticoid feedback to keep the levels high enough to get this longer-term effect.

Lightman: We have looked at that. In our initial studies in adjuvant-induced arthritis, we found that activation of the hypothalamo-pituitary-adrenal axis was associated with a fall in the amount of hypothalamic CRF mRNA at the same time as the amount of POMC mRNA in the pituitary increased.

Experimental allergic encephalomyelitis is another immunologically mediated stress which can activate the hypothalamo-pituitary-adrenal axis. When we studied this we also found that at the time of peak inflammation and corticosteroid secretion there is a fall in the amount of CRF mRNA, which promptly returns to normal when the animal recovers.

Robinson: Paul, you have described many of these co-localized factors. What evidence is there for an equal litany of peptide receptors in these regions?

Presumably, they are present in rather small amounts but they could be present at relatively high concentrations at a confined site. If the litany of peptides present can be changed under different conditions, can one also change the litany of receptors for that peptide? Are we assuming that because the peptides are present at a site they must be active there?

Sawchenko: For many of the co-localized principles in the magnocellular system appropriate receptors or binding sites have been demonstrated in the pituitary. Our knowledge of the manner in which they are regulated under some of these circumstances is pretty fragmentary.

Lightman: There is evidence from Young et al (1990) that glucocorticoid receptors are down-regulated in chronic stress situations, probably at the level of the hippocampus. That's at a different level in the central nervous system, but it does demonstrate that a chronic stress can change receptor complements. There is also some evidence that in chronically stressed animals the anterior pituitaries are hyperresponsive to vasopressin.

Robinson: But that's a clearer situation. I am more interested in these plastic boutons or synapses which may change their complement of peptide. If the peptide concentration is low, presumably the peptide is looking for a relatively local target. If that target is another cell type, how does the second cell type know to produce the relevant receptors? Or are those receptors already latently expressed on that cell? It is OK to produce multiple signals but you also need a set of multiple receptors to receive those signals.

Clarke: The presence of these peptides in neurosecretory terminals does not mean that they are secreted into the portal blood or whatever and that the peptides actually get to the pituitary. There's a whole range of examples of peptides that are in the external zone of the median eminence, which you can demonstrate quite satisfactorily with immunocytochemistry, but which you can't pick up in portal blood even with very sensitive techniques.

Theodosis: If we assume that there is differential control of gene expression, is there then differential control of translation of the gene products? Also, if we assume there is co-localization in the same subcellular organelle, the secretory granule, does that mean that the final processing of the peptides is the same? Is it possible that peptide release at the terminal is also differentially controlled?

Clarke: Two peptides may even be released together at the terminal, for example enkephalin with CRF, but then enzymes might attack specific peptides in the extracellular vascular space between the terminal and the blood vessel.

Sawchenko: There is another point to consider. In isolated neurointermediate lobe, application of some of these low abundance magnocellular peptides can clearly influence oxytocin or vasopressin secretion. Glenn Hatton (1988) has shown that in the posterior pituitary, in particular, pituicytes have a dynamic capability of allowing or restricting the access of some of these terminals to the basal laminae surrounding the perivascular space. This is another kind of mechanism that might be a target for some of these co-localized principles and another way by which these many co-localized peptides could be exerting their effects.

Hatton: Celia Sladek and Huda Akil have tried to measure dynorphin in the blood of animals, which everybody agrees is secreted with vasopressin in the portal blood, but they can't find it. Dynorphin is present in such small quantities that it might be directed towards these local cells. Pituicytes do bind [^{3}H]dynorphin and there is some evidence that dynorphin itself and vasopressin each increase the concentration of free intracellular calcium in these pituicytes in culture (Hatton et al 1990a,b). So a lot of these peptides could have very local effects.

Charlton: To move things one step higher, do we know enough about promoter regions for the genes of these co-localized peptides to hypothesize that they could be sharing a transcription factor? The rates of expression of their genes may depend on some stoichiometric relationship in the unwinding of the DNA. When the transcription factors are present at certain levels, one gene may be turned on and then at a higher level another gene is expressed. Do we know enough about the upstream gene elements to answer that sort of question?

Young: Knowledge of promoter elements for these genes is pretty rudimentary. The CRF gene and the enkephalin gene can both be turned on by stress, for example, in the parvocellular neurons in the PVN. The promoter regions for each gene contain cAMP response elements, so the stress signal could be mediated via cAMP.

Charlton: cAMP is probably also what is turning on expression of c-*fos* in these systems.

Zoeller: Most of the studies that characterize regulatory regions of genes are done in transient expression systems or other systems that optimize the ability to look at those particular regions. When you try to apply those conclusions to an *in vivo* neuron in which several neuropeptide genes may be regulated simultaneously as well as protein products being secreted, in addition to the cellular machinery that has to be regulated commensurate with neuronal activity, then it gets difficult. The concept of transcription factor competition might have some validity but maybe there are other kinds of mechanisms, such as synergism among second messenger systems that converge on a single transcription factor.

Mellon: It is going to be difficult to ask any of these questions by studying genes expressed transiently in cells that aren't appropriately regulated or are not the appropriate cell type. To understand the regulation of these genes, we need model systems that represent the neurons of the hypothalamus.

Gainer: Dr Susan Wray in our lab has developed a model organotypic culture of hypothalamic systems. Amongst the neurons that we can study are the CRF neurons, so we have a model. Given this, what transmitter would you recommend to stimulate these cells? What do we know about the circuitry on the afferent side? Obviously, something important is driving the CRF cells and preventing the glucocorticoid effect in primary stress. What is that drive? Is it NMDA or α-adrenergic input or what?

Poulain: I would like to consider the physiology. We use various stress situations to enhance CRF activity or AVP activity; these do not all have the same effects. For example, if you dehydrate an animal, oxytocin cells apparently are very resistant to osmotic stimulation for quite a while. If you record the electrical activity of vasopressin cells and oxytocin cells, you find that the vasopressin cells react very quickly to dehydration whereas oxytocin cells react much more slowly. Conversely, if you inject cholecystokinin into the third ventricle of an animal, you obtain a massive electrical reaction in oxytocin cells but not in vasopressin cells. So I presume the chain of events in oxytocin cells differs in terms of synthesis of CRF or AVP or oxytocin, depending on the stimulus. Many effects on transcription will depend entirely on the afferent input, i.e. on all the messages resulting from the stimulation.

Gainer: But what are the afferents?

Sawchenko: Clearly, the aminergic ones would be a good bet. Zsolt Liposits' studies (1986) with electron microscopy and our work at the light microscopic level suggest there are at least 3–4 adrenergic or noradrenergic cell groups in the brainstem that interact and target quite specifically these CRF neurons. A monkey wrench has been thrown into the works, however, by the recent proposition that perhaps the dominant neurotransmitters contained in afferents to the endocrine hypothalamus are excitatory amino acids. One often wonders what those small clear vesicles are doing in so many of the terminals on neuroendocrine neurons. This raises the possibility that the fast signalling properties in many of these circuits, even in the long ascending projections, may use one of the excitatory amino acids. But there is some uncertainty as to whether we are talking about glutamate or aspartate *per se* or perhaps the dipeptide, *N*-acetyl aspartylglutamate, which has also been localized in some brainstem catecholamine neurons. Just how dominant or non-dominant these excitatory amino acids are is not yet clear. It may turn out that the amines and the myriad other peptides that are co-localized within afferent neurons perform slower, gating kinds of roles. We need to know more about how these multiple molecules that are co-localized within afferents to neuroendocrine cells are regulated and how they interact.

Gainer: We know a lot more about the anterior pituitary, because we can actually put things on it *in vitro*. What types of transmitters would we put on, if we had such a preparation of CRF neurons *in vitro*?

Clarke: If we use the secretion into portal blood as an index, you can get profound release of CRF and AVP with noradrenaline and neuropeptide Y at very low doses. I suspect there is some physiological relevance to that because in response to a short-term stressor in the sheep model that we use, there is a prolonged activation of the CRF–AVP axis. When we put neuropeptide Y or noradrenaline into the lateral ventricles (admittedly it is not a specific injection into the PVN), we get the same type of response—it is prolonged over 3–4 hours. That makes us believe those two afferents might be important.

Makara: For the rat CRF cells, my bet would be on excitatory amino acids and noradrenaline via α_1-adrenergic receptors as the major stimulatory inputs. Noradrenaline is a dual agent and most likely it also has an α_2 receptor-mediated inhibitory effect on CRF cells.

Palkovits: There are many peptidergic terminals in the PVN that arise in the lower brainstem or limbic neurons. They may make synaptic contacts with CRF cells. Many of these peptides are candidate neurotransmitters because they are present in the PVN in presynaptic terminals. The peptides from the nucleus of the solitary tract include enkephalin, somatostatin, substance P and cholecystokinin. These have been localized in the PVN by immunocytochemistry and retrograde tract tracing. Neurotensin and neuropeptide Y may co-localize with catecholamines or may exist in PVN terminals without them. Fibres containing inhibin β, dynorphin or bombesin from this autonomic nucleus of the brainstem (nucleus tractus solitarii, NTS) run directly up to the PVN. Double immunostaining examined using electron microscopy shows noradrenaline- and neurotensin-containing neurons making direct synaptic contacts with PVN neurons.

Makara: I would like to change the mood of the discussion. Can we enumerate examples of proven co-localized, co-existing, co-acting substances in the central nervous system? Physiologically, I think we will have a hard time to find many.

Gainer: Dr Joan Summy-Long has done some beautiful work on the selective effect of dynorphin secretion from magnocellular neurons that also secrete vasopressin on oxytocin-secreting terminals. She has put all the work that others have done, starting with Bicknell & Leng (1982) who showed that dynorphin selectively inhibits oxytocin cells, in context physiologically (Summy-Long et al 1984). She makes this one of the best examples of a co-localized peptide in a physiological context.

Zoeller: As we move away from neuroendocrine groups of cells, we have to look at the functional consequences of co-secreted peptides on behaviour. In this regard, Dr Elliott Albers has looked at cells of the suprachiasmatic nucleus that contain VIP (vasoactive intestinal peptide), PHI (peptide histidine isoleucine) and gastrin-releasing peptide (Albers et al 1991). When Albers' group injects these three peptides into the suprachiasmatic nucleus, there is a large shift in circadian running wheel behaviour. However, when they inject each peptide singly or in any combination of two, they do not see this effect. In addition, cellular levels of the mRNAs encoding VIP and PHI and that encoding gastrin-releasing peptide exhibit opposite 24-hour profiles of changes (Zoeller et al 1992). Hence, this is a nice system in which to study the functional consequences of co-localized peptides in neurons of the CNS within a specific physiological context.

Theodosis: There is strong anatomical evidence for co-localization of substance P, calcitonin gene-related peptide and glutamate in terminals from primary sensory afferents in the superficial layers of the dorsal horn (Merighi et al 1991). Physiological studies show differential actions on the postsynaptic structure of the neurotransmitter and the neuropeptides, depending on the receptors present. For example, there are two distinct phases of depolarization in dorsal horn neurons; a rapid phase, due to glutamate, and a slower phase, due to substance P (Urban & Randic 1984). Substance P can also modulate the excitatory action of glutamate. This is a beautiful example of physiology and co-localization of peptides/neurotransmitters.

Gainer: Another thing to keep in mind is Hökfelt's idea that one co-released peptide may act as a decoy, i.e. a preferential substrate for proteolysis, leaving the other co-released peptide to function.

Theodosis: Another concept is that two peptides may be released at the same time, but one would have an effect further away than the postsynaptic target, a type of paracrine effect or so-called 'volume transmission' (see Fuxe & Agnati 1991).

Gainer: Can I just come back to the quest on of circuitry? How are we going to decide what the input is on a given cell? The answer I would hear in most circumstances is, when we inject into the PVN we get such a result. That does not give us the postsynaptic response of a cell. The injected peptide may be affecting an interneuron and thus acting indirectly. How are we going to find out the details of the circuitry in these neuroendocrine systems?

Lightman: Although cell culture has problems of its own, obviously we could ask Pamela Mellon to give us cell lines for all the different types of hypothalamic cells we want to study, it would certainly be a great help.

Mellon: We are certainly trying to develop additional cell lines.

References

Albers HE, Liou S-Y, Stopa EG, Zoeller RT 1991 Interaction of colocalized neuropeptides: functional significance in the circadian timing system. J Neurosci 11:846–851

Alper RH, Demarest KT, Moore KE 1980 Dehydration selectively increases dopamine synthesis in tuberohypophyseal dopaminergic neurons. Neuroendocrinology 31:112–115

Bicknell RJ, Leng G 1982 Endogenous opiates regulate oxytocin but not vasopressin secretion from the neurohypophysis. Nature (Lond) 298:161–164

Björklund A, Moore RY, Nobin A, Stenevi U 1973 The organization of tuberohypophyseal and reticuloinfundibular catecholamine neuron systems in the rat brain. Brain Res 51:171–191

Fuxe K, Agnati LF (eds) 1991 Volume transmission in the brain. Raven Press, New York

Goudreau JL, Lindley SE, Lookingland KJ, Moore KE 1990 Functional and neurochemical evidence that tuberohypophysial dopaminergic neurons in the rat project from the periventricular nucleus to the intermediate lobe of pituitary. Soc Neurosci Abstr 16:1178

Hatton GI 1988 Pituicytes, glia, and the control of terminal secretion. J Exp Biol 139:67–79

Hatton GI, Mason WT, Bicknell RJ 1990a Dynorphin A^{1-8} mobilizes intracellular calcium ($[Ca^{++}]_i$) in cells cultured from adult rat neural lobe via an action that does not involve the classical opioid receptors. J Physiol (Lond) 426:79P (abstr)

Hatton GI, Mason WT, Bicknell RJ 1990b Arginine vasopressin mobilizes intracellular calcium in pituicytes cultured from adult rat neural lobes. Neuroendocrinology 52:115 (abstr)

Kiss JZ, Mezey E 1986 Tyrosine hydroxylase in magnocellular neurosecretory neurons. Response to physiological manipulations. Neuroendocrinology 43:519–525

Liposits Z, Phelix C, Paull WK 1986 Adrenergic innervation of corticotropin releasing factor (CRF)-synthesizing neurons in the hypothalamic paraventricular nucleus of the rat. A combined light and electron microscopic immunocytochemical study. Histochemistry 84:201–205

Manzanares J, Lookingland KJ, Moore KE 1990 Atrial natriuretic peptide-induced suppression of basal and dehydration-induced vasopressin secretion is not mediated by hypothalamic tuberohypophysial or tuberoinfundibular dopaminergic neurons. Brain Res 527:103–108

Merighi A, Polak JM, Theodosis DT 1991 Ultrastructural visualization of glutamate and aspartate immunoreactivities in the rat dorsal horn, with special reference to the co-localization of glutamate, substance P and calcitonin gene-related peptide. Neuroscience 40:67–80

Summy-Long JY, Miller DS, Rosella-Dampman LM, Hartman RD, Emmert SE 1984 A functional role for opioid peptides in the differential secretion of vasopressin and oxytocin. Brain Res 309:362–366

Urban L, Randic M 1984 Slow excitatory transmission in rat dorsal horn. Possible mediation by peptides. Brain Res 290:336–342

Whitnall MH 1989 Stress selectively activates the vasopressin-containing subset of corticotropin-releasing hormone neurons. Neuroendocrinology 50:702–707

Whitnall MH, Gainer H 1988 Major pro-vasopressin-expressing and pro-vasopressin-deficient subpopulations of corticotropin-releasing hormone neurons in normal rats. Differential distributions within the paraventricular nucleus. Neuroendocrinology 47:176–180

Whitnall MH, Smyth D, Gainer H 1987a Vasopressin coexists in half of the corticotropin-releasing factor axons present in the external zone of the median eminence in normal rats. Neuroendocrinology 45:420–424

Whitnall MH, Key S, Gainer H 1987b Vasopressin-containing and vasopressin-deficient subpopulations of corticotropin-releasing factor axons are differentially affected by adrenalectomy. Endocrinology 120:2180–2182

Young EA, Akara S, Dallman MF 1990 Decreased sensitivity to glucocorticoid feedback in chronically stressed rats. Neuroendocrinology 51:536–542

Young III WS, Warden M, Mesey Y 1987 Tyrosine hydroxylase mRNA is increased by hyperosmotic stimuli in the paraventricular and supraoptic nuclei. Neuroendocrinology 46:439–444

Zoeller RT, Broyles B, Earley J, Anderson ER, Albers HE 1992 Cellular levels of mRNAs encoding vasoactive intestinal peptide (VIP) and gastrin-releasing peptide (GRP) exhibit different 24-hour rhythms in neurons of the suprachiasmatic nucleus. J Neuroendocrinol 4:119–124

The relative importance of hypothalamic neurons containing corticotropin-releasing factor or vasopressin in the regulation of adrenocorticotropic hormone secretion

G. B. Makara

Institute of Experimental Medicine, Hungarian Academy of Sciences, PO Box 67, H-1450 Budapest, Hungary

Abstract. Corticotropin-releasing factor (CRF-41) and arginine vasopressin (AVP) are the two major factors that regulate adrenocorticotropic hormone (ACTH) secretion. The two neurohormones are co-localized in the parvocellular neurons of the paraventricular nuclei (PVN) of the hypothalamus and are capable of potentiating each others' action on freshly excised anterior pituitary fragments or cells *in vitro*. Transection of all axons entering the medial basal hypothalamus from anterior and lateral directions blocks ACTH release induced by either adrenalectomy or ether-surgery stress. Adrenalectomy-induced ACTH release is almost completely suppressed by a long-term lesion of the PVN. Stress-induced ACTH release is blocked for only a few days after PVN lesion and the pituitary–adrenal response to ether-surgery stress returns to a large extent by a few weeks after PVN lesioning. This remarkable plasticity can be observed also in the homozygous Brattleboro rat, therefore it is not dependent on mediation by AVP. When parvocellular CRF-41- and AVP-containing cells are present, and the anterior lobe ACTH cells are desensitized to the stimulating effects of AVP, the ACTH response to haemorrhage and immobilization is markedly decreased. This indicates that AVP may partially mediate ACTH release under normal conditions. The hypothalamic control of the pituitary–adrenocortical system has a remarkable degree of redundancy which may compensate, at least under stressful conditions, for disruption of the function of CRF-41-containing cells of the paraventricular nucleus, the major source of CRF-41 in the stalk–median eminence.

1992 Functional anatomy of the neuroendocrine hypothalamus. Wiley, Chichester (Ciba Foundation Symposium 168) p 43–53

The factors that influence the secretion of adrenocorticotropic hormone (ACTH) and related pro-opiomelanocortin (POMC)-derived peptides from the anterior pituitary gland are of great importance for the homeostasis of the body. In the

1980s there was an explosive increase in the knowledge of these factors (Makara et al 1984).

More than a dozen factors have been proposed as potential regulators or modulators of ACTH secretion. All of them have well defined actions *in vitro* under favourable conditions. Only a few have been shown to have a definite influence as an endogenous factor in an animal.

There is no doubt that corticotropin-releasing factor (CRF-41) (Vale et al 1981) and arginine vasopressin (AVP) have ACTH-releasing actions *in vivo* and that their levels in the hypophysial portal blood bathing the POMC cells of the anterior pituitary reach concentrations which may alter ACTH secretion (Plotsky 1991). Similar information is not (yet) available for any other factors.

Early studies using peptide antagonists and passive or active immunoneutralization indicated that both of the two major factors may play a role in the regulation of ACTH secretion under conditions where ACTH release is stimulated, but it is not yet clear how much relative weight should be assigned to each. Fairly large doses of antibodies to CRF-41 or AVP failed to counteract the ACTH-releasing effect of stressful stimuli completely but the combination of CRF-41 and AVP antibodies appeared to result in a blockade of ACTH secretion (Rivier & Vale 1983, Linton et al 1985).

Recently, Van Oers et al (1988) have shown that a large dose of monoclonal antibody to CRF-41 completely blocks the response to ether-induced stress in rats. They suggested that on theoretical grounds only a large excess of the

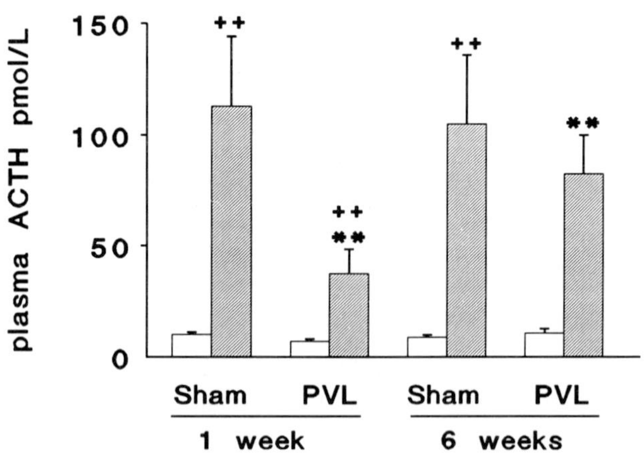

FIG. 1. The effect of a surgical lesion of the hypothalamic paraventricular nucleus (PVL) on the stress-induced release of adrenocorticotropic hormone (ACTH) one and six weeks after the lesion. Cross-hatched bars refer to rats subjected to venesection under ether anaesthesia; open bars represent sham-operated rats. Values are means ± SEM, $n = 8$–14. ++$P<0.01$ compared to resting value; **$P<0.01$ compared to sham-operated group.

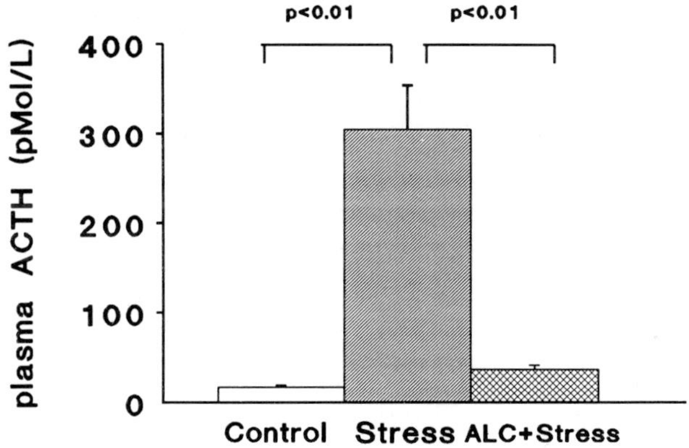

FIG. 2. The effect of an anterolateral cut (ALC) around the medial basal hypothalamus on the stress-induced release of ACTH six weeks later. Cross-hatched bars refer to rats subjected to laparotomy under ether anaesthesia. Values are means ± SEM, $n = 8–17$.

antibody may be expected to block completely the actions of an endogenous factor secreted into the portal circulation.

CRF-41 and AVP are coexpressed in the medial parvocellular neurons of the hypothalamic paraventricular nucleus (PVN), which are known to project to the external zone of the median eminence and the pituitary stalk (Antoni et al 1983, Kiss et al 1984, Sawchenko et al 1984a,b).

Practically all fibres carrying CRF-41 or AVP to the stalk–median eminence region or the neural lobe can be transected by placing an anterolateral cut with a Halász knife around the medial basal hypothalamus (see Makara et al 1984), and the content of CRF and AVP drops to unmeasurable levels in these regions (Dohanics et al 1987). Such a cut blocks the plasma ACTH rise normally elicited by a number of stressful stimuli or by adrenalectomy one week after the cut.

Surgical or electrolytic lesions of the PVN inhibited plasma ACTH rises normally seen during the circadian cycle, and those elicited by adrenalectomy or stressful stimulation (Makara 1979, Makara et al 1981, Ixart et al 1982). The inhibition is not complete and a portion of the response is clearly seen even after a PVN lesion which eliminates all CRF-41-containing cells from this nucleus (Fig. 1). The inhibition of stress-induced ACTH release is strongest 2–5 days after the operation then weakens gradually (Dohanics et al 1986). By 4–6 weeks after the lesion, the stress-induced ACTH release is again considerable.

It should be emphasized that glucocorticoid (dexamethasone) implants near the PVN can block the adrenalectomy-induced rise in plasma ACTH and the suppression of CRF-41 immunostaining in the parvocellular neurons of the PVN (Kovács et al 1986, Kovács & Makara 1988). The ACTH response to adrenalectomy remains blocked even six weeks after lesioning of the PVN (Makara et al 1986). Therefore the PVN is indispensable for ACTH release

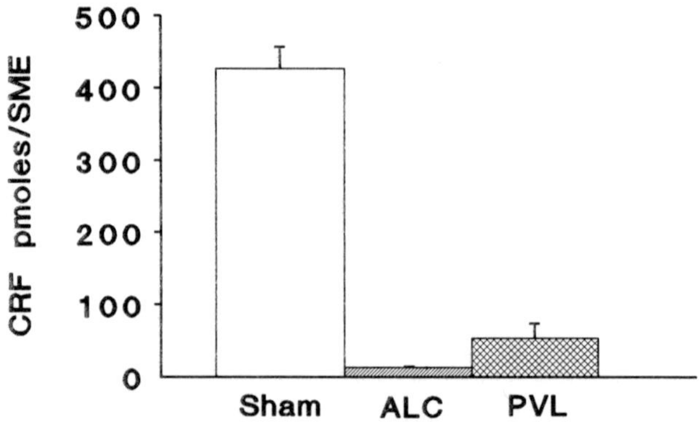

FIG. 3. The CRF-41 content in the stalk–median eminence (SME) region six weeks after a sham operation (Sham, $n=25$), an anterolateral cut around the medial basal hypothalamus (ALC, $n=8$) or a lesion in the paraventricular nucleus (PVL, $n=8$). Values are means ± SEM.

induced by decreased negative glucocorticoid feedback, but the deficit in PVN-lesioned animals can be partially compensated during exposure to stressful stimulation.

What is the mediator that can compensate for the deficit caused by lesioning the CRF-41-containing cells of the PVN? Is it a hypothalamic factor or does it come from outside the hypothalamus?

When searching for answers to these questions, we should remember that an anterolateral cut around the medial basal hypothalamus can completely block the ACTH response normally induced by the powerful stressful stimulus of laparotomy under ether anaesthesia for at least six weeks (Fig. 2). The comparison of the effects of PVN lesions and an anterolateral cut suggests that the factor(s) mediating stress-induced ACTH release after PVN lesion may come from (or through) the hypothalamus to reach the anterior pituitary gland.

We measured the immunoreactive CRF-41 content in the stalk–median eminence of rats and found that after either a PVN lesion or an anterolateral cut there is a marked depletion of the neurohormone, reaching at least 90% (Fig. 3).

This finding and the fact that AVP can strongly potentiate CRF-41-induced ACTH release suggested that AVP may be an important factor in stress-induced ACTH release in the PVN-lesioned rat. Because the homozygous Brattleboro rat is deficient in vasopressin both in the magnocellular and the parvocellular hypothalamic neurons, we studied the stress-induced ACTH release in PVN-lesioned heterozygous (controls capable of producing AVP) and homozygous rats (Makara et al 1989). An interesting finding was that within one week of suffering a lesion in the PVN, both the heterozygous and homozygous rats

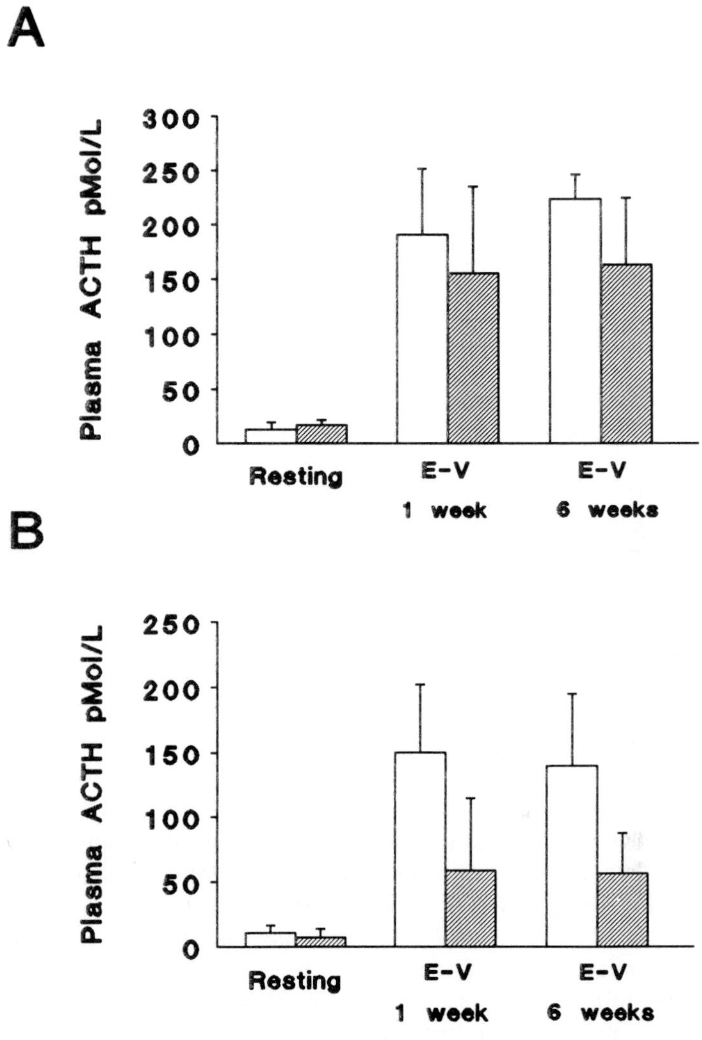

FIG. 4. The effect of stress caused by venesection under ether anaesthesia (E-V) on the plasma ACTH level. (A) Sham-operated rats ($n = 9-14$); (B) rats with paraventricular lesions ($n = 8-11$). Open bars refer to heterozygous Brattleboro rats; hatched bars to homozygous Brattleboro rats. Values are means ± SEM.

showed a marked ACTH release following ether-venesection stress (Fig. 4B). In the heterozygotes, this response was only slightly less than that seen in sham-operated rats (Fig. 4A). These findings, when compared to our previous results in Sprague-Dawley or Wistar rats, suggest that the recovery of stress-induced ACTH release following PVN lesion may differ among various strains of laboratory rat.

The homozygous rats tended to have smaller plasma ACTH responses to ether-venesection stress, and there was no further increase in the responsiveness to stress at six weeks after the lesion, regardless of whether the rats had the ability to synthesize AVP or not. This suggests that (1) AVP may not play an important role in the partial return of the hormonal response to stress following PVN lesions, and (2) a subnormal ACTH response to stress is maintained even in rats with no hypothalamic AVP or paraventricular CRF-41.

After an anterolateral cut, CRF disappears from the stalk–median eminence and, as expected, stress-induced ACTH release is abolished. However, after PVN lesioning, there is a marked discrepancy between the CRF content in the stalk–median eminence (about 10% of the control) and the stress-induced ACTH release, which reaches at least 50% and sometimes 80–90% of the control level. To localize the origin of the sporadic CRF-containing terminals in the stalk–median eminence, we placed a second cut in the retrochiasmatic area six weeks after lesioning the PVN. In sections immunostained for CRF, axons of two groups of CRF-containing cells were traced back to their perikarya in the supraoptic nucleus and the perifornical region (Palkovits et al 1991). With the anterior pituitary ACTH cells, which are possibly hyperresponsive to injected CRF (Bruhn et al 1984), the relatively small number of CRF-containing terminals may be enough to re-establish almost fully the ACTH-releasing ability of the anterior pituitary.

All this evidence is consistent with a major role for CRF-41 and a minor or potentiating role for AVP in the regulation of ACTH secretion. However, vasopressin seems to have some role in mediating, amongst other things, the effect of adrenergic stimulants on ACTH release (Al-Damluji et al 1990).

Neurolobectomy, which eliminates the majority of stored AVP, has also been shown to interfere with stress-induced ACTH release (see Fagin et al 1985, Raff et al 1985). Because the neural lobe has only a small contingent of CRF-containing fibres these effects may also point to a role of AVP.

Recently, we have developed a stereotaxic surgical method for denervating the posterior pituitary gland (neural and intermediate lobes) by compressing the pituitary stalk at a site near its junction with the neural lobe (see Makara 1989). Immediately after compression of the stalk, the operated rats show a marked diabetes insipidus with water intake over 100 ml/day, which subsides in a few days and stabilizes at about twice the normal water intake (50–70 ml/day). Plasma prolactin and corticosterone levels increase sharply within minutes of the compression but return to normal after a day or two.

The vasopressin normally stored in the neural lobe disappears by the 7th day after the operation, while the severed fibres in the stalk start sprouting and innervate the portal capillaries in the stalk just rostral to the site of compression, as shown by the classical neurosecretory stains or immunocytochemical staining for oxytocin, neurophysin or vasopressin. In stalk-median eminence extracted one, four or seven days after stalk compression a remarkable increase in AVP

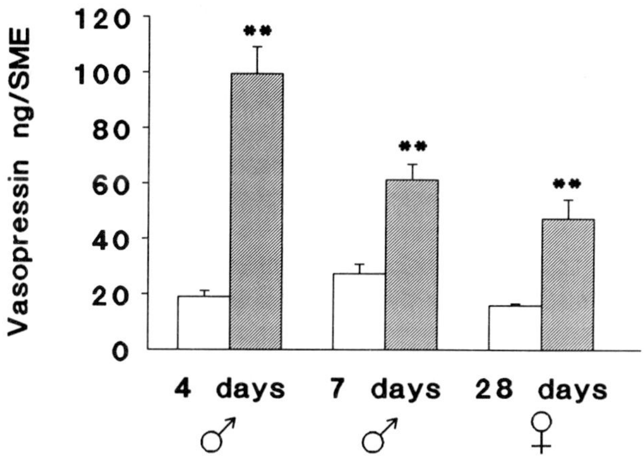

FIG. 5. Vasopressin content in extracts of the stalk–median eminence (SME) made at various times after compression of the pituitary stalk. Open bars, sham-operated rats; cross-hatched bars, rats with stalk compression. Values are means ± SEM; $n = 4$–6.

content is seen (Fig. 5). In preliminary studies in rats one week after posterior pituitary denervation (Plotsky P. M., Makara G. B., unpublished data), we have seen a marked (5–10-fold) rise in the portal plasma AVP level, which is consistent with the accumulation of vasopressin-containing fibres around the portal capillaries.

Such fibres may be the source of inappropriate (ectopic) secretion of AVP that may reach the anterior pituitary ACTH-secreting cells and may affect their responsiveness to AVP. To test this idea, we took anterior lobes from rats which had their pituitary stalk compressed a week before, incubated them in Krebs-Ringer solution and challenged them with graded doses of CRF-41 or AVP. The release of ACTH from the pituitary fragments challenged with CRF was similar in the controls and the rats with posterior pituitary denervation. In contrast, the fragments taken from rats with stalk compression had a markedly diminished responsiveness to AVP compared to the sham-operated controls. This finding is consistent with desensitization due to constant inappropriate secretion of AVP into the portal circulation following stalk compression.

When rats with posterior pituitary denervation were subjected to stressful stimulation through 25% haemorrhage or immobilization (Makara G. B., Jezova D., Jurcovicova J., unpublished), the resulting ACTH release was much less than in the sham-operated controls. The decrease of stress-induced ACTH release in parallel with the desensitization of ACTH release induced by AVP strongly suggests that AVP plays a role in the ACTH release that is normally induced by haemorrhage or immobilization. Thus, AVP may also participate as an

Intrahypothalamic neurohormonal interactions in the control of growth hormone secretion

J. Epelbaum

Unite de dynamique des Systèmes Neuroendocriniens, U159 INSERM, Centre Paul Broca, 2ter rue d'Alésia, 75014 Paris, France

Abstract. The secretion of growth hormone (GH, somatotropin) is regulated by two neurohormones: one inhibitory, somatotropin release-inhibiting hormone (SRIH) or somatostatin, and one stimulatory, GH-releasing hormone (GHRH). There are several lines of evidence for reciprocal interactions between SRIH and GHRH neuronal networks. Anatomically, GHRH terminals contact SRIH-containing neurons in the periventricular nucleus and SRIH-containing fibres innervate GHRH-containing neurons in the arcuate nucleus. Physiologically, SRIH and GHRH are secreted into the portal blood in complementary oscillating patterns. Results from immunizations with anti-SRIH antisera suggest that endogenous SRIH blocks GHRH release from the median eminence. Intracerebroventricular injections of SRIH stimulate secretion of GHRH indirectly, probably via autoinhibition of SRIH neurons in the anterior periventricular region. High resolution autoradiography allowed us to visualize high affinity, specific [^{125}I]SRIH receptors on 30% of GHRH mRNA-containing cells in the ventrolateral portion of the arcuate nucleus. The functional importance of these receptors was demonstrated by blocking endogenous SRIH action with cysteamine, which resulted in an increase in GHRH mRNA levels and desaturation of SRIH receptors in the ventrolateral part of the arcuate nucleus. Alterations in GH production caused by hypophysectomy or acute and chronic GH hypersecretion have opposite effects on the synthesis of SRIH and GHRH, giving further evidence for reciprocal interactions between these two neurohormonal systems.

1992 Functional anatomy of the neuroendocrine hypothalamus. Wiley, Chichester (Ciba Foundation Symposium 168) p 54–68

In humans, as in experimental animals, the ultradian pattern of growth hormone (GH, somatotropin) secretion from the pituitary is pulsatile (for review see Martin 1976). This is particularly striking in the male rat in which the pulses of GH secretion show a regular 3–4 hour periodicity throughout the day with peak values above 100 ng/ml and undetectable trough values (Tannenbaum & Martin 1976). This is the result of the complex interactions between two neurohormones: one inhibitory, somatotropin release-inhibiting hormone

(SRIH) or somatostatin (Brazeau et al 1973), and the other stimulatory, growth hormone-releasing hormone (GHRH) (Guillemin et al 1982, Rivier et al 1982). SRIH-containing neurons are prominent in the hypothalamus, including many interneurons in most nuclei and a neurohumoral system that arises in the anterior periventricular nucleus and projects to the median eminence (see Johansson et al 1984 for review). In the periventricular region, frequent synaptic and non-synaptic connections between SRIH-containing elements have been described (Epelbaum et al 1986). In contrast to the multiplicity of SRIH neuronal systems, GHRH-containing neurons are present only in the mediobasal hypothalamus, mainly in the arcuate nucleus from which tuberoinfundibular neurons reach the neighbouring median eminence (Sawchenko et al 1985). From the median eminence, GHRH and SRIH are released into the portal vasculature, whence they reach their respective receptors on the somatotrophs to elicit mutually antagonistic effects on these cells. In this paper, we shall review the evidence for intrahypothalamic interactions between SRIH- and GHRH-synthesizing neurons that lead to coordinated secretion of these neurohormones in the portal system.

Interactions between SRIH- and GHRH-containing neurons

Neuroanatomical evidence

SRIH- and GHRH-containing neurons are interrelated in the hypothalamus. GHRH terminals contact hypophysiotropic SRIH-containing neurons in the periventricular nucleus (Horvath et al 1989), while SRIH-containing fibres innervate GHRH-containing neurons in the arcuate nucleus (Daikoku et al 1988, Liposits et al 1988, Horvath et al 1989). The arcuate origin of the GHRH fibres that innervate the SRIH neurohumoral system is widely accepted, but the origin of the SRIH innervation of the GHRH neurons is controversial. Using retrograde tracing techniques, Willoughby et al (1989a) were not able to label periventricular SRIH-containing neurons after fluorogold injection around the arcuate nucleus. In contrast, transections of periventricular SRIH afferents to the arcuate nucleus increased the number of GHRH immunoreactive neurons therein (Daikoku & Tsuruo 1990). Similar transections also increased GHRH release from a median eminence–arcuate nucleus complex *in vitro* (Katakami et al 1988).

Physiological evidence

Direct measurements of neurohormonal levels in the hypothalamo-pituitary portal vessels in male rats anaesthetized with ketamine:xylazine (Plotsky & Vale 1985) also support the reciprocal control of GHRH and SRIH secretion: SRIH secretion into the portal circulation occurs in an oscillatory fashion 180° out of phase with GHRH secretion. Thus, GHRH secretion occurs only when SRIH

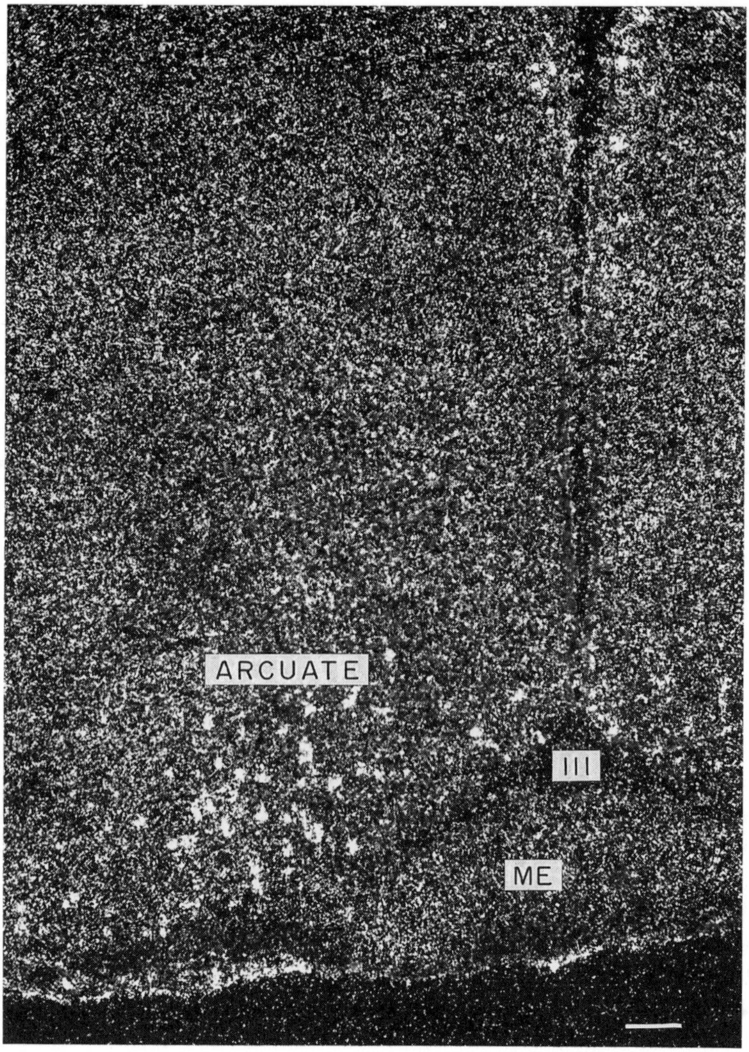

FIG. 1. Cells in the arcuate nucleus of the male rat labelled with [^{125}I]somatotropin release-inhibiting hormone (SRIH). [^{125}I]SRIH binding sites appear as small clusters above neuronal perikarya in the ventrolateral portion of the arcuate nucleus. Note another group of labelled cells dorsally along the wall of the third ventricle. ME, median eminence; III, third ventricle. Dark field illumination; scale bar, 80 μm.

levels are low, which suggests that SRIH directly or indirectly inhibits GHRH secretion. Conversely, intracerebroventricular (icv) injections of GHRH increase the release of SRIH into the portal blood (Mitsugi et al 1990). Additional evidence for reciprocal interactions arises from indirect studies where GH release was measured after passive immunization with anti-SRIH and/or anti-GHRH antibodies (Thomas et al 1985, Katakami et al 1986, Miki et al 1988). Sequential icv injections with highly specific antisera against SRIH or GHRH demonstrated that an acute withdrawal of SRIH triggers a discharge of GHRH from the median eminence and it was concluded that intrahypothalamic SRIH could block GHRH release. Finally, the paradoxical elevation of GH secretion elicited by icv injections of SRIH (Abe et al 1978, Lumpkin et al 1981) is mediated through GHRH release into the portal system (Murakami et al 1987) and abolished by neonatal treatment with monosodium glutamate, which selectively destroys arcuate neurons, including the tuberoinfundibular GHRH system. However, the SRIH stimulation of GHRH secretion is not direct (Willoughby et al 1989b) but likely to be mediated through autoinhibition of SRIH neurons in the anterior periventricular region (Epelbaum et al 1986, Willoughby et al 1989b). Such an hypothesis agrees with recent *in vitro* studies which demonstrated that SRIH perifusion inhibits depolarization-induced GHRH release from the median eminence–arcuate complex (Yamaushi et al 1991, M. T. Bluet-Pajot, personal communication).

Biochemical evidence

High resolution radioautography allowed us to visualize high affinity, specific [^{125}I]SRIH binding sites on a subpopulation of the arcuate perikarya (Epelbaum et al 1989) (Fig. 1). These binding sites are coupled to guanine nucleotide-binding proteins (Bertherat et al 1991a), suggesting that they are functional trans-membrane receptors. On adjacent 5 μm-thick sections, the proportion of [^{125}I]-SRIH-labelled cells co-identified as GHRH mRNA-containing cells by *in situ* hybridization (Bertherat et al 1991c) reached 30% in the ventrolateral portion of the nucleus (Fig. 2). Similar figures were obtained by comparing [^{125}I]SRIH-labelled cells and GHRH-containing cells as visualized by immunohistochemistry (McCarthy et al 1991). This suggested that SRIH could regulate directly GHRH synthesis and/or release from arcuate neurons. Although a selective pharmacological receptor antagonist is not available yet, it is possible to block some of the actions of SRIH by using cysteamine. This compound selectively interacts with the disulphide bond of the SRIH molecule, thereby reducing both its biological and immunological activities. Cysteamine treatment increased both [^{125}I]SRIH binding and the number of cells staining for GHRH mRNA in the ventrolateral arcuate nucleus (Bertherat et al 1991b) (Fig. 3). Cysteamine also enhanced the number of GHRH-containing cells as visualized by immunocytochemistry (Tannenbaum et al 1990). Finally, peripherally administered SRIH strongly diminished GHRH mRNA levels in the arcuate nucleus (Katakami et al 1991).

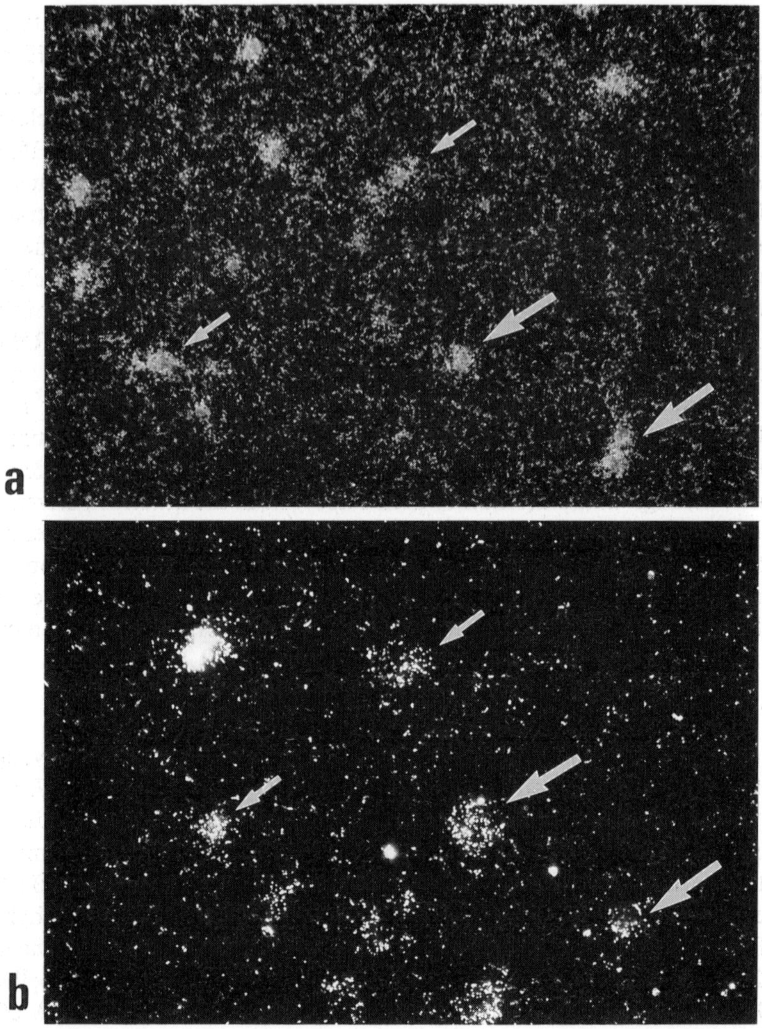

FIG. 2. Two consecutive 5 μm-thick sections in the ventrolateral part of the rat arcuate nucleus labelled with (a) [^{125}I]SRIH and (b) a [^{35}S]oligoprobe complementary to GHRH mRNA. Arrows indicate the 30% of the cells that are labelled with both [^{125}I]SRIH and the oligoprobe. Dark field illumination; magnification ×500.

Reciprocal regulation of SRIH- and GHRH-synthesizing neurons by growth hormone

Additional evidence for reciprocal interactions between SRIH and GHRH hypothalamic neurons is provided by the effects of GH deprivation through hypophysectomy and short- or long-term increases in the levels of GH.

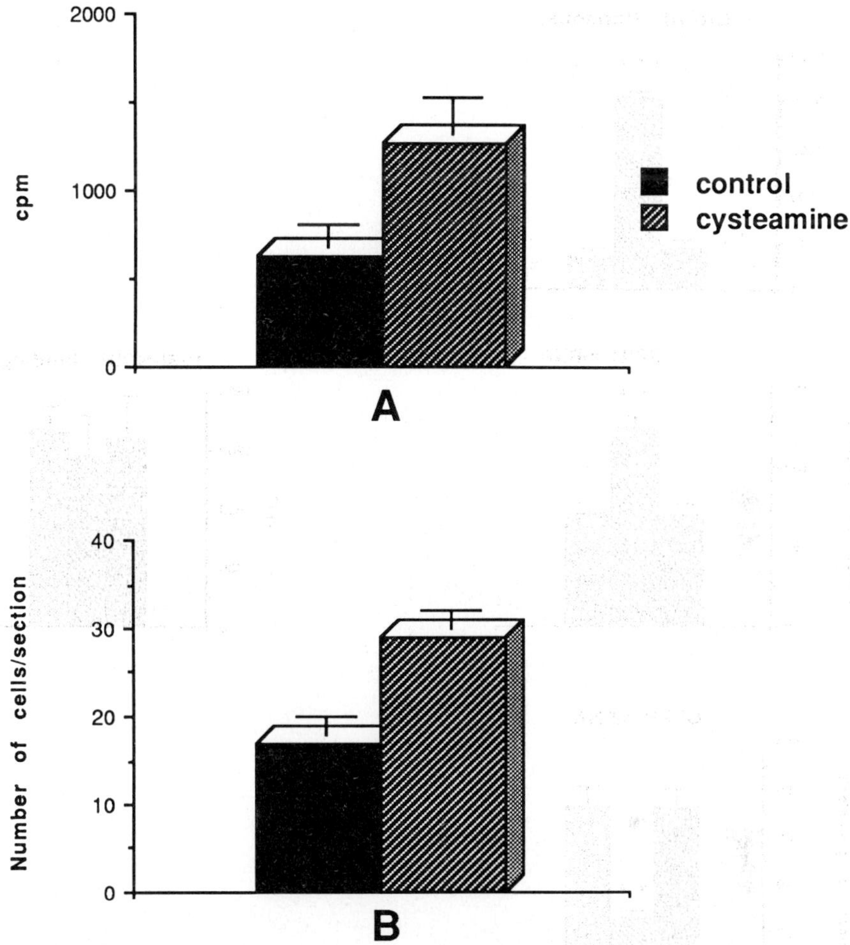

FIG. 3. Parallel effects of cysteamine treatment on (A) the amount of [^{125}]SRIH binding on arcuate nucleus perikarya and (B) the number of cells expressing GHRH mRNA. Adult male rats were injected with cysteamine subcutaneously (300 mg/kg body weight) six hours before they were killed. The SRIH-depleting effect of cysteamine was assessed by measuring the concentrations of the peptide in frontal cortex, which were decreased by 49%. Data are mean ± SEM of four animals in each group. Control vs cysteamine, $P<0.01$.

After hypophysectomy, the hypothalamic levels and release of GHRH and SRIH are decreased, while the amount of GHRH mRNA in the hypothalamus is increased (Chomczinski et al 1988) and SRIH mRNA in the periventricular nucleus is decreased (Rogers et al 1988). Short-term (up to 15 days) treatment with twice daily injections of GH partially reverses these modifications. In adult male rats, GH acutely stimulates hypothalamic SRIH synthesis and release while having the opposite effect on GHRH (see Maiter et al 1990 for review). A model

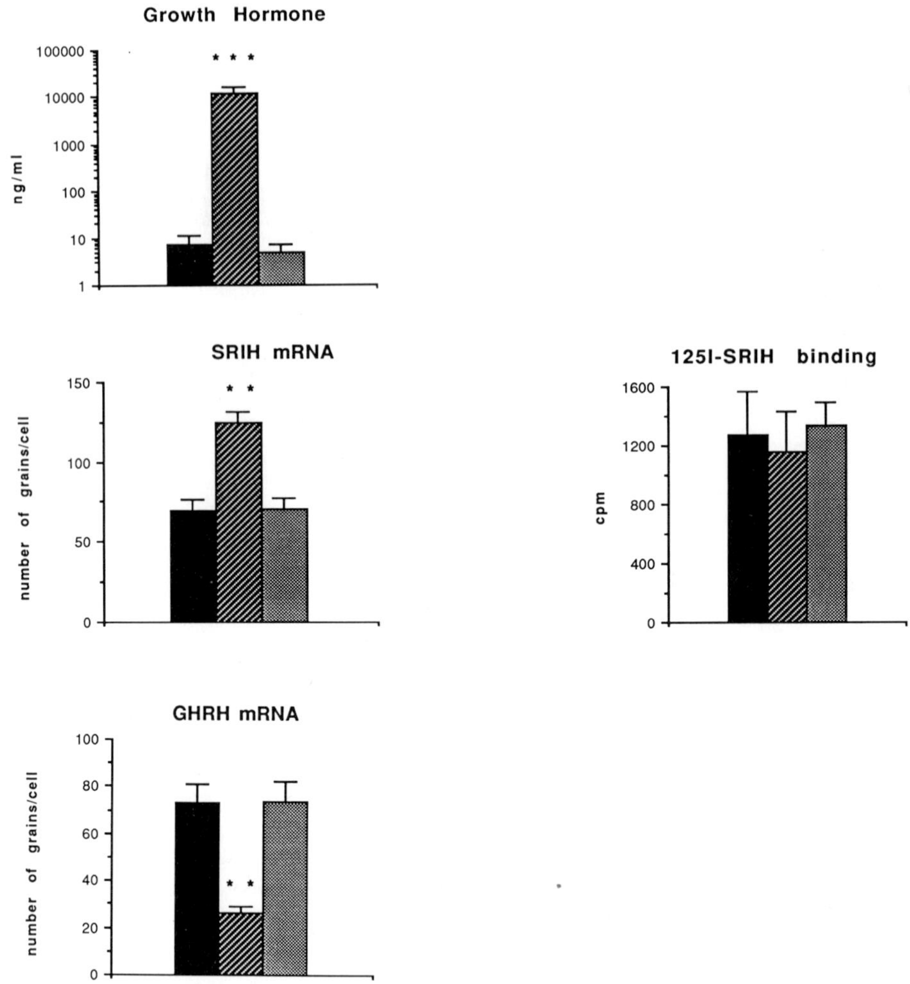

FIG. 4. Effects of chronic growth hormone (GH) hypersecretion on GH levels, the amount of SRIH mRNA in periventricular nucleus neurons and the amount of GHRH mRNA and [^{125}I]SRIH binding in arcuate nucleus neurons. Young adult female rats were sham implanted (control, ■) or implanted with $10–15 \times 10^6$ GC cells then left for 18 weeks (tumour bearing, ▨) or for twelve weeks, when the tumour was resected then left for the remaining six weeks (tumour resected, ▩). Data are mean ± SEM of five animals in each group. $^{**}P<0.01$; $^{***}P<0.001$.

of chronic GH hypersecretion can be created through subcutaneous implantation of GC cells in rats (Tashjian et al 1968). These clonal cells originated from an immortalized rat pituitary tumour and secrete only GH. Six weeks after implantation of the cells, SRIH mRNA levels are specifically raised in

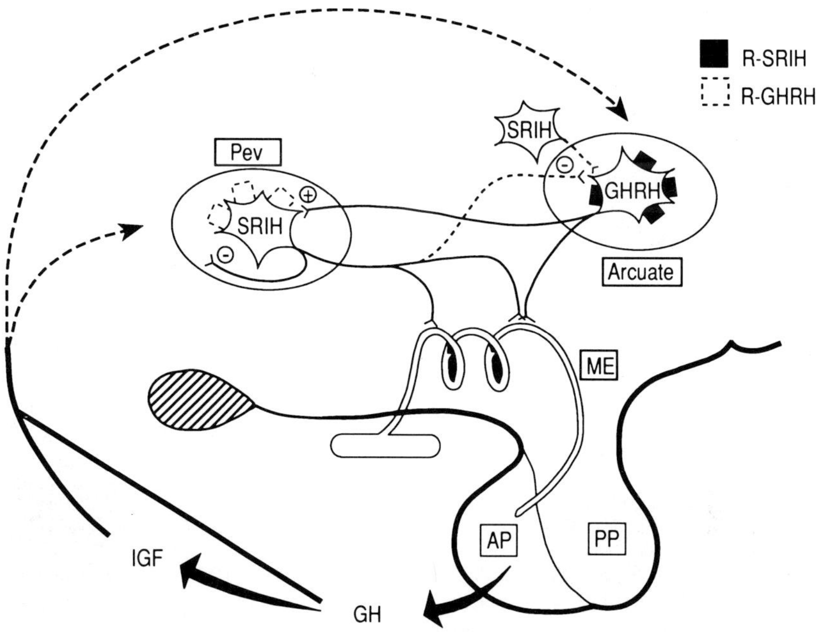

FIG. 5. Schematic representation of the interactions between SRIH and GHRH neurons. Tuberoinfundibular GHRH neurons are located in the arcuate nucleus. SRIH neurons from the periventricular nucleus (Pev) project to the median eminence (ME). Dotted lines indicate postulated connections and full lines verified ones. GHRH stimulates SRIH release through nerve fibres that connect with SRIH neuronal elements in the Pev but there is no evidence as yet for GHRH receptors in this nucleus. SRIH inhibits its own release in the Pev through SRIH/SRIH connections; it inhibits GHRH synthesis and release through specific receptors on arcuate perikarya innervated by SRIH terminals. The origin of the SRIH terminals remains unclear. Circulating GH originating from the anterior pituitary (AP) modulates both GHRH and SRIH neurons either directly or through insulin-like growth factors (IGF). PP, posterior pituitary.

periventricular neurons (Fig. 4). Reciprocally, the levels of GHRH mRNA in the arcuate nucleus are strongly diminished in the presence of high concentrations of GH. These effects persist for at least 18 weeks and are completely reversed after removal of the GH-secreting cells. The numbers of [^{125}]SRIH binding sites on arcuate perikarya are not modified by chronic GH hypersecretion. Therefore SRIH could be involved in the inhibition of the activity of GHRH neurons elicited by chronic GH hypersecretion. It is not yet known whether GH affects SRIH- and GHRH-synthesizing neurons directly or through the mediators of GH action, the insulin-like growth factors, but the recent demonstration of GH receptor mRNA expression in arcuate and periventricular nuclei (Burton et al 1991) strongly favours the first hypothesis.

Conclusion

The rhythmic pattern of GH secretion results from the remarkably precise interactions of two antagonistic neurohormonal systems, as summarized in Fig. 5. There is now convincing evidence that these systems are strongly interrelated and that extrahypothalamic neural influences resulting from metabolic and homeostatic perturbation are likely to affect both the stimulatory and the inhibitory components of the network.

Acknowledgements

The author thanks D. Gourdji for the gift of the GC cells. The collaboration of Jérome Bertherat and Marie Thérèse Bluet-Pajot was of major importance in the author's studies reported herein. These studies were supported by INSERM.

References

Abe H, Kato Y, Iwasaki Y, Chihara K, Imura H 1978 Central effect of somatostatin on the secretion of growth hormone in the anesthetized rat. Proc Soc Exp Biol Med 159:346–357

Bertherat J, Slama A, Kordon C, Videau C, Epelbaum J 1991a Characterization of pericellular ^{125}I-Tyr0-DTrp8 somatostatin binding sites in the rat arcuate nucleus by a newly developed method: quantitative high resolution light microscropic radioautography. Neuroscience 41:571–579

Bertherat J, Berod A, Normand E et al 1991b Somatostatin depletion by cysteamine increases somatostatin binding and growth hormone releasing factor messenger ribonucleic acid in the rat arcuate nucleus. J Neuroendocrinol 3:115–118

Bertherat J, Dournaud P, Berod A et al 1991c GRF neurons in the arcuate nucleus are a subpopulation of ^{125}I-SRIF labeled cells in the mediobasal hypothalamus. Abstracts of 73rd Endocrine Society annual meeting, June 19–22, Washington, DC (abstr 1280)

Brazeau P, Vale W, Burgus R et al 1973 Hypothalamic polypeptide that inhibits the secretion of immunoreactive pituitary growth hormone. Science (Wash DC) 179:77–79

Burton KA, Kabigting EB, Clifton DK, Steiner RA 1991 Distribution and regulation of GH receptor mRNA in the adult male rat brain. Meeting programme of 73rd Endocrine Society annual meeting, June 19–22, Washington DC (abstr 1555)

Chomczynski P, Downs TR, Frohman LA 1988 Feedback regulation of growth hormone (GH)-releasing hormone gene expression by GH in rat hypothalamus. Mol Endocrinol 2:236–241

Daikoku S, Tsuruo Y 1990 Neuron systems regulating growth hormone-releasing hormone-containing neurons in the rat hypothalamus. Neuroendocrinology 52:S_1:121 (abstr 358)

Daikoku S, Hisano S, Kawano H et al 1988 Ultrastructural evidence for neuronal regulation of growth hormone secretion. Neuroendocrinology 47:405–415

Epelbaum J, Tapia Arancibia L, Alonso G, Astier H, Kordon C 1986 The anterior periventricular hypothalamus is the site of somatostatin inhibition of its own release: an in vitro and immunocytochemical study. Neuroendocrinology 44:255–259

Epelbaum J, Moyse E, Tannenbaum GS, Kordon C, Beaudet A 1989 Combined autoradiographic and immunohistochemical evidence for an association of somatostatin binding sites with growth hormone-releasing factor-containing nerve cell bodies in the rat arcuate nucleus. J Neuroendocrinol 1:109–116

Guillemin R, Brazeau P, Böhlen P, Esch F, Ling N, Wehrenberg WB 1982 Growth hormone releasing factor from a human pancreatic tumor that caused acromegaly. Science (Wash DC) 218:585–587

Horvath S, Palkovits M, Görcz T, Arimura A 1989 Electron microscopic immunocytochemical evidence for the existence of bidirectional synaptic connections between growth hormone releasing hormone and somatostatin containing neurons in the hypothalamus of the rat. Brain Res 481:8–15

Johansson O, Hökfelt T, Elde RP 1984 Immunohistochemical distribution of somatostatin-like immunoreactivity in the central nervous system of the adult rat. Neurosciences 13:265–339

Katakami H, Arimura A, Frohman LA 1986 Growth hormone (GH)-releasing factor stimulates hypothalamic somatostatin release: an inhibitory feedback effect on GH secretion. Endocrinology 118:1872–1877

Katakami H, Downs TR, Frohman LA 1988 Inhibitory effect of hypothalamic medial preoptic area somatostatin on growth hormone releasing factor in the rat. Endocrinology 123:1103–1109

Katakami H, Hidaka H, Romanelli RJ, Mayo KE, Matsukura S 1991 Inhibition by somatostatin of growth hormone-releasing factor gene expression in the hypothalamic arcuate nucleus. Meeting programme of 73rd Endocrine Society annual meeting, June 19–22, Washington DC (abstr 1608)

Liposits Z, Merchenthaler I, Paull WK, Flerkó B 1988 Synaptic communication between somatostatinergic axons and growth hormone-releasing factor (GRF) synthesizing neurons in the arcuate nucleus of the rat. Histochemistry 89:247–252

Lumpkin MD, Negro-Vilar A, McCann SM 1981 Paradoxical elevation of growth hormone by intraventricular somatostatin: possible ultrashort loop feedback. Science (Wash DC) 211:1072–1074

Maiter DM, Gabriel SM, Koenig JI, Russell WE, Martin JB 1990 Sexual differentiation of growth hormone feedback effects on hypothalamic growth hormone-releasing hormone and somatostatin. Neuroendocrinology 51:174–180

Martin JB 1976 Brain regulation of growth hormone secretion. In: Martini L, Ganong WF (eds) Frontiers in Neuroendocrinology. Raven Press, New York vol 4:129–168

McCarthy GF, Beaudet A, Tannenbaum GS 1991 Co-localization of somatostatin receptors and growth hormone-releasing factor immunoreactivity in neurons of the arcuate nucleus. Meeting programme of 73rd Endocrine Society annual meeting, June 19–22, Washington DC (abstr 1287)

Miki N, Ono M, Shizume K 1988 Withdrawal of endogenous somatostatin induces secretion of growth hormone releasing factor in rats. J Endocrinol 117:245–252

Mitsugi N, Arita J, Kimura F 1990 Effects of intracerebroventricular administration of growth hormone-releasing factor and corticotropin-releasing factor on somatostatin secretion into rat hypophysial portal blood. Neuroendocrinology 51:93–96

Murakami Y, Kato Y, Kabayama Y, Inoue T, Koshiyama H, Imura H 1987 Involvement of hypothalamic growth hormone (GH)-releasing factor in GH secretion induced by intracerebroventricular injection of somatostatin in rats. Endocrinology 120:311–316

Plotsky PM, Vale W 1985 Patterns of growth hormone releasing factor and somatostatin secretion into the hypophysial portal circulation of the rat. Science (Wash DC) 230:461–463

Rivier J, Spiess J, Thorner M, Vale W 1982 Characterization of a growth hormone releasing factor from a human pancreatic islet tumor. Nature (Lond) 300:276–278

Rogers KV, Vician L, Steiner RA, Clifton DK 1988 The effects of hypophysectomy and growth hormone administration on pre-prosomatostatin messenger ribonucleic acid in the periventricular nucleus of the rat hypothalamus. Endocrinology 122:586–591

Sawchenko PE, Swanson LW, Rivier J, Vale W 1985 The distribution of growth hormone releasing factor (GRF) immunoreactivity in the central nervous system of the rat: an immunohistochemical study using antisera directed against rat hypothalamic GRF. J Comp Neurol 237:100–115

Tannenbaum GS, Martin JB 1976 Evidence for an endogenous ultradian rhythm governing growth hormone secretion in the rat. Endocrinology 98:562–570

Tannenbaum GS, McCarthy GF, Zeitler P, Beaudet A 1990 Cysteamine-induced enhancement of growth hormone-releasing factor (GRF) immunoreactivity in arcuate neurons: morphological evidence for putative somatostatin/GRF interactions within hypothalamus. Endocrinology 127:2551–2560

Tashjian AH, Yasumura Y, Levine L, Sato GH, Parker ML 1968 Establishment of clonal strains of rat pituitary tumor cells that secrete growth hormone. Endocrinology 82:342–352

Thomas CR, Groot K, Arimura A 1985 Antiserum to rat growth hormone (GH)-releasing factor suppresses but does not abolish antisomatostatin-induced GH release in the rat. Endocrinology 116:2174–2178

Willoughby JO, Brogan M, Kapoor R 1989a Hypothalamic interconnections of somatostatin and growth hormone releasing factor neurons. Neuroendocrinology 50:584–591

Willoughby JO, Brogan M, Kapoor R 1989b Intrahypothalamic actions of somatostatin and growth hormone releasing factor on growth hormone secretion. Neuroendocrinology 50:592–596

Yamaushi N, Shibasaki T, Ling N, Demura H 1991 In vitro release of growth hormone releasing factor (GRF) from the hypothalamus: somatostatin inhibits GRF release. Regul Pept 33:1171–1178

DISCUSSION

Robinson: The most striking thing about GH in the rat is the enormous sexual dimorphism in the pattern of its secretion and there is much evidence to suggest that this is due to a difference in the somatostatin (or SRIH) secretory patterns. Have you looked carefully at males versus females to see whether the distribution or the abundance of somatostatin-binding sites on GHRH neurons differs?

Epelbaum: We have only preliminary data for females. We see the same kind of pattern, namely pericellular binding of somatostatin in the arcuate nucleus. We don't know whether this varies in different conditions of oestrogen treatment or during the oestrous cycle.

Robinson: What is the evidence that there are no somatostatin receptors on terminals of somatostatin projections to the GHRH neurons? Somatostatin analogues suppress somatostatin secretion: can you exclude the possibility that the somatostatin-binding sites are on terminals apposed to the GHRH cells rather than on the cell bodies of the GHRH neurons themselves?

Epelbaum: We treated animals with monosodium glutamate and thereby destroyed the GHRH cell bodies; this abolished somatostatin binding.

Robinson: But what happens to the terminals on arcuate cells in this experiment?

Epelbaum: We don't know whether there is retrograde degeneration of the terminals. We should look at this by electron microscopy but there are technical problems at the moment—we don't have a fixative that retains the binding of [^{125}I] somatostatin.

Liposits: Have you seen somatostatin binding over GHRH neurons in the perifornical region, more laterally than the arcuate nucleus?

Epelbaum: No.

Liposits: That means there are two distinct GHRH populations—one is inhibited by somatostatin and the other is probably not. Could the somatostatinergic cells that are present in the dorsomedial part of the arcuate nucleus contribute to the inhibition of the GHRH neurons or does the inhibition come exclusively from the periventricular hypophysiotropic somatostatin cells?

Epelbaum: That's a very good question, I would like to know a way to answer it. There have been two kinds of studies, one by Willoughby et al (1989) trying to do retrograde tracing to find the origin of the somatostatin neuron. He couldn't find any in the periventricular nucleus. On the other hand, Daikoku & Tsuruo (1990) transected the periventricular afferents to the arcuate nucleus and saw an increase in the immunoreactivity for GHRH after these transections.

Lightman: Does this mechanism help to explain the pulsatile nature of growth hormone secretion?

Epelbaum: In the rat, I think so, but maybe not for other animals. In the rat everything is based on the paper by Plotsky & Vale (1985), which shows that there must be a lowering of the somatostatin concentration in the portal blood before the GHRH levels increase. Such results are in keeping with those of our cysteamine study: we show that a short-term treatment with cysteamine increases GHRH mRNA levels in the arcuate hypophysiotropic neurons. Somatostatin can also decrease the release of GHRH from the mediobasal hypothalamus *in vitro*. Finally, Dr Oliver has some results from work on sheep.

Oliver: In our laboratory, E. Magnan and others (1992) have observed that an intravenous injection of a somatostatin analogue (BIM 23014) into intact male sheep causes a rapid decrease in the concentration of GH in the plasma and also a rapid decrease in the amount of GHRH in hypophysial portal plasma. This is an acute experiment.

Epelbaum: That could be an additional way for somatostatin to block GH release directly at the level of the GHRH neuron.

Clarke: The studies of Plotsky & Vale (1985) need to be carefully interpreted. The inverse relationship we are talking about is not profound. In their paper, the data that are presented for growth hormone are from different animals than those which were used for the collection of the GHRH and somatostatin data. Their argument for an inverse relationship between GHRH and somatostatin was based on the proposition that, in the rat at least, the growth hormone release is very accurately locked into the time clock. They are the only data we have from the rat. Personally, I'm cautious about the interpretation.

Robinson: We showed (Clark & Robinson 1988) that if you give prolonged intermittent infusions of somatostatin to a female rat which secretes GH continuously, you can convert that pattern into a pulsatile pattern and stimulate growth. More importantly, the pulses induced by somatostatin withdrawal can be 90% obliterated by GHRH immunoneutralization. Rapid withdrawal of natural somatostatin, which has a short half-life, does generate pulsatile GH secretion that is mostly due to GHRH release—it is not just the passive removal of somatostatin inhibition of the pituitary gland. Whether that's the primary drive for the pulsatile secretion of GH is another question, because blocking GHRH with a neutralizing monoclonal antibody abolished GH pulses. Both GHRH and somatostatin must play a role. The question is whether somatostatin acts centrally to control its own release (or that of GHRH) or whether its inhibitory action at the pituitary gland is more important.

GH can feedback to inhibit its own release; you have suggested this may occur in the hypothalamus. How do you think that GH gets to this site of action? We have shown that GH can inhibit the pulsatile release pattern quite quickly, within 30 minutes of administration of a bolus of GH (Carlsson et al 1990). We are still puzzled about what the target of that peripherally administered GH is.

Epelbaum: The arcuate nucleus is definitely outside the blood–brain barrier.

Robinson: But does the GH act directly on the GHRH cells or on the somatostatin input to those GHRH cells?

Epelbaum: We don't have any data on that.

Palkovits: In the animals treated with monosodium glutamate, 80% of GHRH and somatostatin immunoreactivity has been eliminated from the arcuate nucleus. Are the expression patterns of any other peptides in the arcuate nucleus influenced in these animals? Especially, those that co-localize with GHRH and somatostatin, for example, galanin?

Epelbaum: It has been shown that galanin stimulates GH secretion by a central mechanism, not at the pituitary level. We have no direct evidence for galanin action on GHRH or somatostatin expression or secretion. We have been looking at galanin receptors in the hypothalamus. We find that if you desaturate the receptors of endogenous galanin, you can find many galanin receptors in the median eminence, so there could be an interaction at this level with galanin. But galanin is present in most cells in the hypothalamus, so whether the galanin we see is co-localized in the GHRH neuron, I don't know.

Liposits: We have recently found that there are galaninergic synapses on hypophysiotropic somatostatin cells (Liposits et al 1991), so probably galanin does drive somatostatin cells. It might also influence GHRH neurons, but we don't have ultrastructural evidence for that yet. Both systems are probably influenced by galanin and, furthermore, galanin is co-synthesized with GHRH, so it's a very complex interaction.

Barraclough: Circulating GH causes an increase in the amount of insulin-like growth factor (IGF) in the plasma. Clearly, this is how GH exerts its peripheral effects. Does IGF feed back into the brain to regulate the release of GHRH and/or somatostatin?

Epelbaum: It could, because there are IGF_1 receptors in the median eminence.

Robinson: I would like to comment on the central GH. Hollingshead et al (1989) produced a transgenic mouse that expressed human GH only in the brain. The mouse was dwarf rather than giant because secretion of the endogenous mouse GH was suppressed. A dramatic reduction in GHRH synthesis was found, so central GH can have a profound effect on synthesis of GHRH and presumably on its endogenous GH-releasing effects on the pituitary. That experiment was in a transgenic animal; I don't know whether natural GH does the same thing.

Charlton: What promoter was used to drive expression of the human GH?

Robinson: A weird one unrelated to the GH promoter (it contains elements of the long terminal repeat of mouse mammary tumour virus) but in this line the ectopic expression of human GH was restricted to the CNS (Hollingshead et al 1989).

Pfaff: Dr Palkovits, what do you think about GH in the brain? Do you believe it is produced there?

Palkovits: I have no idea. From the literature, the GH in brain is highly concentrated in a special part of the amygdala, mainly in the central amygdaloid nucleus. It would be interesting to know whether the cognate receptors are there.

Sawchenko: Jacques, you found these pericellular somatostatin-binding sites on roughly 30% of the GHRH cells. Is that an accurate estimate? Was there a clear topographic order in their distribution or do you think your values may represent an underestimate owing to the technical difficulties associated with trying to compare autoradiographic images on different tissue sections?

Epelbaum: I feel that it is an underestimate because statistically we have about a 55% chance to cut the same cells on two consecutive sections, which means we are underestimating by almost a factor of two the number of cells that express both [^{125}I]somatostatin-binding sites and GHRH mRNA.

Theodosis: Can you be sure that this pericellular distribution is on the somata? Could it not be on something else in the neuropil, like glial cells?

Epelbaum: The only way to be sure is to look by electron microscopy.

Sawchenko: Is there evidence to suggest, in any aspect of this system, that the somatostatin cells which talk to the GHRH cells, or vice versa, are hypophysiotropic neurons? Are the intracerebral projections known to arise as collaterals of neuroendocrine neurons or is that just an assumption?

Epelbaum: We don't know from which somatostatin cells these terminals are coming. They could be short interneurons and terminate in the arcuate nucleus because we know there are some there.

Sawchenko: Micky, do we know if the GHRH cells in the arcuate nucleus are responsible for the projection to the somatostatin cells in the anterior periventricular nucleus?

Palkovits: All I know is that the GHRH terminals or fibres disappear from the anterior periventricular area almost completely after a knife cut between the arcuate and the periventricular nuclei.

Sawchenko: Is it known whether they persist in animals treated neonatally with monosodium glutamate, which would be expected to destroy the GHRH neurons in the arcuate nucleus?

Palkovits: Most of the GHRH terminals disappeared from the periventricular nucleus three months after monosodium glutamate treatment of newborn rats.

References

Carlsson LMS, Clark RG, Robinson ICAF 1990 Sex difference in growth hormone (GH) feedback in the rat. J Endocrinol 126:27–35

Clark RG, Robinson ICAF 1988 Paradoxical growth promoting effects induced by patterned infusions of somatostatin in female rats. Endocrinology 122:2675–2682

Daikoku S, Tsuruo Y 1990 Neuron systems regulating growth hormone-releasing hormone-containing neurons in the rat hypothalamus. Neuroendocrinology $52S_1$:121 (abstr 358)

Hollingshead PG, Martin L, Pitts SL, Stewart TA 1989 A dominant phenocopy of hypopituitarism in transgenic mice resulting from central nervous system synthesis of human growth hormone. Endocrinology 125:1556–1564

Liposits Zs, Valenca MM, Lopez FJ, Merchenthaler I, Negro-Vilar A 1991 Galanin neurons innervate anterior periventricular somatostatin cells and modulate somatostatin secretion from arcuate nucleus-median eminence fragments. Neurosci Abstr 17:434

Magnan E, Cataldi M, Guillaume V et al 1992 Acute changes in growth hormone-releasing hormone secretion after injection of BIM 23014, a long acting somatostatin analog, in rams. Life Sci

Plotsky PM, Vale W 1985 Patterns of growth hormone releasing factor and somatostatin secretion into the hypophysial portal circulation of the rat. Science (Wash DC) 230:461–463

Willoughby JO, Brogan M, Kapoor R 1989 Hypothalamic interconnections of somatostatin and growth hormone releasing factor neurons. Neuroendocrinology 50:584–591

Regulation of prolactin transformation in the rat pituitary

Clark E. Grosvenor and *Flavio Mena

*Department of Molecular and Cell Biology, Eberly College of Science, Paul M. Althouse Laboratory, The Pennsylvania State University, University Park, Pennsylvania 16802, USA and *Institute of Biomedical Investigations, National University of Mexico, 04510 Mexico City D.F., Mexico*

Abstract. We have investigated the mechanisms involved in the inhibition by dopamine of the transformation of prolactin within the anterior pituitary of the lactating rat. The degree of inhibition depends on the intracellular age of prolactin, being greater in newly synthesized (<1 hour) and in older (>12 hours since biosynthesis) hormone and lesser in prolactin synthesized 4–8 hours earlier. Transformation occurs in prolactin granules and involves an increase in oligomeric forms of prolactin at the expense of the monomeric form. A reversible disulphide-linked mechanism may be involved in dopamine inhibition of prolactin transformation; it is dependent upon the intracellular and/or intragranular pH via a Na^+/H^+ exchange mechanism. Transient suppression of dopamine inhibition may lower the intracellular/intragranular pH and subsequently cause transformation of the hormone. Developmentally, dopamine secretion by tuberoinfundibular neurons is seriously impaired and the response of pituitary lactotrophs to dopamine is reduced in adult rats deprived of milk prolactin during Days 2–5 post partum. These results suggest milk prolactin ingested during a critical post partum period may exert an organizational effect upon dopamine secretion and its function on the pituitary lactotroph during adulthood.

1992 Functional anatomy of the neuroendocrine hypothalamus. Wiley, Chichester (Ciba Foundation Symposium 168) p 69–86

In the first part of this paper we examine possible mechanisms in the regulation by dopamine of the transformation of prolactin within the lactotrophic cell in the anterior pituitary of the lactating rat. In the second, we present evidence that the ingestion of milk prolactin by the neonatal rat during a critical post partum period influences the secretion of dopamine by hypothalamic neurons as well as the inhibitory effectiveness of dopamine upon prolactin secretion in the adult rat.

Transformation and release of prolactin

The concept is presented that prolactin is stored in a pre-releasable form within the lactating rat's anterior pituitary, which, upon secretory activation, is

TABLE 1 Comparison of the characteristics of secretion of prereleasable and transformed prolactin in the lactating rat

Characteristics of suckling-induced transformation of prereleasable to releasable prolactin in lactotroph	*Characteristics of suckling-induced release of transformed prolactin into the circulation*
Occurs rapidly (within the first five minutes of suckling)	Occurs in a steady, minute-by-minute manner
15–60 μg of prolactin are transformed	400–800 ng/min (RP-1 Standard) are released
Occurs periodically (every 1.5–2 hours)	Prolactin is released only when suckling is applied
Has a high threshold of activation	Has a lower threshold of activation than prolactin transformation
Is inhibited by dopamine	Is stimulated by thyrotropin-releasing hormone and possibly other prolactin-releasing factors
Prereleasable prolactin is soluble at pH 7.0–7.2	Releasable prolactin is insoluble at pH 7.0–7.2
Prolactin synthesized either <1 hour or >12 hours beforehand is *not* transformed; that synthesized >12 hours earlier is destroyed by lysosomes	Prolactin transformed >12 hours earlier is not released but is lysed

See Grosvenor & Mena (1982).

transformed into a releasable form before exocytotic release (Grosvenor et al 1979, Grosvenor & Mena 1982, Lawson et al 1987). This model implies that processing and transformation of the hormone occur sequentially after biosynthesis (Mena et al 1984, 1989a) and that hypothalamic factors may differentially affect transformation and release, depending on the age of the hormone (Mena et al 1989b).

The transformation process is inhibited by the prolactin-inhibiting hormone, dopamine (Grosvenor et al 1980), and is stimulated by the dopamine antagonists, haloperidol and domperidone (Grosvenor et al 1984). Evidence (see Grosvenor & Mena 1982 for review) suggests that suckling causes prolactin transformation by transiently suppressing the release of dopamine from the hypothalamus. A striking concomitance of the brief removal of the dopaminergic tone is a marked increase in the sensitivity of the release of transformed prolactin to putative prolactin-releasing factors, such as thyrotropin-releasing hormone. Releasable prolactin within the anterior pituitary gland has a lower threshold for activation and is less soluble than prolactin that has not undergone transformation. It was this difference in the solubility of transformed prolactin (Mena et al 1982) that, in the early studies, made it seem as though it had disappeared from the pituitary gland with the onset of suckling. (See Table 1 for a comparison of the characteristics of the two forms of prolactin.)

In the lactating rat, the storage time of prolactin within the pituitary gland, i.e. the degree of intracellular maturation of the hormone since biosynthesis, determines its availability for transformation into releasable forms and release (Mena et al 1984). Mature prolactin (that synthesized 4–8 hours previously) is preferentially secreted over younger (synthesized one hour previously) or older (synthesized 16–24 hours previously) hormone: the older hormone is degraded by lysosomes within the lactotroph (Smith & Farquhar 1966, DeMarco et al 1982). These results, together with those from previous studies (Mena et al 1984, 1989a), suggest that processing of transformation, release and degradation of prolactin occur sequentially, rather than preferentially, after biosynthesis and in parallel with maturation of prolactin in granule form (Farquhar 1977, Walker & Farquhar 1980).

Functional heterogeneity within the lactotroph population has been suggested to account for the preferential release of new prolactin (Farquhar 1977). However, it is likely that the different methods employed to define newly synthesized and mature prolactin, plus the fact that releasability and secretory patterns of prolactins of different ages change as a function of time during incubation (Mena et al 1984, 1989a,b and present study), contribute to the various interpretations concerning secretion of new, mature and old hormone. On the other hand, heterogeneity of lactotrophs may reflect the ability of cell subpopulations to change according to physiological requirements for prolactin release and such heterogeneity may be low in the anterior pituitary of the lactating rat (see Nagy & Frawley 1990, Mena et al 1989a for further discussion).

The respective inhibitory and stimulatory influences of dopamine and thyrotropin-releasing hormone on prolactin secretion are clearly influenced by the intracellular age of prolactin (Fig. 1). Lesser inhibition by dopamine and greater stimulation by thyrotropin-releasing hormone of prolactin secretion occurred when prolactin was labelled *in vivo* with [^{3}H]leucine four or eight hours before removal and incubation than when labelling was at 0.5, 1 or 16–24 hours before incubation (Mena et al 1989b). These results confirm previous results which showed that thyrotropin-releasing hormone stimulates the secretion of primarily mature prolactin (Mena et al 1989a), whereas dopamine more effectively inhibits the release of newly synthesized prolactin (Macleod & Lehmeyer 1974, Cheung et al 1981). An interesting possibility is that in the lactating rat, hypothalamic regulation of prolactin transformation and release is synchronized with intracellular regulatory mechanisms in the lactotroph and thus has different effects according to the time after biosynthesis of the hormone. For example, mature prolactin may be transformed and released preferentially because intracellular mechanisms (Mena et al 1986, Martinez-Escalera et al 1986) facilitate its release and/or because it is less susceptible to inhibition by dopamine and more sensitive to secretagogues such as thyrotropin-releasing hormone than is newly synthesized or older, stored prolactin.

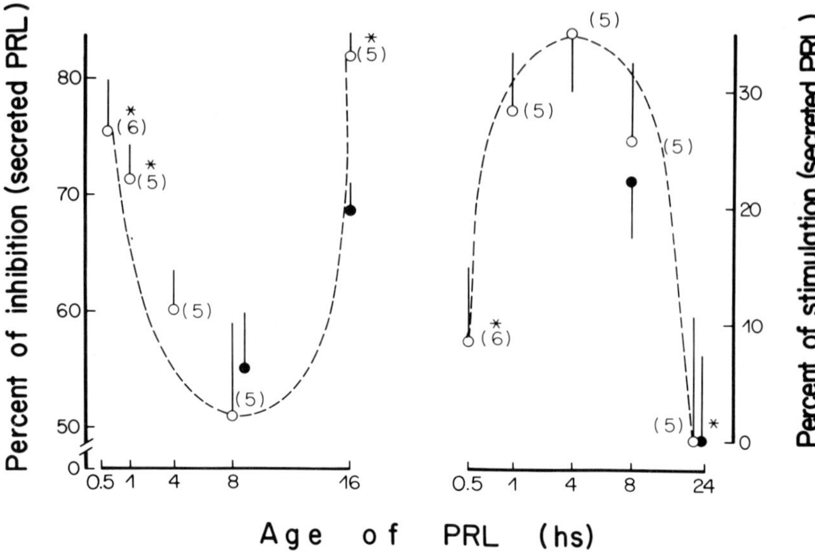

FIG. 1. Effects of dopamine (left panel) and thyrotropin-releasing hormone (right panel) on prolactin release in lactating rats. Prolactin was labelled *in vivo* with [^{3}H]leucine for 0.5, 1, 4, 8, 16 or 24 hours, then the anterior pituitary gland was removed, cut into four pieces and each piece was cultured for four hours. Dopamine (17 μM) or thyrotropin-releasing hormone (0.35 μM) was added at the onset of incubation; ascorbic acid (0.1%) was added with dopamine to prevent its oxidation. The total amount of prolactin released (●) was compared with the amount of newly synthesized [^{3}H]prolactin released (○) during the four hour incubation. Data represent the percent change in secretion compared to control values (without dopamine or thyrotropin-releasing hormone). Data are expressed as the mean + SEM; the number of rats per group is shown in parentheses. *$P<0.025$ versus unstarred values.

Recent studies suggest that intragranular pH and a disulphide-linked mechanism (Mena et al 1986) may be involved in prolactin transformation and that hypothalamic inhibition of prolactin transformation may be exerted by interaction with these systems (Martinez-Escalera et al 1986). During *in vitro* transformation, oligomeric forms of prolactin increase in number at the expense of the monomeric form of the hormone, which may account for the loss of detectability by polyacrylamide gel electrophoresis (PAGE) (7.5%) during transformation. Oligomerization occurs after thiol/disulphide interactions; dopamine may inhibit prolactin transformation by depressing such interactions. Moreover, there was a higher proportion of monomeric prolactin in the secreted hormone than in the anterior pituitary glands, which indicates that prolactin transformation is a reversible phenomenon (Mena et al 1992).

We determined that changes in prolactin detectability during transformation occurred in the granular fraction of the hormone (Mena et al 1992). The involvement of disulphide-linked mechanisms was shown to be pH dependent. In acid (pH 6.5) or alkaline (pH 8.2) conditions, changes in prolactin detectability were similar in all respects to those observed during physiological transformation or repletion of prolactin in the anterior pituitary, respectively (Mena et al 1986). To determine whether intracellular and/or intragranular pH played an important role in prolactin secretion, we analysed the *in vitro* effects of agents known to influence intracellular pH by different mechanisms (for example dinitrophenol, amiloride, NH_4Cl and nigericin) on basal and dopamine-regulated prolactin transformation and release by incubated fragments of anterior pituitary or on prolactin detectability by PAGE in anterior pituitary homogenates.

Fragments of anterior pituitary from lactating rats (eight hours after nursing) were incubated (in Earle's medium pH 7.2 for 30–240 minutes following standard procedures) either alone or in the presence of the pH-regulating agents. Prolactin in the tissue and medium was analysed by PAGE (7.5%) and densitometry. In some instances, tissue and medium samples were also run on 12.5% polyacrylamide gels, with or without SDS (sodium dodecyl sulphate), followed by Western blotting.

Dinitrophenol (200 µM), nigericin (5 mM) and NH_4Cl (35 mM) selectively counteracted inhibition of prolactin transformation by dopamine but showed little or no effect upon dopamine inhibition of prolactin release (Fig. 2). On the other hand, amiloride (1–4 mM) increased prolactin transformation and inhibited prolactin release in a dose-dependent fashion (Fig. 2). When co-incubated with dopamine, the effect of amiloride on prolactin transformation was no longer observed, but the inhibition of prolactin release was stronger than with either agent alone (Fig. 2). This suggests that dopamine and amiloride regulate prolactin transformation via the same mechanism but use different mechanisms to inhibit its release.

When anterior pituitary homogenates were incubated with amiloride at pH 7.5, prolactin detectability by PAGE decreased strongly in a dose- and time-dependent fashion (Fig. 3). Amiloride specifically inhibits Na^+/H^+ exchange (Benos 1982) and its inhibitory effects upon prolactin release are counteracted in part by veratrine and ouabain (Mena F, Aguayo D, Morales MT, unpublished), compounds that increase intracellular Na^+ and stimulate prolactin release (Scammell & Dannies 1983). This suggests that a Na^+/H^+ exchange mechanism exists within lactotrophs and/or prolactin granules, which regulates internal pH. This is important for the transformation and release of prolactin by the anterior pituitary in lactating rats. The effects of amiloride upon anterior pituitary homogenates are prevented by N-ethylmaleimide (Fig. 3), which supports the idea that polymerization via disulphide links is involved in amiloride-induced loss of detectable prolactin.

Thus, intracellular acidification may be an event normally associated with prolactin transformation and release in lactotrophs. Because transient interruption of dopamine tone upon lactotrophs is considered at present to

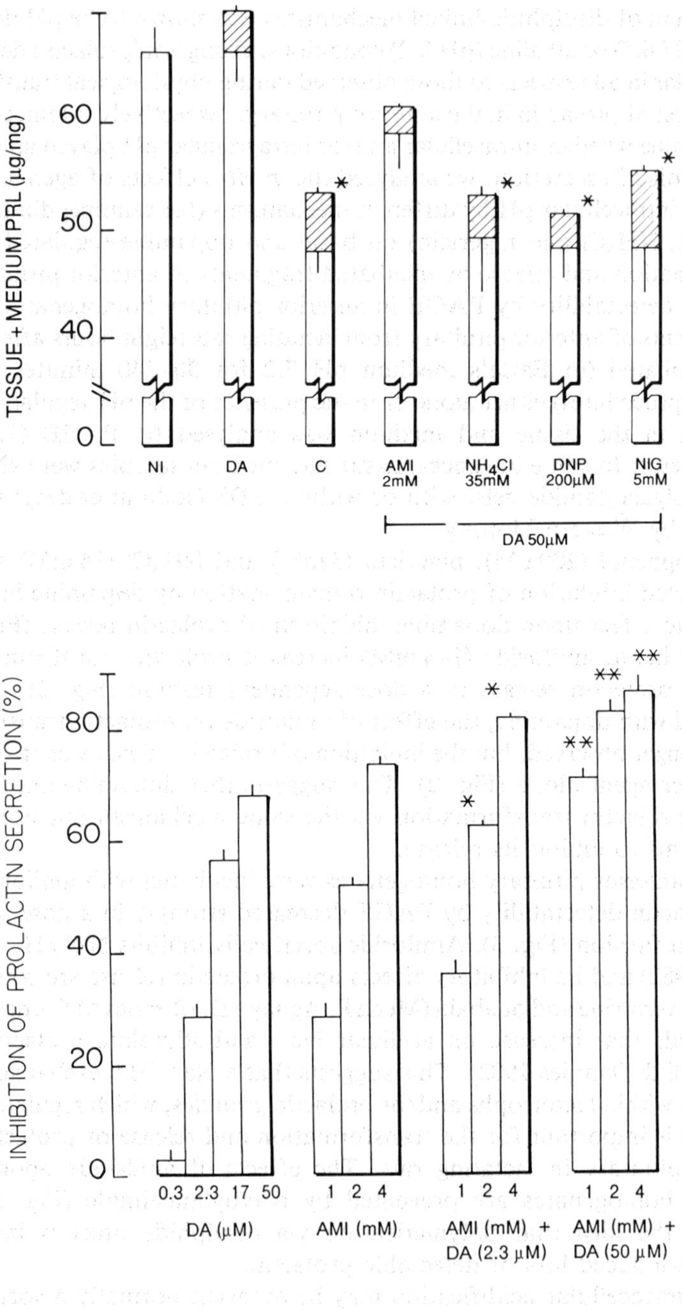
TISSUE + MEDIUM PRL (μg/mg)
60
50
40
0
NI
DA
C
AMI 2mM
NH4Cl 35mM
DNP 200μM
NIG 5mM
DA 50μM
INHIBITION OF PROLACTIN SECRETION (%)
80
60
40
20
0
0.3 2.3 17 50
DA (μM)
1 2 4
AMI (mM)
1 2 4
AMI (mM) + DA (2.3 μM)
1 2 4
AMI (mM) + DA (50 μM)

FIG. 2

be the cause of prolactin transformation and the subsequent increase in releasable prolactin (Grosvenor et al 1984), intracellular and/or intragranular acidification may result from this interruption of dopamine tone. Moreover, it is likely that the inhibition of prolactin transformation by dopamine is mediated, at least in part, by an amiloride-sensitive Na^+/H^+ antiport system. Additional studies are required on the interaction of these mechanisms with other, well established mechanisms involved in the physiology of prolactin secretion.

Neuroendocrine effects of a neonatal deficiency in milk-derived prolactin

Prolactin is present at high concentrations in the milk of lactating rats and other species (see Koldovsky 1980, Grosvenor & Whitworth 1983). Radiolabelled prolactin has been traced from the maternal circulation into the milk within the mammary gland, then into the stomach of the suckling offspring and finally into their systemic circulation (Whitworth & Grosvenor 1978). Through the episodic feeding and absorption from the gastrointestinal tract, neonatal rats may be exposed to a total of 300 ng of prolactin reaching the circulation per 24 hours (Whitworth & Grosvenor 1978). It is interesting that this occurs during a period when the neonatal pituitary gland secretes relatively little prolactin (Döhler et al 1977, Döhler & Wuttke 1975, Ojeda & McCann 1974).

Injection of mother rats with bromocriptine (250 µg/kg body weight, s.c.) daily from Days 2–5 post partum reduced secretion of prolactin from the anterior pituitary and thus the amount of prolactin in milk (Grosvenor & Whitworth 1983). The pups of these rats had significantly decreased steady-state concentrations and turnover rates of dopamine in the median eminence at 33–35 days of age (Shyr et al 1986, Shah et al 1988) (Table 2). Consistent with these results, there was a significant increase in the serum prolactin concentration.

These effects were completely abolished by prolactin replacement in rats deprived of prolactin neonatally (Table 2). Neonatal prolactin deficiency did not reduce tuberohypophysial dopamine secretion, nor did it alter noradrenergic

---

FIG. 2. Top: The influence of dopamine in the absence or presence of different acidificants on prolactin transformation in the anterior pituitary glands from lactating rats after these were removed and placed in incubation medium. The empty bars represent prolactin in the tissue; hatched bars represent secreted prolactin; transformed prolactin is represented by the difference. The values are the amount of prolactin present after incubation of fragments of anterior pituitaries with the reagents for 30 minutes. Reagents were added at the onset of incubation. NI, non-incubated control, DA, dopamine; C, incubated control; AMI, amiloride; DNP, dinitrophenol; nig, nigericin. Bottom: Effects of dopamine and amiloride, alone or in combination, on 4 h *in vitro* release of prolactin. Data represent the percent change in secretion compared to control values.

*$P<0.05$ vs. unstarred values. **$P<0.01$ vs unstarred values. Comparisons are among the four pieces of individual pituitaries. $n=5$.

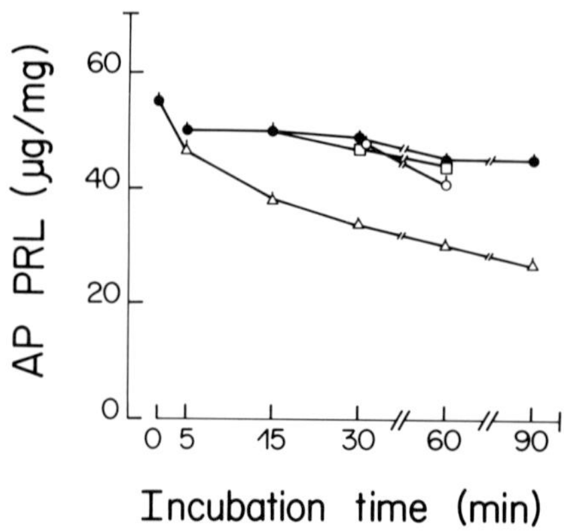

FIG. 3. The effect of amiloride and N-ethylmaleimide on the concentration of prolactin in the anterior pituitary. Neonatal rats were prevented from suckling for eight hours. The anterior pituitaries were removed and incubated with 2 mM amiloride, 1 mM N-ethylmaleimide or both at pH 7.2, 37 °C. The pituitaries were homogenized at the indicated times in 0.05 M Tris-HCl buffer (pH 7.2) and prolactin was separated by PAGE and determined by densitometry. $n = 4$.

activity in either tuberoinfundibular or tuberohypophysial neurons. That there may be a critical period for the effects of neonatal prolactin deficiency is suggested by the finding that neither the reduction in median eminence dopamine turnover nor the elevation of serum prolactin occurred if the reduction in milk prolactin caused by injection of the mother with bromocriptine was delayed until Days 9–12 of lactation (Shyr et al 1986). The hyperprolactinaemia resulting from neonatal prolactin deficiency persists beyond the onset of puberty until at least 90–100 days of age (Shah et al 1988). The onset of puberty is not altered by this treatment, but the offspring display oestrous acyclicity for 20–30 days during the immediate postpubertal period and exhibit prepubertally raised levels of serum oestradiol. In a subsequent study, female rats injected subcutaneously with 1 μg oestradiol benzoate on post partum Days 1–3 also displayed hyperprolactinaemia as young adults compared to matched controls injected with oil vehicle (Crowley et al 1990). In these rats, as well as in neonatally prolactin-deprived female rats, ovariectomy abolished the hyperprolactinaemia. The rats injected with oestradiol benzoate also exhibited reduced steady state-levels and turnover of dopamine in the median eminence when 35 days of age (Table 2) and these were still significantly reduced on post partum Day 60, though they had returned to normal by Day 60 in neonatally prolactin-deprived rats (Table 2).

These results suggest that one function of milk-derived prolactin in the neonate is to influence indirectly the maturation of the inhibitory tuberoinfundibular dopaminergic controls of prolactin secretion via ovarian secretion, and that

TABLE 2 Effects of different maternal and neonatal treatments on serum prolactin concentrations and rates of dopamine turnover in the median eminence in 35 and 60 day old rats

	Day 35		*Day 60*	
	Dopamine turnover pg/μg/h	*Serum prolactin ng/ml*	*Dopamine turnover pg/μg/h*	*Serum prolactin ng/ml*
Treatment of mother				
Days 2–5 post partum				
Saline control	101 ± 16	12.5 ± 2.3	105 ± 8	3.2 ± 0.6
Bromocriptine	19 ± 5*	34.1 ± 2.0*	102 ± 4	30.6 ± 3.2*
Bromocriptine + ovine prolactin	65 ± 7	9.9 ± 3.0	ND	ND
Days 9–12 post partum				
Saline control	66 ± 12	10.5 ± 1.6	ND	ND
Bromocriptine	82 ± 11	12.3 ± 2.1	ND	ND
Treatment of offspring				
Days 1–3 post partum				
Oil control	89 ± 13	12.8 ± 2.1	45 ± 4	2.2 ± 1.0
Oestradiol	30 ± 4*	32.3 ± 6.2*	25 ± 7*	8.3 ± 0.9*

*$P < 0.01$ versus control. Values are mean ± SEM; $n = 5–8$. ND, not determined.

a deficiency in milk-derived prolactin during a brief postnatal critical period leads to a reduction in the dopaminergic inhibition of prolactin secretion that persists into adulthood.

The functional development of the tuberoinfundibular system occurs postnatally in rats (see Grosvenor et al 1990 for review). Although present on the day of birth, neurons in the arcuate nucleus do not begin to synthesize or release dopamine until approximately postnatal Day 5, as the processes of these neurons grow into the median eminence to meet the vessels of the primary capillary plexus, which also form at this time. The dopaminergic inhibition of prolactin secretion develops gradually during the first and second postnatal weeks, even though dopamine receptors are demonstrable in the pituitary gland on postnatal Day 1. Thus, milk-derived prolactin, perhaps acting in concert with endogenously secreted prolactin, stimulates the synthesis and/or release of dopamine from developing tuberoinfundibular neurons. In the mature animal, higher concentrations of circulating prolactin increase tuberoinfundibular activity by enhancing dopamine synthesis and release. This action has been observed as early as postnatal Day 15, but it is not known whether it occurs before then.

Hypoprolactinaemia decreases tuberoinfundibular activity in adulthood. Because prolactin deficiency on Days 2–5 post partum results in decreased dopamine turnover some time later (Shah et al 1988, Shyr et al 1986), prolactin may also exert a stimulatory 'organizational' effect during the immediate postnatal period. A precedent for the hormonal regulation of dopamine neuronal development comes from work showing that the postnatal onset of dopaminergic inhibition of α-melanocyte-stimulating hormone secretion is stimulated by the hormone itself (Lichtensteiger et al 1983). Thus, the two neuroendocrine dopamine systems may be similar in that their postnatal functional development is, in part, regulated by the hormones they are destined to inhibit in the mature animal. Hyperprolactinaemia persists in neonatally prolactin-deficient rats in spite of a return to normal of tuberoinfundibular activity by Day 60 (see Table 2). Additional studies suggested that neonatal prolactin deficiency alters other aspects of the hypothalamic control of prolactin secretion in adulthood. A degree of pituitary hyperplasia is present—there is an approximately 20% increase in cells per pituitary obtained during enzymic dispersal (C. E. Grosvenor, B. Kacsoh, W. Hymer, unpublished observations). Second, in contrast to normal anterior pituitary cells maintained in culture, pituitary cells obtained from 100-day-old neonatally prolactin-deficient rats are almost completely unresponsive to the effects of a dopamine agonist *in vitro*, which usually decreases the release of prolactin and reduces the cytoplasmic levels of prolactin messenger RNA (Shaf et al 1988). In addition, these cells are somewhat more responsive to thyrotropin-releasing hormone on these parameters (Shah et al 1988). However, prolactin release occurs to a similar extent in cultured anterior pituitary cells from neonatal oestrogen-treated and control rats, suggesting normal responsiveness of dopamine receptors on lactotrophic cells in both groups (Crowley et al 1990). The altered responses of the lactotroph to dopamine and thyrotropin-releasing hormone in neonatally prolactin-deficient rats, unlike alterations in tubero-infundibular activity, therefore, do not appear to involve the ovary. Thus, much like neonatal deprivation of androgen (Harris 1964, MacLusky & Naftolin 1981, McEwen 1981), neonatal deprivation of prolactin alters sensitivity to hormones in adulthood. Because tuberoinfundibular activity in neonatal prolactin-deficient offspring eventually returns to normal (see Table 2), the persistently elevated serum prolactin may, in part, be maintained by the depressed responses of the pituitary lactotroph to dopamine (Shah et al 1988).

Acknowledgement

Supported by USPHS Grant HD 04358 to Clark E. Grosvenor and DGAPA-UNAM Grant IN206991 to Flavio Mena.

References

Benos DJ 1982 Amiloride: a molecular probe of sodium transport in tissues and cells. Am J Physiol 242:C131–C135

Cheung CY, Kuhn RW, Weiner RI 1981 Increased responsiveness of the dopamine-mediated inhibition of prolactin synthesis after destruction of the medial basal hypothalamus. Endocrinology 108:747–751

Crowley WR, Shah GV, Watanobe H, Grosvenor CE 1990 Effects of neonatal exposure to estradiol on prolactin secretion and activity of the tuberoinfundibular dopamine system in young adulthood: comparison with neonatal prolactin deficiency. Neuroendocrinology 2:19–24
De Marco L, Mashiter K, Peters TJ 1982 Suckling withdrawal increases pituitary lysosomal enzyme activities and prolactin protease in lactating rats. Endocrinology 110:1178–1182
Döhler KD, Wuttke W 1975 Changes with age in levels of serum gonadotropins, prolactin and gonadal steroids in prepubertal male and female rats. Endocrinology 97:898–907
Döhler KD, von zur Mühlen A, Döhler U 1977 Pituitary luteinizing hormone (LH), follicle stimulating hormone (FSH) and prolactin from birth to puberty in female and male rats. Acta Endocrinol 85:718–728
Farquhar MG 1977 Secretion and crinophagy in prolactin cells. In: Dellman HD, Johnson JA, Klachko DM (eds) Comparative endocrinology of prolactin. Plenum Press, New York p 37
Grosvenor CE, Mena F 1982 Regulating mechanisms for oxytocin and prolactin secretion during lactation. Neuroendocr Perspect 1:69–102
Grosvenor CE, Whitworth NS 1983 Accumulation of prolactin by maternal milk and its transfer to the circulation of the neonatal rat—a review. Endocrinol Exp 17:271–282
Grosvenor CE, Mena F, Whitworth NS 1979 The secretion rate of prolactin in the rat during suckling and its metabolic clearance rate following increasing intervals of non-suckling. Endocrinology 104:372–377
Grosvenor CE, Mena F, Whitworth NS 1980 Evidence that the dopaminergic-PIF mechanism regulates the depletion-transformation phase and not the release phase of prolactin secretion during suckling in the rat. Endocrinology 106:481–485
Grosvenor CE, Goodman GT, Mena F 1984 Control of the multiphase secretion of prolactin in the lactating rat. In: Mena F, Valverde C (eds) Prolactin secretion: a multidisciplinary approach. Academic Press, Orlando, FL, p 275–284
Grosvenor CE, Terry LC, Meyers JS, Crowley WR 1990 In: Krasnegor N, Bridges R (eds) Mammalian parenting: biochemical, neurobiological and behavioural determinants. Oxford University Press, New York p 324–343
Harris GW 1964 Sex hormones, brain development and brain function. Endocrinology 75:627–647
Koldovsky O 1980 Hormones in milk. Life Sci 26:1833–1836
Lawson DM, Haisenleder DJ, Gala RR, Moy JA 1987 The occurrence of pituitary prolactin depletion-transformation in lactating rats: dependence on strain of rats, homogenization conditions and method of assay. J Endocrinol 113:71–80
Lichtensteiger W, Schlumpf M, Davis MD 1983 Catecholamine and nicotine in early neuroendocrine organization. Monogr Neural Sci 9:213–224
MacLeod RM, Lehmeyer JE 1974 Studies on the mechanism of the dopamine-mediated inhibition of prolactin secretion. Endocrinology 94:1077–1085
MacLusky NJ, Naftolin F 1981 Sexual differentiation of the central nervous system. Science (Wash DC) 211:1294–1303
Martinez-Escalera G, Clapp C, Morales MT, Lorenson MY, Mena F 1986 Reversal by thiols of dopamine-, stalk-median eminence- and zinc-induced inhibition of prolactin transformation in adenohypophyses of lactating rats. Endocrinology 118:1803–1807
McEwen BS 1981 Neural gonadal steroid actions. Science (Wash DC) 211:1303–1311
Mena F, Martinez-Escalera G, Clapp C, Aguayo D, Forray C, Grosvenor CE 1982 A solubility shift occurs during depletion-transformation of prolactin within the lactating rat pituitary. Endocrinology 111:1086–1091
Mena F, Martinez-Escalera G, Clapp C, Grosvenor CE 1984 In vivo and in vitro secretion of prolactin by lactating rat adenohypophyses as a function of intracellular age. J Endocrinol 101:27–32

Mena F, Clapp C, Aguayo D, Lorenson MY, Martínez-Escalera G 1986 Thiol regulation of depletion-transformation and release of prolactin by the pituitary of the lactating rat. Endocrinology 118:1795–1802
Mena F, Clapp C, Aguayo D, Martínez-Escalera G 1989a Differential effects of thyrotropin-releasing hormone on in vitro release of in vivo or in vitro newly synthesized and mature prolactin by lactating rat adenohypophyses. Further evidence for a sequential pattern of hormone release. Neuroendocrinology 49:207–214
Mena F, Clapp C, Aguayo D, Morales MT, Grosvenor CE, Martínez de la Escalera G 1989b Regulation of prolactin secretion by dopamine and thyrotropin-releasing hormone in lactating rat adenohypophyses: influence of intracellular age of the hormone. Endocrinology 125:1814–1820
Mena F, Aguayo G, Hummelt C, Clapp G, Martínez de la Escalera G, Morales MT 1992 Prolactin secretion during lactation. In: Negro-Vilar A, Perez-Palacios G (eds) Serono symposium on reproduction, growth and development. Acapulco, Mexico. Raven Press, New York, in press
Nagy GM, Frawley LS 1990 Suckling increases the proportion of mammotrophes responsive to various prolactin-releasing stimuli. Endocrinology 127:2079–2084
Ojeda SR, McCann SM 1974 Development of dopaminergic and estrogenic control of prolactin release in the female rat. Endocrinology 95:1499–1505
Scammell JG, Dannies PS 1983 Veratridine and ouabain stimulate calcium-dependent prolactin release. Endocrinology 113:1228–1235
Shah GV, Shyr SW, Grosvenor CE, Crowley WR 1988 Hyperprolactinemia after neonatal prolactin (PRL) deficiency in rats: evidence for altered anterior pituitary regulation of PRL secretion. Endocrinology 122:1883–1890
Shyr SW, Crowley WR, Grosvenor CE 1986 Effect of neonatal prolactin deficiency in prepubertal tuberoinfundibular and tuberohypophyseal dopaminergic neuronal activity. Endocrinology 119:1217–1221
Smith RE, Farquhar MG 1966 Lysosome function in the regulation of the secretory process in cells of the anterior pituitary gland. J Cell Biol 31:319–347
Walker AM, Farquhar MG 1980 Preferential release of newly synthesized prolactin granules is the result of functional heterogeneity among mammotrophs. Endocrinology 107:1095–1104
Whitworth NS, Grosvenor CE 1978 Transfer of milk prolactin to the plasma of neonatal rats by intestinal absorption. J Endocrinol 79:191–199

DISCUSSION

Hatton: You said dopamine had the ability to lower the intragranular pH. Do you know what the mechanism of that is?

Grosvenor: No, except that a Na^+/H^+ pump is involved. When dopamine is absent, the pH of the granules tends to become more acidic. When dopamine is present, the pH in the granules becomes more alkaline. So during transformation, presumably the intragranular pH falls, which allows these events we have observed to occur. *In vitro*, we can alter prolactin from one form to another reversibly, simply by adjusting the pH.

Zoeller: You assay the transformation of prolactin by PAGE. Is it possible that prolactin 'matures' at the same time as the vesicle 'matures', i.e. the vesicle itself becomes able to be released? I am thinking that in order for the prolactin to be released, there has to be some mechanism that allows fusion of the vesicle to the plasma membrane and subsequent exocytosis. If there is, this mechanism may be an important point of regulation.

Grosvenor: We don't know to what extent vesicle maturation is involved in prolactin transformation. The granules migrate to the exocytotic pouch where the pH is more alkaline. Presumably, transformed prolactin, which is relatively acidic, becomes more alkaline and thus more soluble in the exocytotic pouch, which may facilitate its release into the intracellular space.

Charlton: In terms of targeting secretion within cells, is the 'youngest' hormone targeted towards the endothelial side of the cell? Hormone is apparently preferentially released from the endothelial side.

Grosvenor: I can't answer that. We have done the same intracellular age manipulations with granule preparations as with pituitary homogenates and obtained the same results: prolactin synthesized 2–8 hours previously is transformed and released, younger or older prolactin, in terms of time since biosynthesis, is not. Some argue that there is a pool of prolactin that is not packaged into granules but moves rapidly through the cytoplasm to the cell membrane where it is released. We have no evidence for this route of prolactin secretion in the lactotroph of the lactating rat.

Prolactin transformation also occurs during the oestrous cycle of the rat, peaking during pro-oestrus and oestrus when oestrogen secretion is maximal. One can induce prolactin transformation in ovariectomized rats by oestrogen replacement, so prolactin transformation is not confined to the lactating animal. We have looked at the effect of intracellular age using labelling *in vitro* and found exactly the same results as *in vivo*.

Barraclough: Do you believe in the existence of a prolactin-releasing factor which increases prolactin secretion even in the presence of normal or raised levels of portal blood dopamine?

Grosvenor: It is possible. Dopamine suppression of prolactin transformation in the pituitary is transient, it occurs in response to a sensory activation such as suckling. It takes only a few seconds for transformation to occur and then dopamine secretion resumes. It is not that dopamine can't inhibit the release of prolactin, but its ability to do so is greatly reduced, because of transformation, such that thyrotropin-releasing hormone or vasoactive intestinal peptide or other prolactin-releasing agents (because I think there are several) can easily release the transformed form of prolactin.

Nagy: How does the recent evidence that quite a low dose of dopamine, 10^{-10} M, which is close to the concentration in the hypophysial portal blood, can stimulate tremendously prolactin secretion in rats both *in vivo* and *in vitro* fit with your data? Denef et al (1980) first reported that a dose of dopamine 1000-fold lower than that required for maximal inhibition could actually stimulate prolactin secretion from pituitary cells *in vitro*. The most convincing *in vivo* data have recently come from Marc Freeman's lab. When female rats were treated with α-methyl-*p*-tyrosine, plasma prolactin levels increased, as expected (Arey 1991). Infusion of 10 ng/kg of dopamine per minute resulted in a further increase in plasma prolactin. These effects could be blocked with domperidone (a specific D_2 dopamine receptor blocker), which suggests the D_2 receptor is involved in the stimulatory effect.

Grosvenor: I don't know how these results fit, but they should certainly be taken into account.

Nagy: We also know there is a translational block in prolactin synthesis during the first 4–5 days after birth in rats. Should we consider post-transcriptional regulation, I mean before translation? And do you think a milk prolactin is one of the factors which can regulate prolactin translation? There are several examples, like acetylcholine receptors, of a block after transcription before translation.

Grosvenor: The translational mechanism to make prolactin is present at birth, but the rat doesn't make prolactin for 4–5 days after. This suggests that some mechanism is inhibiting translation. The hyperprolactinaemia that we see may in part be due to suppression of this inhibition mechanism. If the inhibition is related to milk prolactin, then the translation of prolactin mRNA might be turned on earlier in rats deprived of milk prolactin neonatally.

Nagy: That's one possibility, but we have recently investigated the acute oestradiol-induced prolactin secretion *in vitro* (Ellekmann et al 1991). We treated anterior pituitary cells from ovariectomized rats in culture with oestradiol and there was no increase in prolactin secretion. We used the reverse haemolytic plaque assay that enabled us to quantify the percentage of cells that released prolactin. When we co-cultured anterior lobe cells with intermediate lobe cells and treated them with oestradiol for three hours, we found about a 10–15% increase in the number of prolactin-secreting cells. Even when we co-cultured these cells in the same dish at separate sites then removed the neurointermediate lobe cells (*in vitro* neurointermediate lobectomy) before the oestradiol treatment, we saw no effect of oestradiol. So the intermediate lobe cells are really necessary for the oestradiol-induced increase in the number of prolactin-secreting cells in the anterior lobe. This may be similar to the example you described, in that there is an increase in the amount of oestradiol that parallels the increase in prolactin levels. It would be really interesting to look at the number of prolactin-producing cells in your system.

The other thing is that the intermediate lobe is tonically inhibited by dopamine, so another possible site of action of bromocriptine is the intermediate lobe.

Grosvenor: You and Steve Frawley have shown that at the time prolactin transformation occurs there is a mobilization of cells from the central part to the periphery of the pituitary. So prolactin release is not just mobilization of prolactin from one form to the other, there is also a mobilization of cells.

Nagy: Only 10% of the anterior lobe cells located exclusively surrounding the neurointermediate lobe supplied by short portal vessels may participate in the acute release of prolactin. 10 minutes suckling results in a tremendous change in the responsiveness of these cells measured *in vitro* in what we call the inner zone. For example, one cannot detect any dopamine inhibition in these cells after 10 minutes suckling and the responsiveness to thyrotropin-releasing hormone and angiotensin II increases tremendously. If one does a range of dose-response curves, one can see stimulation of prolactin secretion by 10^{-10} M dopamine in suckled rats. This relates to the possibility that the anterior lobe, especially in the case of prolactin, can be regulated from the neurointermediate lobe direction. We tend to neglect the possible role of the short portal blood vessel system in the regulation of anterior pituitary hormone secretion. It gives the peripheral hormones a chance to affect directly a certain region of anterior pituitary glands because the blood washes through the posterior lobe and intermediate lobe, then arrives at this inner zone of the anterior lobe. So you may have to consider these possibilities for a physiologically relevant regulation.

Charlton: In the blood there is very little glycosylated prolactin, yet in the milk it is glycosylated. Is the glycosylated prolactin a product of the gene in the mammary gland epithelial cells or have the cells taken in the prolactin and transported it to the Golgi for glycosylation?

Grosvenor: It is probably a little of both. The prolactin may even become glycosylated in the milk itself. The phosphorylating and glycosylating enzymes are present in milk, but they are undoubtedly synthesized in the secretory cells of the mammary gland. Of the different bioactive forms of prolactin in milk, most are phosphorylated and about 15% are glycosylated. We are now determining which sugars are present and we are separating the glycosylated variants.

Gainer: You suggest the prolactin may be glycosylated in the lumen of the mammary gland. The lumen is the extracellular side. Is this an N-asparaginyl glycosylation? I have never heard of that reaction happening outside the Golgi.

Grosvenor: I don't know. We are going to take fresh rat milk, add the 24 kDa form of prolactin and see if we get the glycosylated form. I doubt we will. We are also going to label the sugars and try to follow the synthesis of glycosylated prolactin through the typical glycosylation pathways.

Gainer: What are the frequencies of glycosylation and phosphorylation of prolactin in the pituitary?

Grosvenor: Prolactin in the pituitary is about 99% non-glycosylated in the rat. In the pig there is a fairly high amount of glycosylated prolactin. In the rat, the concentration is highest in early post partum milk and decreases towards the end of lactation.

Gainer: Is the glycosylated form the active form?

Grosvenor: It is one of the bioactive forms; there are others. The phosphorylated forms of prolactin are biologically active and present in high concentrations in milk and, of course, there is a fair amount of unconjugated, monomeric prolactin that is bioactive and immunoreactive.

Nagy: When you talk about bioactivity, do you mean only prolactin that is active in the Nb_2 cell assay (Tanaka et al 1980)?

Grosvenor: Not exclusively. We have also used the rabbit mammary gland explant bioassay, which is very laborious and very expensive, to confirm the results obtained with the Nb_2 cell assay.

Nagy: The Nb_2 cell line is a lymphoma cell line, so why should it respond to prolactin and what type of bioactivity of prolactin would it detect?

Grosvenor: Nb_2 cells have prolactin receptors. They proliferate in a sensitive, dose-dependent manner in response to all mammalian prolactins. The only non-prolactin hormone I know of that causes proliferation of Nb_2 cells is human growth hormone.

Halasz: In the experiments on neonatal deprivation of milk-derived prolactin, do you have any evidence that bromocriptine was acting directly on the offspring?

Grosvenor: We cooperated with Sandoz (Basel, Switzerland) on these studies. You can find a metabolically inactive product of bromocriptine in milk, but bromocriptine itself does not appear in milk or in the offspring in an active form. We also injected bromocriptine directly into rat offspring and did not achieve any of the effects we obtained with neonatal milk prolactin deficiency.

Halasz: Did you try feeding the pups with prolactin-deprived milk?

Grosvenor: We are doing experiments right now utilizing a technique which has the cute name of 'pup in the cup'. It is commonly used by behavioural psychologists and nutritionists. Young rats are left with the mother for one day to give them the necessary antibodies, then we place them individually in Styrofoam cups in a water bath that closely approximates body temperature. Diets with or without milk prolactin are delivered through a gastric catheter via a constant fusion pump. After four days we take out the catheter and put the pup back with the mother.

Another tactic we will use is to infuse rat pups with milk infranatant, which contains the prolactin variants plus many other hormones and growth factors. Then we will add specific monoclonal antibodies raised against the various forms of prolactin and look at dopamine secretion, lactotroph response to dopamine and to thyrotropin-releasing hormone, etc, during adulthood. We hope to determine which forms of prolactin are involved in the defects we have observed in adult rats that were deprived of milk prolactin during the critical first few days after birth.

Lightman: This is somewhat worrying for bottle-fed babies. Are bottle-fed babies hyperprolactinaemic?

Grosvenor: I am working on this with some clinicians at Penn State Medical School. I am told that babies who are fed milk formulae which do not contain hormones have more problems, particularly immunological ones, than do babies who are breast fed. We know there is a lot of prolactin in human colostrum and there is also a lot in amniotic fluid which the baby absorbs during the last stages of pregnancy. We are initiating some studies to look at these immunological signs in the hope that we can tie those two things together. There was a retrospective study which showed that a substantial number of children who come to the clinics with precocious puberty are without any known aetiology and of those, most are hyperprolactinaemic.

Lightman: It would be easy to study the prolactin levels in breast-fed and bottle-fed babies.

Grosvenor: Yes, but to study this retrospectively, by asking mothers whether or not they breast fed, is impossible. We are undertaking a prospective study. We have three groups of women: one will feed by breast, one will feed a formula off the shelf which contains no prolactin, and one will feed milk containing a cow-based prolactin. About 10% of the bioactive prolactin in bovine milk escapes pasteurization.

Plant: The hyperprolactinaemia seen in precocious puberty could be a consequence of puberty rather than its cause, i.e. a direct action on the lactotroph of ovarian oestrogen secreted in response to the initiation of puberty. In this regard, are the children you refer to hyperprolactinaemic with respect to a normal age-matched prepubertal child or are they hyperprolactinaemic with respect to a normal pubertal child? Precocity is produced by many causes. The only one of significance here is a central precocity, a premature reawakening of the hypothalamic gonadotropin-releasing hormone pulse generator.

Grosvenor: These case histories were reviewed by a paediatrician, who told me that for a significant number of the children who showed precocious puberty, both male and female, she couldn't track down the cause of the hyperprolactinaemia.

Plant: What type of precocious puberty?

Grosvenor: I don't know. These children were patients at the University of Tennessee Medical School. I had no direct contact with them. I don't mean to imply that the precocious puberty resulted from prolactin deprivation. I don't know whether the oestradiol levels were checked, for example.

Plant: What does a lactotroph of an infant rat look like compared with a lactotroph in an adult rat?

Grosvenor: They look pretty similar to me.

Plant: As I understand it, the hypophysial portal communication isn't fully differentiated in a very young rat.

Grosvenor: That's true. I don't know when that connection matures. There is more to this than simply having a window during which substances can be absorbed because in the rat the gastrointestinal tract is porous to macromolecules up to 18 days of age. These effects occur only in the first 4–5 days of life. This is also the time the rat brain is relatively plastic.

Barraclough: You attribute some of these changes to the ovarian secretion of oestradiol. Normally, neonatal rats have high plasma levels of oestradiol, but this steroid is bound to α-fetoprotein and therefore biologically inactive.

When you deprive new-born rats of prolactin, it is possible that you are affecting the synthesis or availability of α-fetoprotein. This would make free oestradiol available to the brain to produce the subsequent syndrome you describe. Most likely, when you inject oestradiol, the α-fetoprotein is unable to bind all the exogenous steroid and the unbound steroid would have the same defeminizing effects within the brain. These animals, in adulthood, develop polyfollicular ovaries that make excessive oestradiol because follicular rupture never occurs. The high plasma oestradiol levels produced by these ovaries also lead to hyperprolactinaemia in the adult.

Grosvenor: We can produce vaginal opening in rats 3–4 days earlier than normal if we reduce the milk prolactin levels ingested by the rat on Days 2–5 post partum. You are probably correct. We did not consider the α-fetoprotein, all we did was measure the oestradiol levels by radioimmunoassay and they were elevated. When we ovariectomized the neonatally prolactin-deprived rats on Day 10, we prevented the hyperprolactinaemia and we prevented the alterations in tuberoinfundibular dopamine turnover. Putting those results together, we concluded that the ovary was involved in the induction of hyperprolactinaemia through oestradiol secretion affecting dopamine secretion.

Gainer: So are you proposing that there is a down-regulation of dopamine receptor in rats that were deprived of prolactin neonatally? Is that part of the reason the response to dopamine is missing in the adults?

Grosvenor: From preliminary studies it appears that two things are occurring in the neonatally prolactin-deprived rat. First, there seem to be fewer dopamine neurons, although that's subject to interpretation. Secondly, in collaboration with Bill Crowley, we are using a tyrosine hydroxylase probe and preliminary studies suggest synthesis of dopamine is impaired. But after puberty, this recovers.

Gainer: But the response of the lactotroph to dopamine does not, which is why I thought it could be a dopamine receptor problem.

Grosvenor: We don't yet know whether down-regulation (or alteration) of the dopamine receptor in the lactotroph is involved or whether there is a defect in the second messenger signal transduction system. The intriguing thing to me is that this is an inductive effect which occurs early on and is maintained during adult life.

References

Arey BJ 1991 Infusion of low concentrations of dopamine stimulate prolactin secretion in vivo. 73rd Annual Meeting of the Endocrine Society, Washington DC

Denef C, Manet D, Dewals R 1980 Dopaminergic stimulation of prolactin release. Nature (Lond) 285:243–246

Ellekmann E, Nagy G, Frawley LS 1991 Rapid augmentation of prolactin cell number and secretory capacity by an estrogen-induced factor released from the neurointermediate lobe. Endocrinology 129:838–842

Tanaka T, Shiu RPC, Gout PW, Beer CT, Noble RL, Friesen HG 1980 A new sensitive and specific bioassay for lactogenic hormones: measurement of prolactin and growth hormone in human serum. J Clin Endocrinol Metab 51:1058–1063

What can we learn from sampling hypophysial portal blood?

Iain J. Clarke

Prince Henry's Institute of Medical Research, P.O. Box 152, Clayton, Victoria 3168, Australia

Abstract. Since it has become possible to sample hypophysial portal blood from sheep without totally compromising pituitary function, several important features of the secretion of hypothalamic hormones have been elucidated. The secretion of gonadotropin-releasing hormone (GnRH) has been detailed most thoroughly with the important observation that each pulsatile discharge of luteinizing hormone (LH) is the direct result of a large secretory episode of GnRH from the hypothalamus. There is high fidelity in the GnRH relationship in terms of frequency and amplitude. During the LH surge, additional factors such as an alteration in the degree of enzymic degradation of GnRH may be important physiological mechanisms. The secretion of factors that control the release of growth hormone (GH), prolactin and adrenocorticotropic hormone (ACTH) have also been studied. Hypothalamic factors controlling GH and ACTH release do not bear such an explicit relationship to the secretory episodes of pituitary hormone as seen with the GnRH/LH axis. The factor involved in the acute stress-induced release of prolactin has not yet been identified in sheep.

1992 Functional anatomy of the neuroendocrine hypothalamus. Wiley, Chichester (Ciba Foundation Symposium 168) p 87–103

Several significant advances have been made in our understanding of the way in which the hypothalamus controls the endocrine system since it became possible to sample hypophysial portal blood. In the early 1980s we set out to develop a model in which one could monitor hypothalamic secretions in conscious animals. This was achieved in sheep and has been the model of choice for the last decade.

Methodological considerations

The earliest models developed by Worthington (1966) allowed collection of portal blood using a transoral approach in anaesthetized rats; the pituitary stalk was cut then blood was aspirated into tubing. Porter & Smith (1967) also developed a transpharangeal approach in which a polyethylene cannula was fitted over the cut stump of the stalk–median eminence. In assessing these two methods, Gibbs (1984) found that the levels of hypothalamic hormones measured were similar in both, but these techniques do have considerable limitations. Both use anaesthetized animals and, since it is well known that most anaesthetics compromise hypothalamic function, this is a serious problem. Fink and colleagues (Sarkar et al 1976) overcame this by using althesin, which does not compromise gonadotropin releasing-hormone (GnRH) secretion, but it is short acting and not suitable for use in larger animals.

Another method for the assessment of neuropeptide/neurotransmitter secretion from the median eminence is that of push–pull perfusion. This has been used successfully in rats (Levine & Ramirez 1980), monkeys (Levine et al 1985), sheep (Levine et al 1982) and rabbits (Ramirez et al 1986). With this method, GnRH secretion has been found to be reasonably well correlated with luteinizing hormone (LH) secretion, but substantial tissue damage occurs at the tip of the cannula (see Pau et al 1989). This raises the question of the extent to which the perfusate contains material from damaged cells.

We have developed a method to sample hypophysial portal blood from a site downstream from the median eminence in conscious sheep (Clarke & Cummins 1982, Caraty & Locatelli 1988), with which we can monitor hypothalamic secretions in a more physiologically relevant manner. Our method involves transphenoidal surgery with the placement of two 12G needles through the nasal bone, directed towards the pituitary gland. One needle is directed at the face of the pituitary gland, such that an appropriately fashioned stillete will cut through the portal vessels on the face of the gland; the other needle is used to aspirate blood. An artificial sinus is created in the sphenoid bone and no apparatus is left in contact with the pituitary. This is important because touching the vessels causes vasoconstriction and any pressure on the gland must therefore alter blood flow. A method developed by Moenter et al (1990) is slightly different to ours in that a cone-shaped apparatus is placed in front of the pituitary gland and the cutting of the portal vessels is not as definitively predetermined as in our method. The advantage of both methods is that sheep are prepared for sampling and then allowed to recover from surgery for days or weeks before sampling. Thus, samples can be collected from conscious, resting animals in an interactive environment.

The reason I have belaboured the issue of methodology is that it has considerable bearing on what data may be obtained regarding biologically relevant material in hypophysial portal blood. For example, methods that involve cutting the stalk or perfusing tissue can cause artifacts because cells within the median eminence may burst and release material that is not normally secreted into portal blood.

The relationship between the secretion of GnRH, luteinizing hormone and follicle-stimulating hormone

Basal secretion in gonadectomized animals

The sampling of conscious sheep indicated quite clearly that the pulsatile pattern of LH secretion is due primarily to the episodic discharge of GnRH into hypophysial portal blood (Clarke & Cummins 1982). A similar pattern of secretion was seen using push–pull perfusion in ovariectomized ewes (Levine et al 1982), although the levels measured in perfusate were much lower than those seen in portal blood. In addition to the major Gaussian peaks of GnRH activity in hypophysial portal blood, minor peaks of GnRH activity can also be detected (Clarke & Cummins 1982, Levine et al 1982); these often correlate with minor perturbations in LH secretion.

The effect of castration on GnRH secretion was nicely demonstrated by Caraty & Locatelli (1988). The pulsatile secretion of GnRH increased, indicating a release from negative feedback regulation by gonadal hormones at the hypothalamic level in male sheep. In long-term castrated males, the appearance of a greater number of 'silent' GnRH pulses (unaccompanied by LH pulses) was noted and the authors suggested that the GnRH/LH relationship becomes uncoupled in these animals. On the other hand, we have observed 'silent' GnRH pulses in ovary-intact ewes (Clarke et al 1987), so this characteristic may not be confined to castrated males.

The effect of castration on GnRH secretion in the rat has been more controversial. Studies using push–pull perfusion of the median eminence indicated that there was no increase (Levine & Ramirez 1982). Levine & Duffy (1988) did, however, demonstrate a modest increase in GnRH pulse frequency after castration of male rats. The issue was resolved when Dluzen & Ramirez (1987) used push–pull perfusion of the pituitary gland and found a large increase in GnRH pulse frequency and amplitude after castration of male rats. These results cause one to question the relevance of the absolute concentrations recorded by perfusion of the median eminence in rats.

Oestrous cycle

Our original studies in sheep (Clarke et al 1987) showed that GnRH pulses were less frequent and of greater amplitude during the luteal phase of the oestrous cycle than during the follicular phase. These results have been confirmed by Moenter et al (1991), who demonstrated that the frequency of GnRH pulses increases and the amplitude of pulses decreases across the follicular phase. The change in amplitude may simply be a function of increasing pulse frequency, since releasable stores of GnRH in the terminals may be reduced with more frequent discharges. In support of this notion, it is interesting that the excitatory amino acid *N*-methyl aspartate causes more LH release in rams during the non-breeding season (when pulse frequency is low) than during the breeding season (when pulse frequency is high) (Lincoln & Wu 1991). Whilst it is necessary to measure GnRH pulse amplitudes in these two situations before a definitive statement can be made, it seems likely that the different effects of aspartate are due to the greater build-up of GnRH stores in the non-breeding season.

The prepubertal period

Rodriguez and Wise (1989) have sampled bull calves during the transition from infancy to the prepubertal period (Weeks 2–12 of age). GnRH pulses were detected throughout this period whereas LH pulses began to appear in jugular blood only after eight weeks of age. These findings suggest that GnRH may not act as a secretagogue in the pituitary of the infant, and it would be interesting to investigate stimulus–secretion coupling at this time.

Seasonal and post-partum anoestrus

In ewes, GnRH pulse frequency is low during seasonal anoestrus (Clarke 1988). There is also a low GnRH pulse frequency in the early post partum period, with a gradual increase in frequency over a period of weeks (Wise 1990).

The LH surge

In all species examined, it is clear that a rise in the secretion of GnRH occurs at the time of the oestrogen-induced preovulatory LH surge. This was first demonstrated in the rat by Sarkar et al (1976). Our early results in sheep (Clarke et al 1987) suggested that variable patterns of GnRH secretion are seen at the time of the LH surge in animals experiencing oestrous cycles. More recent evidence from Moenter et al (1991) indicates that a definite rise in GnRH occurs at this time, although the magnitude of the rise varies considerably among animals. In ovariectomized ewes given oestrogen injections, a GnRH rise can also be seen at the time of the induced LH surge (Clarke & Cummins 1985, Caraty et al 1989, Moenter et al 1990). The most profound increase in GnRH secretion that we have seen is in seasonally anoestrous ewes given oestradiol (Clarke 1988).

The variability in the GnRH surge between animals provokes discussion as to what is the most important feature of the rise in GnRH secretion. Is it important to have high frequency GnRH pulses at this time or is there an increase in GnRH pulse amplitude or is there some mechanism that causes a rise in the baseline levels of GnRH in portal blood? Currently, we favour the proposition that a rise in baseline levels of GnRH in portal blood is crucially important in the generation of the LH surge. This could be brought about by a local reduction in the levels of enzyme that degrade GnRH and we have preliminary results to support this proposition (Smith et al 1982). In ovariectomized animals given oestradiol benzoate and then killed at the time of their LH surge, we found that the activity of enzymes degrading GnRH in the median eminence was substantially curtailed (Smith et al 1992). This offers a mechanism for the prolonged secretion of GnRH at this time. Furthermore, it may be that enzyme activity in the extracellular space of the median eminence usually degrades GnRH at all other times, ensuring that secretory episodes are shortlived.

Secretion of GnRH and follicle-stimulating hormone

The peripheral concentration of follicle-stimulating hormone (FSH) in plasma rarely indicates a pulsatile pattern of release and the relationship between secretion of FSH and GnRH is not as distinct as that between secretion of LH and GnRH (Li et al 1989). Results from a range of species suggest that GnRH promotes FSH synthesis and that secretion does not require the secretagogue function of GnRH (Clarke 1987).

Factors in portal blood that control prolactin secretion

On the basis of various criteria it is clear that prolactin secretion is controlled by the hypothalamus. Thus, acute prolactin responses to stressful stimuli or centrally acting drugs (e.g. serotonin) can be prevented by hypothalamo–pituitary disconnection (HPD) (Thomas et al 1987). In spite of this, we have been unable to identify any single putative prolactin-releasing factor in sheep except thyrotropin-releasing hormone, which is probably a non-specific stimulator (Thomas et al 1988). In addition, we have attempted to identify, in portal blood, peptides that may be prolactin-releasing factors but have not succeeded. It is possible that factors controlling prolactin secretion (and indeed, the secretion of other pituitary hormones) originate from the posterior lobe of the pituitary gland and reach the anterior lobe via short portal vessels (Thomas et al 1989a). The concept that factors controlling prolactin secretion arise from the posterior lobe has been promoted by Ben-Jonathan & Peters (1982). This hypothesis could be tested if one developed a method to sample from the short portal vessels, as opposed to the long vessels draining from the median eminence.

Early studies indicated that dopamine could be a physiological inhibitor of prolactin secretion, since it was found at higher concentrations in hypophysial portal blood than in peripheral blood (Ben-Jonathan 1985). We re-examined this issue in the sheep. Whilst dopamine was found to inhibit prolactin secretion, we were surprised that there was no portal/jugular differential in dopamine levels (Thomas et al 1989b,c). On the other hand, we found that noradrenaline was equally efficient in lowering prolactin levels in sheep in which the hypothalamo–pituitary connection had been cut and the levels of this amine in portal blood were three times higher than in peripheral blood (Thomas et al 1989b,c).

Factors that control the secretion of adrenocorticotropic hormone

After the isolation of corticotropin-releasing factor (CRF), it was possible to demonstrate that this neuropeptide was secreted into portal blood and stimulated the secretion of adrenocorticotropic hormone (ACTH) (Gibbs & Vale 1982, Rivier et al 1982). Another neuropeptide, arginine vasopressin (AVP), potentiates the action of CRF on ACTH secretion in rats (Gillies et al 1982) and is also found in portal blood (Zimmerman et al 1973). It was interesting, therefore, that when we investigated this axis in sheep, we found that AVP concentrations in hypophysial portal blood far exceeded those of CRF and that stress caused a more profound response in the former (Engler et al 1988). Furthermore, the ovine pituitary releases more ACTH in response to AVP than CRF (Familari et al 1989).

A more difficult issue that relates to the control of ACTH secretion is the possible existence of a corticotropin release-inhibiting factor (CRIF). This has been proposed by Engler et al (1990) on the basis that disconnection of the pituitary gland from the hypothalamus causes ACTH levels to increase. The identity of such a factor remains to be determined.

Factors that control growth hormone secretion

It has been well accepted for many years that somatostatin inhibits the secretion of growth hormone but the role of growth hormone-releasing factor (GHRH) has been more difficult to delineate. There have been major problems in the assay of GHRH in hypophysial portal blood, but we recently overcame this and were able to record GHRH, somatostatin and GH secretion in conscious sheep (Frohman et al 1990). Concordance between GHRH and GH secretion rates was not strict, nor were there obvious decrements in levels of somatostatin at the time of GH pulses. Reduced food intake, which elevates GH levels in plasma, also reduced somatostatin levels in portal blood (Thomas et al 1991), suggesting a functional link. So far, we have not been able to manipulate levels of GHRH in portal blood.

Conclusions

In the past decade there has been a quantum leap in our knowledge of the secretions from the hypothalamus. The secretion of GnRH has been well described but more study is required of the factors controlling ACTH, GH and prolactin release. It is apparent that some regulators, e.g. prolactin-releasing factors and CRIF, have not yet been identified. Also the role of the short portal vessels linking the posterior and anterior lobes of the pituitary gland is not well defined.

References

Ben-Jonathan N 1985 Dopamine: a prolactin-inhibiting hormone. Endocr Rev 6:564–589

Ben-Jonathan N, Peters LL 1982 Posterior pituitary lobectomy: regulation of anterior pituitary prolactin secretion. Science (Wash DC) 213:659–661

Caraty A, Locatelli A 1988 Effect of time after castration on secretion of LHRH and LH in the ram. J Reprod Fertil 82:263–269

Caraty A, Locatelli A, Martin GB 1989 Biphasic response in the secretion of gonadotrophin-releasing hormone (GnRH) in ovariectomized ewes injected with estradiol. J Endocrinol 123:375–382

Clarke IJ 1987 New concepts in GnRH action on the pituitary gland. Semin Reprod Endocrinol 5:345–352

Clarke IJ 1988 Gonadotrophin releasing hormone secretion in anoestrous ewes and the induction of GnRH surges by oestrogen. J Endocrinol 117:355–360

Clarke IJ, Cummins JT 1982 The temporal relationship between gonadotropin-releasing hormone (GnRH) and luteinising hormone (LH) secretion in ovariectomised ewes. Endocrinology 111:1737–1739

Clarke IJ, Cummins JT 1985 Increased GnRH pulse frequency associated with estrogen-induced LH surges in ovariectomised ewes. Endocrinology 116:2376–2383

Clarke IJ, Thomas GB, Yao B, Cummins JT 1987 GnRH secretion throughout the ovine estrous cycle. Neuroendocrinology 46:82–88

Dluzen EE, Ramirez VD 1987 In vivo activity of the LHRH pulse generator as determined with push–pull perfusion of the anterior pituitary gland of unrestrained intact and castrate male rats. Neuroendocrinology 45:328–332

Engler D, Pham T, Fullerton MJ, Funder JW, Clarke IJ 1988 Studies on the regulation of the hypothalamic–pituitary–adrenal axis in sheep with hypothalamic–pituitary disconnection. I. Effect of an audio visual stimulus and insulin-induced hypoglycemia. Neuroendocrinology 48:551–560

Engler D, Pham T, Liu JP et al 1990 Studies of the regulation of the hypothalamic–pituitary–adrenal axis in sheep with hypothalamic–pituitary disconnection. II. Evidence for in vivo ultradian hypersecretion of proopiomelanocortin peptides by the isolated anterior and intermediate pituitary. Endocrinology 127:1956–1966

Familari M, Smith A, Smith R, Funder JW 1989 Arginine vasopressin is a much more potent stimulus to ACTH release from ovine anterior pituitary cells than ovine corticotropin releasing factor. 1 In vitro studies. Neuroendocrinology 50:152–157

Frohman LA, Downs TR, Clarke IJ, Thomas GB 1990 Measurement of growth hormone-releasing hormone and somatostatin in hypothalamic–portal plasma of unanaesthetized sheep: spontaneous secretion and response to insulin-induced hypoglycemia. J Clin Invest 86:17–24

Gibbs DM 1984 Collection of pituitary portal blood: a methodologic analysis. Neuroendocrinology 38:97–101

Gibbs DM, Vale W 1982 The presence of corticotropin releasing factor-like immunoreactivity in hypophysial portal blood. Endocrinology 111:1418–1420

Gillies GE, Linton EA, Lowry PJ 1982 Corticotropin releasing activity of the new CRF is potentiated several times by vasopressin. Nature (Lond) 299:355–357

Levine JE, Duffy MT 1988 Simultaneous measurement of luteinizing hormone (LH)-releasing hormone, LH and follicle-stimulating hormone release in intact and short-term castrate rats. Endocrinology 122:2211–2221

Levine JE, Ramirez VD 1980 *In vivo* release of luteinizing hormone-releasing hormone estimated with push–pull cannulae from the mediobasal hypothalamus of ovariectomized, steroid-primed rats. Endocrinology 107:1782–1790

Levine JE, Ramirez VD 1982 Luteinizing hormone-releasing hormone release during the rat estrous cycle and after ovariectomy, as estimated with push–pull cannulae. Endocrinology 111:1439–1448

Levine JE, Pau K-Y, Ramirez VD, Jackson GL 1982 Simultaneous measurement of luteinizing hormone-releasing hormone and luteinizing hormone release in unanaesthetized ovariectomized sheep. Endocrinology 111:1449–1455

Levine JE, Norman RL, Gliessman PM, Oyama TT, Bangsberg DR, Spies HG 1985 *In vivo* gonadotropin-releasing hormone release and serum luteinizing hormone measurements in ovariectomized, estrogen-treated rhesus monkeys. Endocrinology 117:711–721

Li J-Y, Francis H, Clarke IJ 1989 Effects of bovine follicular fluid on pulsatile secretion of gonadotrophin releasing hormone and gonadotrophins in ovariectomised rats. J Neuroendocrinol 1:61–64

Lincoln GA, Wu FCW 1991 Luteinising hormone responses to N-Methyl-D, L-Aspartate during a photoperiodically-induced reproductive cycle in the ram. J Neuroendocrinol 3:309–318

Moenter SM, Caraty A, Karsch FJ 1990 The estradiol-induced surge of gonadotropin-releasing hormone in the ewe. Endocrinology 127:1375–1384

Moenter SM, Caraty A, Locatelli A, Karsch FJ 1991 Pattern of GnRH secretion leading up to ovulation in the ewe: existence of a preovulatory GnRH surge. Endocrinology 129:1175–1182

Pau K-YF, Hess DL, Kaynard AH, Wei-Zhi J, Gliessman BM, Spies HG 1989 Suppression of mediobasal hypothalamic gonadotropin-releasing hormone and plasma luteinizing hormone pulsatile patterns by phentolamine in ovariectomized rhesus monkeys. Endocrinology 124:891–898

Porter JC, Smith KR 1967 Collection of hypophysial stalk blood in rats. Endocrinology 81:1182–1185

Ramirez VP, Ramirez AD, Slamet W, Nduka E 1986 Functional characteristics of the luteinizing hormone releasing hormone pulse generator in conscious unrestrained female rabbits: activation by norepinephrine. Endocrinology 118:2331–2339

Rivier C, Brownstein M, Spiess J, Rivier J, Vale W 1982 *In vivo* corticotropin-releasing factor-induced secretion of adrenocorticotropin, β-endorphin, and corticosterone. Endocrinology 110:272–278

Rodriguez RE, Wise ME 1989 Ontogeny of pulsatile secretion of gonadotropin releasing hormone in the bull calf during infantile and pubertal development. Endocrinology 124:248–256

Sarkar DK, Chiappa SA, Fink G 1976 Gonadotrophin-releasing hormone surge in pro-oestrous rats. Nature (Lond) 264:461–463

Smith AI, Wallace CA, Tetaz T, Glucksman H, Roberts JL Clarke IJ 1992 Post-secretory processing of gonadotrophin releasing hormone on the ovine hypothalamo–pituitary axis. In: Sheppard KE, Boublik JH, Funder JW (eds) Stress and reproduction. Serono Symposia Publications, Raven Press, New York, p 207–218

Thomas GB, Cummins JT, Clarke IJ 1987 Secretion of prolactin in response to serotonin requires an intact hypothalamo–pituitary axis in the ewe. Neurosci Lett 83:323–326

Thomas GB, Cummins JT, Yao B, Gordon K, Clarke IJ 1988 The release of prolactin is independent of the secretion of thyrotropin-releasing hormone into hypophysial portal blood of sheep. J Endocrinol 117:115–122

Thomas GT, Cummins JT, Canny BJ et al 1989a The posterior pituitary regulates prolactin, but not ACTH or gonadotrophin secretion in the sheep. Endocrinology 125:2204–2211

Thomas GB, Cummins JT, Smythe GA et al 1989b Concentrations of dopamine and noradrenaline in hypophysial portal blood in the sheep and the rat. J Endocrinol 121:141–147

Thomas GB, Cummins JT, Doughton BW et al 1989c Direct pituitary inhibition of prolactin secretion by dopamine and noradrenaline in sheep. J Endocrinol 123:393–402

Thomas GB, Cummins JT, Francis H, Sudbury AW, McCloud PI, Clarke IJ 1991 Effect of restricted feeding on the relationship between hypophysial portal concentrations of growth hormone (GH)-releasing factor and somatostatin, and jugular concentrations of GH in ovariectomized ewes. Endocrinology 128:1151–1158
Wise ME 1990 Gonadotropin-releasing hormone secretion during post-partum anoestrus in the ewe. Biol Reprod 43:719–725
Worthington WC 1966 Blood samples from the pituitary stalk of the rat: method of collection and factors determining volume. Nature (Lond) 210:710–712
Zimmerman EA, Carmel PW, Hussain MK et al 1973 Vasopressin and neurophysin: high concentrations in monkey hypophyseal portal blood. Science (Wash DC) 182:925–927

DISCUSSION

Young: In your method for sampling portal blood, the blood drips into an artificial sinus, then you suck it out and measure it. How long is it from when the blood comes out of the blood vessels to when it is sucked up into your collecting apparatus? How stable are the various peptides during this procedure?

Clarke: This is certainly a problem. One of the ways we attempt to circumvent it is to have a suction apparatus which is hooked up to a vacuum system for rapid removal of blood. It is as quick as we can get it, but I don't know how quick that is.

Young: How often do you remove the blood?

Clarke: It is continuous. The catheter lies on the bottom of the cavity and continuous suction is applied. So as soon as the blood is secreted it's removed, usually into a tube on ice which contains various inhibitors, depending on the peptides we are studying.

Young: What kind of volume is being released?

Clarke: We generally sample every five minutes and over that period we get 2–3 ml of blood. The amount of blood we collect is determined by how big we make the cut. We generally try to cut half of the portal blood vessels in the portal bed, which gives us around that volume. Dr Oliver and Drs Caraty and Karsch have done more prolonged experiments than we have, but I think the volume they sample is less.

Oliver: We collect about 0.5 ml portal blood every 5–10 minutes. We don't make a cut, we make a small lesion of one or two portal vessels with a needle inserted into one cannula.

Sawchenko: In your sampling paradigm, how can you be sure that the portal blood isn't being diluted by local bleeding or drainage from some other sources?

Clarke: One needs to allow the animal to recover for a sufficient time that the wound, for example, is not going to contribute to the sample or that the bleeding from the sphenoid bone is not going to contribute. In general, we inject the animal with heparin three hours before we start collecting blood. During that period we apply suction as if we were taking samples and generally we do not see any bleeding.

Palkovits: When you make the cut, what percentage of the portal blood is collected?

Clarke: Our animals are always prepared under direct vision of the portal bed. It is important to open the dura mater so that one can see the entire portal bed. I have never seen two animals with the same portal bed—the possible array of vessels on the anterior face of the pituitary is endless, some run straight down the midline, some are distributed much more widely. I always attempt to cut through at least 50% of those vessels to a depth of 2 mm.

Palkovits: Do you expect that the portal blood is homogeneous? In the rat, the composition of portal blood is quite different, depending on whether you take samples from the lateral vessels, the dorsal ones or the ventral ones. If you take from the ventral and the lateral vessels, you may detect no dopamine at all because the dopamine is in the dorsal ones. LHRH is already collected from fibres by the beginning of the stalk; other substances, such as CRF and GHRH are still in neuronal fibres on the stalk and are released into the circulation far beyond.

Clarke: I understand the problem. We always try to cut from the midline out to one side.

Palkovits: That is OK for LHRH.

Clarke: And also for dopamine: Porter & Smith (1967) showed that there is a difference between lateral and medial but not between top and bottom. We have monitored catecholamine levels in portal blood. Whereas the noradrenaline levels were threefold higher than in peripheral blood, the dopamine concentrations were the same in portal and peripheral blood.

Moore: In dairy cattle, tuberoinfundibular dopaminergic neurons terminate along the infundibular stalk (Zinn et al 1990). Where do tuberoinfundibular dopaminergic neurons terminate in the sheep?

Clarke: They are certainly not below our cut because our cut is at the level of the pituitary gland. I know from bitter experience that if one makes lesions high in that portal bed, one gets spurious results. We always cut down very low in the pituitary gland. You must keep away from the median eminence to get information that makes sense.

Palkovits: We cut the median eminence and the pituitary stalk in coronal sections and took samples from each 300 μm rostrocaudally. There was a huge difference in dopamine and noradrenaline levels measured in the rostral or the post-infundibular median eminence portions.

Clarke: You are talking about the median eminence!

Palkovits: And the stalk! It is very heterogeneous in terms of the nerve terminals and axon fragments. It might be that the portal blood is homogeneous at the caudal end of the stalk, where it enters the pituitary.

Clarke: That's where we sample—where the blood enters the pituitary gland, below the stalk.

Palkovits: But if you sample only 50% of the portal blood, you may miss the dorsal or more lateral veins, which may not be exactly the same, in terms of blood composition, as the ventral ones. Have you compared them systematically?

Clarke: As I say, you have to keep away from the median eminence to have a sensible model, so I certainly haven't tried that.

Grosvenor: There could be another explanation for the lack of dopamine, which is that the sheep is exquisitely sensitive to stress. Stress turns off secretion of dopamine in the rat. Maybe you have shut off dopamine, which leads to prolactin being transformed into a releasable form (see Grosvenor & Mena, this volume).

Charlton: Have you tried using separate audio and visual sources of stress?

Clarke: No, it would be nice to do that. Dennis Engler did that work. He used what he calls audiovisual stress, whereas I always just say barking dog!

Grosvenor: In the rat the stressful stimulus is olfactory.

Moore: In the rat the activity of the tuberoinfundibular dopaminergic neurons is 2–3 times higher in the female than in the male (Moore et al 1987). I was somewhat disturbed that you find such low values of dopamine in the portal blood of your sheep. Have you looked at gender differences in the dopamine levels in the portal blood?

Clarke: No, we have not looked at males. We have looked at intact and castrated females under a variety of physiological conditions. All those assays were done by George Smythe using gas chromatography and mass spectrometry, which is very complicated and highly specific and he is fastidious when it comes to doing catecholamine measurements. I am confident that what we are seeing is right. I was mystified as to why one can't see dopamine. If the lack of detectable dopamine was anything to do with degradation or whatever, one might expect to see the same things happening to noradrenaline. We can detect noradrenaline, so one should be able to measure dopamine, if it was there.

Moore: It is hard to figure out why there is such a low level of dopamine in the portal blood of sheep in comparison to the rat.

Clarke: We have done hundreds of sheep. When one looks, for example, at GnRH in portal blood in terms of absolute concentrations, the levels are generally fairly consistent from animal to animal and from year to year. So the relative contribution that might be coming from an external source would be low, if there was one at all. One can't stop an experiment half way through, plug up the portal vessels and measure the amount of leakage, so it's a very difficult question to answer. Another factor is that occasionally, because one has to open the dura mater to do these experiments, there is release of some cerebrospinal fluid from around the pituitary gland. So we always do haematocrits on the samples to make sure that hasn't happened. Occasionally, one sample has a low haematocrit and we know we have some cerebrospinal fluid.

Lightman: Fred Bell et al (1991) have shown that metaclopramide doesn't have much effect on prolactin levels in the sheep. Do other people find the same?

Clarke: Metaclopramide has an effect, at least in our hands.

Nagy: What about domperidone in sheep?

Clarke: Domperidone also has the expected effect.

Nagy: I have some information about the possible role of the posterior lobe in the regulation of prolactin secretion in the anterior pituitary. The main regulator of prolactin secretion consists of two dopaminergic systems. One, the tuberohypophysial system, is sensitive to dehydration. In intact lactating rats you can detect a nice suckling-induced prolactin release. If you dehydrate these animals, although dehydration has no effect on the other dopaminergic system, the tuberoinfundibular system, there is a massive depletion of the suckling-induced prolactin release within 24 hours. If you rehydrate these animals, the release response is restored. At the same time, domperidone-induced prolactin secretion does not change. I think we have to alter our ideas about the regulation of prolactin secretion. It is difficult, because the hypothalamic regulation through the median eminence is considered to be the only classical pathway. Dopamine from the posterior lobe, directly or indirectly, can markedly influence prolactin secretion.

Oliver: There are species differences in the relative contribution of the short portal vessels to the vascularization of the anterior pituitary gland. Is this contribution very important in the sheep?

Clarke: I think that is yet to be resolved. We have to see what is in those short portal vessels. The best way to do that may be with a dialysis probe. If, for example, there are things like catecholamines coming across to the anterior lobe through the short portal vessels, you would pick them up with a dialysis probe.

Epelbaum: You said that noradrenaline lowers prolactin levels in sheep in which the hypothalamo–pituitary connection had been cut. Did you see any change in noradrenaline after subjecting the sheep to stress?

Clarke: We haven't looked at that.

Makara: Concerning stress-induced prolactin release, in the rat, after posterior pituitary denervation (i.e. stalk compression), we get a decrease in stress-induced prolactin release, when the stressful stimulus is either haemorrhage or immobilization.

Epelbaum: What is the ratio between the amounts of somatostatin in the peripheral blood and in the portal blood?

Clarke: We don't generally measure any somatostatin in the peripheral blood. We use an extraction assay and the samples are separated chromatographically on C18 columns.

Epelbaum: Is there any somatostatin-binding protein in sheep? Friesen and co-workers (Ogawa et al 1978) described one in rats.

Clarke: I am not aware of one.

Oliver: We have some results on GHRH and somatostatin levels that we have not published yet because we don't understand them very well. We agree with Iain Clarke's results and find also that there is a 70% positive correlation between GHRH pulses in portal blood and GH pulses in the jugular vein. However, we also found a significant percentage of somatostatin peaks occurring in the same samples as GHRH.

Clarke: Our correlation between pulses of GH and GHRH was 50–60%; it is a significant relationship. I don't know what a relationship between GHRH and somatostatin might mean.

Robinson: Iain, you presented the data on GHRH and somatostatin as though they will give the GH physiologists a headache, but GH physiologists look at GHRH not just as releasing GH but also as being trophic to the somatotroph, both in terms of synthesis of GH and in the regulation of the somatotroph population. Frequent GHRH pulses without GH release are not silent because they are doing something else to the pituitary. GHRH in the portal blood is not just teaching us about release, it should also be teaching us about GH synthesis and somatotroph regulation.

Clarke: I accept that. I didn't refer to GHRH as silent pulses, it is just that one expects us to provide clean answers by looking at things in portal blood, and they are not clean.

Poulain: You measure GnRH in the portal blood by radioimmunoassay. Is it biologically active?

Clarke: The assay will detect only GnRH(1–10); it is a very specific antibody. GnRH(1–10) is usually bioactive.

Poulain: There may be a pulse which is extremely brief and then a by-product which is not biologically active that remains for some time and is detected by the radioimmunoassay. I am thinking about oxytocin release: when you measure the amount of oxytocin in blood by radioimmunoassay, there is a high level present for a long time. If you use a bioassay, you realize that the active oxytocin is very short lived.

Clarke: I don't think it is relevant in the case of GnRH. Almost all radioimmunoassays for GnRH will only recognize GnRH(1–10), which is fully processed and stored in the terminal.

Plant: With the GnRH pulses during the oestrogen-induced preovulatory LH surge, I am a little confused because Fred Karsch comes to the opposite conclusion—namely that there are no GnRH pulses (Moenter et al 1991). His reasoning is that since the concentration of GnRH in blood perfusing the median eminence is zero, a discrete discharge of GnRH into the primary plexus will result in a peak in portal blood that will fall to zero again immediately the secretory discharge is terminated.

Clarke: Fred and I have a conceptual difference. He has seen our data and he suggested that what I am seeing are not pulses. He said there are no pulses in that surge, whereas I would say there are pulses. My reason is that there is a single upswing and a period of decrement. They are small but I believe they are pulses. That gets into the whole issue of what you call a pulse, which I suggest we don't do.

I also do not believe that because there is no GnRH being secreted from terminals, levels in portal blood should go back to zero. If there is a GnRH pulse that is prolonged because the enzyme that normally degrades GnRH is no longer present in the median eminence, then the GnRH would be diffusing into portal blood all the time and the GnRH levels in portal blood would not fall back to zero. The period of the LH surge is one of virtually continuous secretion.

Let's take the argument one step further. In an ovariectomized animal not treated with steroid, we see nice clean pulses of GnRH over 5–7.5 minutes. Consider this scenario: the GnRH is secreted from a terminal into an extravascular space, and from there into the portal blood system. A GnRH-degrading enzyme is in the median eminence to mop up whatever doesn't get into the portal blood vessels in a defined period of time, so there will be a Gaussian pulse of GnRH. If that doesn't occur during the LH surge because the enzyme is not present, there will be a smear rather than a sharp pulse.

Nagy: Can you inject GnRH antibody into the portal vessels at the same time?

Clarke: But it would then be hard to do an assay. The way to do that experiment is to use the appropriate enzyme inhibitors, which Ian Smith has information about.

Smith: I think specific peptidases may be very important in the hypothalamic regulation of pituitary function.

For example, when we examine HPLC profiles of immunoreactive CRF in sheep median eminence taken from a normal unstressed animal, the majority of the immunoreactive CRF elutes in the position of the synthetic 1-41 molecule. In sheep that have been chronically treated with the glucocorticoid dexamethasone, however, we see a marked diminution of the amount of CRF 1-41 and an increase in the peaks eluting at the positions of CRF 37-41, 36-41 and 16-41. The cleavages generating these peptide products are not consistent with what we know about the processing enzymes we would expect to find in secretory granules higher up in the biosynthetic pathway. The results are more consistent with peptidases acting post-synaptically.

In studies on immunoreactive angiotensin in portal plasma taken from sheep by Iain Clarke, the levels of different angiotensin peptides were established using a battery of antisera and HPLC to separate the immunoreactive products, then compared with levels in jugular plasma. The amount of circulating angiotensin I is greatly reduced in hypophysial portal plasma compared to the jugular plasma, whereas the amount of angiotensin(1–7) is much increased in the hypophysial portal plasma compared to the jugular plasma. This suggests that as the peptide passes through the capillary vasculature, there is some conversion.

The enzyme that generates angiotensin(1–7) is similar to the one that cleaves GnRH, suggesting these peptide fragments could be generated by the same enzyme.

To test the hypothesis that these peptides were being secreted, I took some jugular and portal plasma from animals that were resting and from animals that were stressed. These samples were incubated with synthetic GnRH. This was a crude experiment to see if there was any difference in the degrading activity in the portal plasma or the jugular plasma. We could not detect any difference in the degrading profiles.

We then made crude membrane preparations from the median eminence and from anterior pituitary, preoptic area and medial basal hypothalamus. These were also incubated with synthetic GnRH. The primary product was GnRH(1–3) which is consistent with cleavage by the enzyme 3.4.24.11, a membrane-associated protease. When we look at the V_{max} of the degradation, there's not a lot of difference between the activities in the medial basal hypothalamus, the preoptic area and the anterior pituitary; there is slightly more activity in the median eminence. However, when we looked at the soluble enzyme activity from the same tissues that generate GnRH(1–5), the activity in the mediobasal hypothalamus was similar to that in the preoptic area and in the anterior pituitary, but the median eminence had about 5–6 times more degrading activity than the other three tissues. This suggests that there is selective expression of the enzyme within that tissue. Interestingly, during the LH surge, the enzyme activity disappears, so we feel that this enzyme may specifically modulate GnRH.

I think therefore that specific peptidases may mediate, in the case of CRF, a negative feedback action, thereby modulating the availability of bioactive CRF. In the case of angiotensin, these peptidases may protect the pituitary from circulating peptides, such as angiotensin, which have stimulatory effects on the pituitary.

Finally, if you look at the curves of GnRH release, especially the rapid sampling, measured by Iain Clarke, those profiles are strictly Gaussian. One might predict that secretion in that system would give a rapid rise then a slow tail-off as the peptide slowly perfuses into the capillary vessels. Our hypothesis is that the enzymes are there to maintain the strictly Gaussian release profiles of these peptides essential for full activation of the pituitary.

Gainer: That's an interesting hypothesis. Could it be tested in this system by delivering exogenous GnRH to the catheter more upstream?

Clarke: We have done that in a limited sense. We have put GnRH into the jugular vein, but it could be done in a range of different endocrine conditions. This whole business of the role of degrading enzymes is something that we stumbled across only recently, we need to go a lot further with it.

Gainer: The hypothesis is that you have to have a short 5–7 minute Gaussian pulse or you would compromise this response. So what if you made the pulse last for 10 minutes?

Clarke: We have done experiments in hypothalamo–pituitary disconnected animals where we presented the pituitary *in vivo* with a range of pulses of different period, amplitude and frequency, or a longer pulse versus a shorter one, etc. Basically, the shorter and sharper the pulse is, the bigger is the LH response. That is confirmed by other studies. For example, Jim and Rosslyn McIntosh in New Zealand did a series of *in vitro* studies where they showed that the leading edge of the pulse of GnRH was what determined the response at the pituitary level (McIntosh & McIntosh 1983).

Gainer: Is the response being measured as the height of the peak or the interval between peaks?

Clarke: Both. You get beautiful curves: with the short sharp pulse you get a nice pulse of LH; as you lengthen the pulse of GnRH, the LH pulses become longer and flatter and smaller.

Zoeller: You could also change the shape of the curve by changing the rate of blood flow. Have you looked at that?

Clarke: If you have a tissue in which catecholamines are released, it is obviously vulnerable to regulation via effects on blood flow. Certainly, when doing surgery on the hypothalamo–pituitary unit, you only have to touch a blood vessel and it clamps up totally—they are very responsive. For that reason, I am very reticent to do anything that involves putting catecholamines into the arcuate median eminence region of an animal because there will be effects on the blood flow. We haven't done formal studies using microspheres, for example; maybe I should.

Zoeller: Are there some cardiovascular peptides like atrial natriuretic factor (ANF) or endothelin that affect the amount of LH in the serum in a pharmacological sense or a crude sense?

Nagy: Endothelin can stimulate secretion of FSH and LH *in vitro* (Kanyicska et al 1991).

Clarke: I don't know of any profound effect of ANF on LH secretion.

Makara: Concerning the inhibitor of ACTH secretion, the elusive CRIF, there is some evidence from the lab of A.F. Antoni (Edinburgh) that brain-derived ANF might be such a factor in the rat. To show the effectiveness of ANF injected intravenously, you have to lesion the paraventricular nucleus because otherwise systemic injection of ANF will reduce blood pressure and may induce CRF secretion. In an animal with a lesioned paraventricular nucleus, it is easy to show that ANF affects the pituitary. As far as I understand it, immunoneutralization with anti-ANF antibody will increase ACTH levels. So there might be a tonic ANF secretion to the anterior lobe in the rat.

References

Bell FR, Lightman SL, Simmonds A 1991 Dopamine modulation of prolactin and vasopressin but not behavior on satiation of sheep. Am J Physiol 260:1194–1197

Grosvenor CE, Mena F 1992 Regulation of prolactin transformation in the rat pituitary. In: Functional anatomy of the neuroendocrine hypothalamus. Wiley, Chichester (Ciba Found Symp 168) p 69–86

Kanyicska B, Burris T, Freeman ME 1991 Endothelin-3 inhibits prolactin and stimulates LH, FSH and TSH secretion from pituitary cell culture. BBRC (Biochem Biophys Res Commun) 174:338–343

McIntosh RP, McIntosh JE 1983 Influence of the characteristics of pulses of gonadotropin releasing hormone on the dynamics of luteinizing hormone release from perifused sheep pituitary cells. J Endocrinol 98:411–421

Moenter SM, Brand RC, Karsch FJ 1991 Pattern of GnRH secretion during the LH surge in the ewe. Proc 73rd Annu Meet Endocr Soc, Washington, DC, abstr 16

Moore KE, Demarest KT, Lookingland KJ 1987 Stress, prolactin and hypothalamic dopaminergic neurons. Neuropharmacology 26:801–808

Ogawa N, Thompson T, Friesen HG 1978 Characteristics of a soluble somatostatin binding protein. Can J Phys Pharm 56:48–53

Porter JC, Smith KR 1967 Collection of hypophysial stalk blood in rats. Endocrinology 81:1182–1185

Zinn SA, Lookingland KJ, Chapin LT, Moore KE, Tucker HA 1990 Prolactin regulation of dopaminergic neurons in the infundibulum pituitary stalk of bull calves. Proc Soc Exp Biol Med 193:98–103

Immortalized hypothalamic gonadotropin-releasing hormone neurons

Pamela L. Mellon*‡, William C. Wetsel§, Jolene J. Windle*, Marcelo M. Valença§, Paul C. Goldsmith†, David B. Whyte*, Satish A. Eraly*, Andres Negro-Vilar§ and Richard I. Weiner†

*The Salk Institute and the School of Medicine, University of California, San Diego, CA 92186-5800 and §Laboratory of Molecular and Integrative Neuroscience, National Institute of Environmental Health Sciences, Research Triangle Park, NC 27709 and †Reproductive Endocrinology Center, University of California, San Francisco, CA 94143, USA

Abstract. The neuroendocrine hypothalamus has been intensively studied using whole animals and tissue slices. However, it has been difficult to approach questions at the molecular and cellular level. By targeting expression of the oncogene product, simian virus 40 T antigen, in transgenic mice using the regulatory domain of the rat gonadotropin-releasing hormone (GnRH) gene, we have produced specific hypothalamic tumours. These tumours have been cultured to produce clonal cell lines (GT-1 cells) that express T antigen, GnRH and many other neuronal markers, but do not express other hypothalamic hormones. These immortal cell lines have a distinctive neuronal phenotype, process the GnRH peptide accurately and secrete GnRH in a pulsatile pattern. Thus, by targeting oncogenesis to a defined population of neurons using the regulatory region of a gene that is expressed late in differentiation of that cell lineage, we have succeeded in immortalizing hypothalamic GnRH neurons. The GT-1 cell lines are an excellent model for future molecular, cell biological, physiological and biochemical investigations into the mechanisms involved in regulation of GnRH and the characteristics of an isolated central nervous system neuron. Their derivation demonstrates the utility of targeting tumorigenesis to specific differentiated neurons of the central nervous system in transgenic mice.

1992 Functional anatomy of the neuroendocrine hypothalamus. Wiley, Chichester (Ciba Foundation Symposium 168) p 104–126

The hypothalamic neurosecretory neurons that produce gonadotropin-releasing hormone (GnRH) control reproductive function in mammals by secreting GnRH in a pulsatile pattern to regulate the secretion and synthesis of gonadotropic hormones from the anterior pituitary. These neurons are regulated by several neuromediators, such as monoamines, steroids and peptide hormones (Weiner et al 1988). Studies of the ultrastructure *in vivo* using double staining techniques have

‡*Present address:* Departments of Reproductive Medicine & Neurosciences 0674, Center for Molecular Genetics, UCSD School of Medicine, 9500 Gilman Drive, La Jolla, CA 92093-0674, USA

yielded information as to the interconnections between various neurons, indicating that opioid GABA-containing and oestrogen-concentrating neurons may communicate directly with the GnRH neurons. Advances in microdialysis, direct sampling of portal blood, hypothalamic slice preparation and *in situ* hybridization have contributed substantial information on the complexity of neurochemical information involved in the integrated control of GnRH pulsatility. However, many questions remain. The present *in vivo* approaches are not adequate to establish the direct action of an agent on the GnRH neuron itself or on the transcription of GnRH mRNA. Unfortunately, the low number and scattered localization of GnRH neurosecretory neurons in the rostral hypothalamus makes study of their cell and molecular biology difficult. For these purposes, isolated GnRH neuronal cultures are crucial. In addition, the study of genetic specification of GnRH gene expression to this precisely restricted neuronal population would be greatly aided by the derivation of cultured hypothalamic neuronal cell lines expressing GnRH.

The scarcity of naturally occurring neuronal tumours of the central nervous system (CNS) and the difficulty of transforming or immortalizing highly differentiated CNS neurons experimentally (Cepko 1989) have inhibited the development of CNS neuronal cell lines in which to apply molecular genetic and biochemical approaches. Several cell lines of peripheral nervous system origin have been established from neuroblastomas and pheochromocytomas; however, previous attempts to obtain cell lines of differentiated CNS neurons by various means have met with little success. Recently, Ronnett et al (1990) reported the derivation of a cell line from a human cerebral cortical tumour which will differentiate to the neuronal phenotype in the presence of a mixture of nerve growth factor, cyclic AMP and IBMX (3-isobutyl-1-methylxanthine). This is an important development, but the approach depends on the rare occurrence of human tumours with the unique ability to differentiate in this manner. Direct transformation of cultured embryonic cerebrocortical cells with retroviruses produced multipotential precursors or glial cells (Cepko 1989, Frederiksen et al 1988, Giotta & Cohn 1981, Giotta et al 1980, McKay 1989). One cell line (Frederiksen et al 1988) obtained by transformation with a temperature-sensitive T antigen will differentiate into a mixed culture with both neural and glial phenotypes upon shift to the non-permissive temperature. Infection of primary cultures of hypothalamic tissue with SV40 (simian virus 40) resulted in peptide-producing cell lines that were not fully differentiated (de Vitry et al 1974).

Transgenic mice have been used successfully to produce tumours in specific tissues (Hanahan 1989, Jenkins & Copeland 1989, Rosenfeld et al 1988), such as the anterior pituitary (Windle et al 1990) and the pancreas (Hanahan 1985), by specifically targeting expression of the potent oncogene, SV40 T antigen, using regulatory domains of the glycoprotein hormone α-subunit and insulin genes, respectively. Cell lines cultured from T antigen-targeted tumours sometimes maintain the differentiated phenotype of the original tissue (Efrat et al 1988a, Ornitz et al 1987, Windle et al 1990) and serve as excellent cell culture model systems for the relevant cell type.

Targeted oncogene expression for the derivation of tumours of neuronal tissues has proved difficult. Tissue-specific expression of T antigen under the control of the glucagon promoter in transgenic mice specifically induced tumours of the α cells of the pancreas. Normally, glucagon is also produced in neurons of the brain, but CNS tumours were not observed (Efrat et al 1988b). T antigen driven by the promoter for the phenylethanolamine N-methyltransferase gene (PNMT, an enzyme in the catecholamine synthetic pathway), was expressed appropriately in the retina but not in brain (Baetge et al 1988). Retinal cell

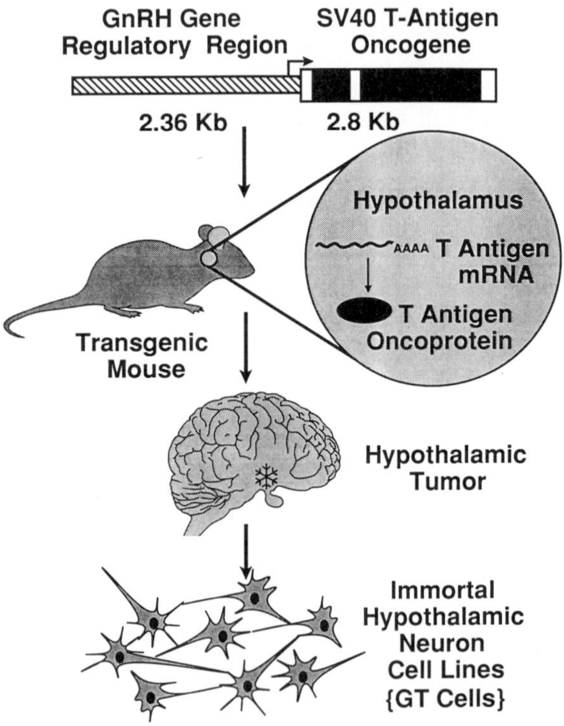

FIG. 1. Targeting of the GnRH–T antigen hybrid gene in transgenic mice. The GnRH-Tag gene contains 2.36 kb of the rat GnRH gene 5′ flanking region, including the mRNA start sites linked to the entire coding region of SV40 T antigen from the *Bgl*1 site to the *Bam*H1 site. The T antigen gene is spliced to produce two mRNAs that give rise to large and small T antigens, both of which are thought to be required for efficient transformation and immortalization (Choi et al 1988). The region of 5′ flanking DNA used in this construction (supplied by KeWen Dong and J. L. Roberts) was later determined to contain a deletion of a 731 bp *Eco*R1 fragment at −441 bp from the start site (J. Adelman, personal communication, Bond et al 1989). It thus contains DNA from −2987 to −1172 appended to −441 to +104.

tumours were obtained; however, neither the tumours nor derived cell lines expressed the endogenous PNMT gene (Hammang et al 1990). In addition, transgenic mice bearing T antigen under the control of the promoter from the growth hormone-releasing hormone (GHRH) gene had non-specific thymic hyperplasia but there was no transformation of cells of the hypothalamus (Botteri et al 1987).

The finding that the mouse GnRH gene was expressed appropriately in transgenic mice (Mason et al 1986b) and entirely reversed the sterile phenotype of hypogonadal mutant mice (which carry a deletion of the GnRH gene) supported the feasibility of using transgenic technology to target oncogene expression to the GnRH neurons. Through introduction of a hybrid gene composed of the GnRH promoter coupled to the coding region for T antigen into transgenic mice (Fig. 1), we have produced specific tumours of GnRH-secreting neurons (Mellon et al 1990). Clonal cell lines derived from these tumours express GnRH mRNA and secrete properly processed GnRH (Wetsel et al 1991) in a pulsatile pattern. They express neuronal markers, including several associated with presynaptic vesicles and membranes, but do not express glial markers. They extend lengthy neurites that end in growth cones or contact other cells (Liposits et al 1991, Mellon et al 1990). Thus, by targeting oncogenesis to a specific, small population of neurons using the regulatory region of a gene that is expressed late in the differentiation of that cell lineage, we have succeeded in immortalizing hypothalamic GnRH neurons which maintain many of the differentiated functions found *in vivo*.

A GnRH–T antigen transgene results in brain tumours and sterility

Transgenic mice carrying the hybrid GnRH–T antigen (Tag) gene (Table 1) were obtained. Neither male nor female founder mice were fertile, though they were normal in size. The gonads and accessory sex organs were underdeveloped, suggesting that expression of T antigen interfered with sexual maturation. Several

TABLE 1 Transgenic mice carrying a gonadotropin-releasing hormone–T antigen transgene

Mouse	*Sex*	*Tumour*	*Age*	*Fertility*
GT-1	Female	Hypothalamic	4.5 months	Infertile
GT-2	Female	None	>1 year	Infertile
GT-3	Female	Other—brain	3.5 months	Infertile
GT-4	Female	Other—brain	2.5 months	Infertile
GT-5	Female	Hypothalamic	2.0 months	Infertile
GT-8	Female	Other—brain	2.5 months	Infertile
GT-12	Male	None	>1 year	Infertile
GT-13	Female	None	>1 year	Infertile

of the mice showed neurological signs consistent with development of brain tumours. In founder mice GT-2, GT-3 and GT-8, tumours were observed in the hindbrain, localized in the subarachnoid space between the cerebellum and brainstem and, in one case, superior to the parietal cortex. In two other transgenic mice, GT-1 and GT-5, anterior hypothalamic tumours were observed. Several of the sterile mice survived with no neurological signs for over one year and were sacrificed for immunohistochemical analysis.

The anterior pituitary of these mice had little, if any, immunocytochemically detectable luteinizing hormone. Immunopositive GnRH neurons were observed in the preoptic area of transgenic mice, although no positive nerve terminals were observed in the median eminence (unpublished 1991). In a developmental study of GnRH–Tag transgenic mice, we observed T antigen expression in the olfactory placode on or before embryonic day 12 (E12). Furthermore, the neurons were capable of migrating along the nervus terminalis into the brain. By E16, neurons immunopositive for GnRH and T antigen were observed in large numbers in the telencephalon and preoptic areas. GnRH cells expressing T antigen extend long axonal processes and form contacts with neighbouring neurons. Thus, T antigen expression may inhibit the ability of GnRH neurons to target the median eminence during development.

Culture of immortalized neuronal cell lines

Tumours isolated from GT-4 and GT-1 mice were cultured as follows. After surgical removal, tumours were dispersed by incubation at 37 °C in Hanks buffered saline with 10 mg/ml collagenase and 10 µg/ml DNase for 30 minutes, followed by trituration and plating in plastic culture plates or Matrigel-coated plates in Dulbecco's modified Eagle's medium (DMEM) with 15% fetal bovine serum. Cells derived from the hindbrain tumour of the GT-4 mouse grew very quickly, adhered strongly to tissue culture plastic, formed coherent monolayers and were epithelial in appearance (Fig. 2A,B). The dispersed GT-1 tumour cells were extremely heterogeneous and included cells with neural and glial phenotypes, both of which proliferated in culture (Fig. 2C–F). The more adherent cells (Fig. 2C,D) grew rapidly and formed extensive networks. Occasionally, balls of cells formed which were not very adherent and radiated extensive neurites that made contact with distant cells (Fig. 2E,F). These could be removed by agitation of the plate and enriched in separate culture dishes. After six months of repeated passage using differential plating on plastic culture dishes to separate glia and other cell types, characteristic cultures of a pure cell population (GT-1) were established from the less adherent cell balls and cloned by serial dilution (GT1-1, GT1-3, GT1-7). The more adherent cells were also separated into different populations, but either failed to persist in culture for over six months or failed to express GnRH. Clonal cell lines are maintained in DMEM with 5% fetal bovine serum, 5% equine serum, 100 units/ml

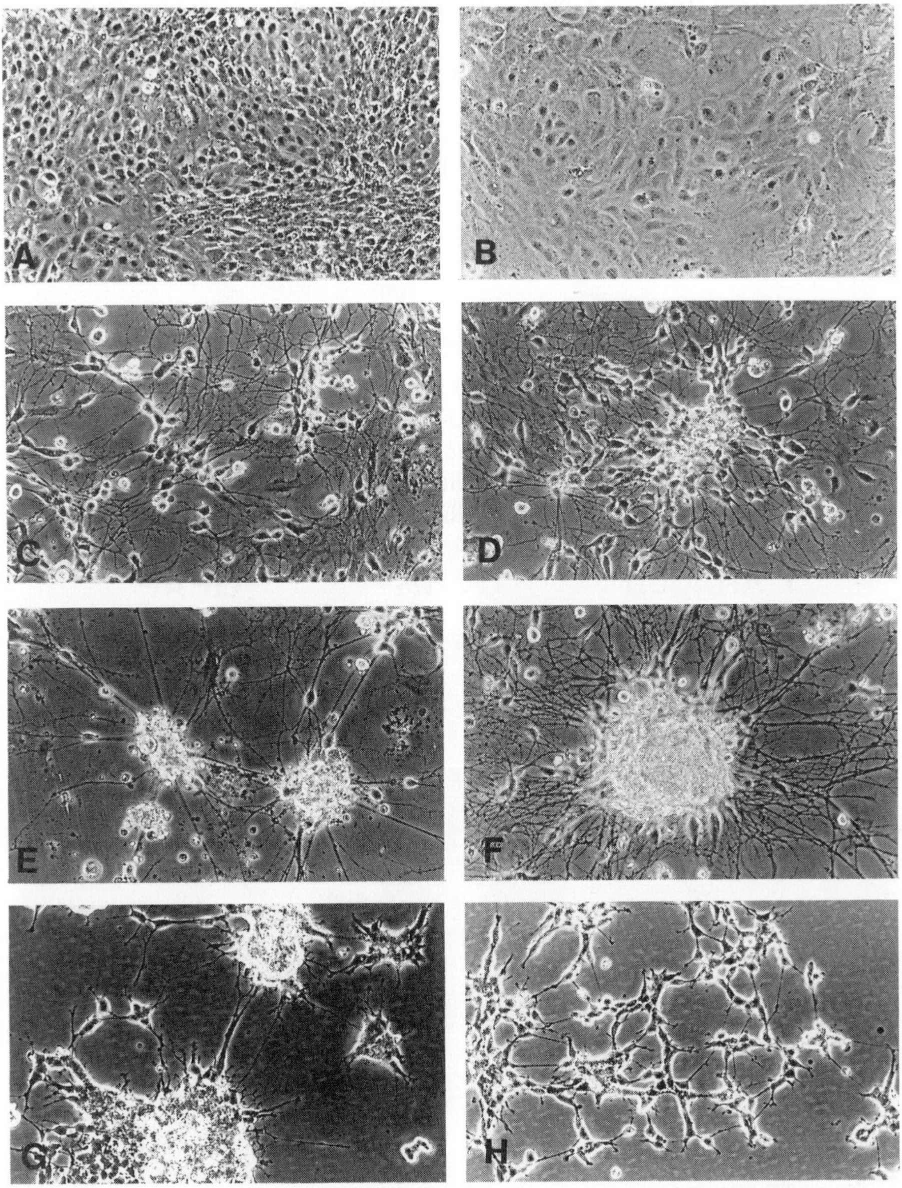

FIG. 2. Morphology of cultured cells from the GT-1 and GT-4 tumours. Phase contrast micrographs demonstrate the phenotypes of (A,B) GT-4 cell cultures, (C–F) complex early GT-1 cultures and (E,H) cloned GT1-3 cells, including neurite formation, growth cones and cell–cell contacts. Cells were cultured on plastic tissue culture plates in (A–F) regular medium or (G,H) OptiMEM (Gibco) without serum for three days. Magnification (A) ×70 and (B–H) ×175.

penicillin, 100 μg/ml streptomycin and 4.5 mg/ml glucose. These clonal cells continue to form ball-like structures (Fig. 2G) but also will spread on the plastic (Fig. 2H) and grow to confluence. Cloned cell lines attach to plastic culture dishes and exhibit a distinct neuronal phenotype, including the extension of multiple lengthy neurites which often contact distant cells or end in apparent growth cones (Fig. 2H). This neuronal phenotype is more profound when cells are cultured in the absence of serum to inhibit cell division. These cell lines have been propagated continuously for two years with no apparent change in phenotype and can be stored in liquid nitrogen.

Expression of GnRH and T antigen in GT cells

The expression of the transgene mRNA was investigated in cell cultures derived from the GT-1 and the GT-4 tumours. Northern blotting (Fig. 3) demonstrated the presence of T antigen mRNA in the purified cultures of both cell types but not in normal brain or in C6 cells, a glioma cell line. The expression of GnRH mRNA was found only in the GT-1 cells. Neither the GT-1 or GT-4 cell culture

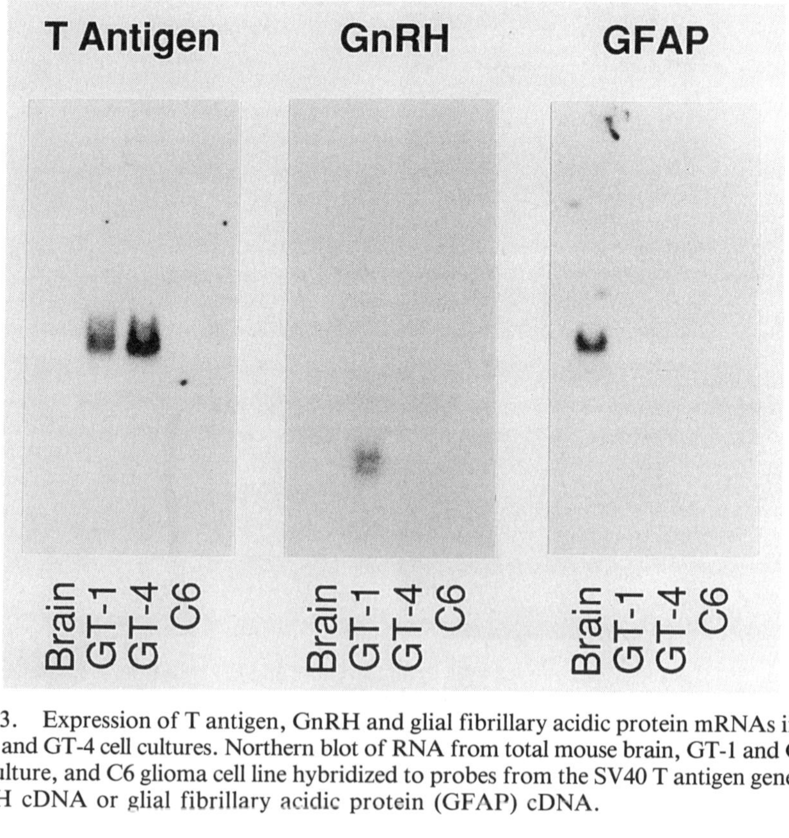

FIG. 3. Expression of T antigen, GnRH and glial fibrillary acidic protein mRNAs in the GT-1 and GT-4 cell cultures. Northern blot of RNA from total mouse brain, GT-1 and GT-4 cell culture, and C6 glioma cell line hybridized to probes from the SV40 T antigen gene, rat GnRH cDNA or glial fibrillary acidic protein (GFAP) cDNA.

expressed glial fibrillary acidic protein, the glial-specific form of intermediate filament (Lewis et al 1984). This confirms the non-hypothalamic derivation of the GT-4 tumour and is consistent with the non-neuronal phenotype displayed by these cells in culture. Thus, the GT-4 tumour may have been derived from choroid plexus, a tumour type observed commonly in transgenic mice carrying the T antigen gene (Brinster et al 1984), while the GT-1 tumour expresses GnRH, indicating its origin as hypothalamic.

GT1-7 Cells do not express genes for other neuropeptides

The neuroendocrine function of GT cells appears to be limited to GnRH expression. Hybridization with probes for somatostatin, corticotropin-releasing hormone, GHRH and pro-opiomelanocortin detected mRNAs corresponding to these hypothalamic neuroendocrine hormones in control brain polyA$^+$ mRNA, but failed to detect mRNAs in GT1-7 polyA$^+$ mRNA, indicating the highly specific nature of the expression of GnRH in GT-1 cells (Fig. 4). The glycolytic enzyme, enolase, occurs in three forms: non-neuronal enolase, expressed in glial cells; a muscle-specific form; and neuron-specific enolase which is strictly localized to central and peripheral neurons. The mRNA for the neuronal form was present in higher concentration in GT1-7 cells than in total brain. Vimentin is an intermediate filament protein which is common to many tissues (Wood et al 1989) and its mRNA is highly expressed in GT1-7 cells (Fig. 4).

Biosynthesis and processing of the GnRH precursor

The GT-1 cell lines (GT1-1, GT1-3 and GT1-7) synthesize nanogram quantities of GnRH and GnRH-associated peptide (GAP) (Wetsel et al 1991). Comparison of the GAP:GnRH molar ratios among the lysates and media revealed at least eightfold more GAP-like than GnRH-like immunoreactive materials. Thus, the GT-1 cell lines secrete more pro-GnRH or GnRH intermediate products that do GnRH nerve terminals in slice preparations (Wetsel & Negro-Vilar 1989). To characterize further the GnRH and GAP immunoreactivities, we separated materials in cell lysates and media using size-exclusion and high pressure liquid chromatography (HPLC). GnRH-like species were found to elute in the void volume and at approximately 1200 relative molecular mass (M_r) (Fig. 5A); the latter co-eluted with synthetic GnRH. GAP antiserum detected four species: in the void volume and at approximately 10 000, 8500 and 6700 M_r (Fig. 5B). The 6700 M_r species co-eluted with synthetic GAP-(1-56) both in the size-exclusion column and in subsequent reversed-phase column analysis (Wetsel et al 1991.) Because the mouse and rat GAP amino acid sequences differ from each other by only five residues (Mason et al 1986a), this GAP-like material is probably authentic mouse GAP-(1-56). By comparison, separation of the GnRH material from the size-exclusion column using HPLC revealed the

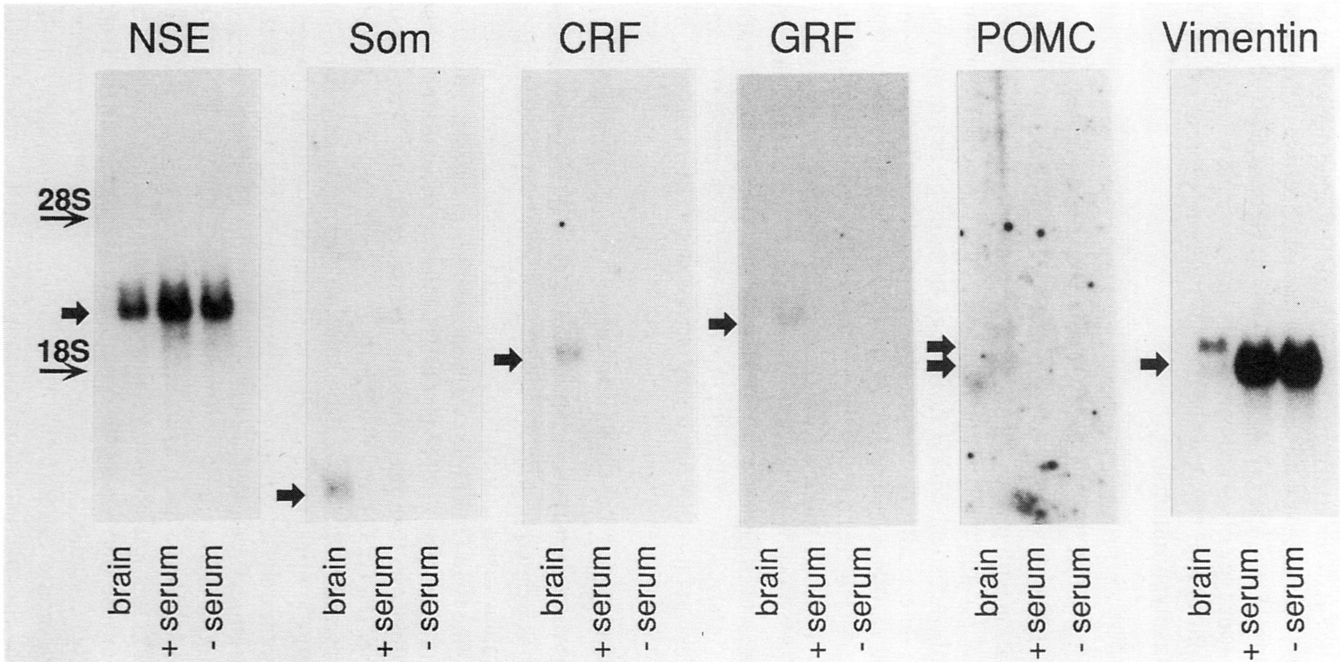

FIG. 4. GT cells express neuron-specific enolase and vimentin and fail to express several neuropeptide mRNAs. Northern blot of 5 μg of polyA$^+$ RNA from GT1-7 cells cultured in Dulbecco's modified Eagle's medium with (+) or OptiMEM (Gibco) without (−) serum is compared to 3 μg of polyA$^+$ RNA from mouse brain. There is expression of neuron-specific enolase (NSE) and vimentin but not somatostatin (Som; Montminy et al 1984), pro-opiomelanocortin (POMC; Eberwine & Roberts 1984), corticotropin-releasing hormone (CRF; Imaki et al 1989) and growth hormone-releasing hormone (GHRH; Mayo et al 1985) in GT1-7 cells. Expression of these neuropeptide genes occurs in very few cells in the brain but is detectable in polyA$^+$ RNA, while the GT1-7 cells are a homogeneous clonal cell line and expression is undetectable.

presence of at least four different peptides: GnRH-(Gln^1, Gly^{11}, Lys^{12}, Arg^{13}), GnRH-(Gln^1, Gly^{11}, Lys^{12}), GnRH-(Gly^{11}) and GnRH (Wetsel et al 1991). Taken together, these results indicate that all the GT-1 cell lines synthesize the pro-GnRH protein and process it to GAP(1-56) and GnRH as the final products. A similar scheme for processing the GnRH precursor has been described in rat hypothalamus (Wetsel et al 1988).

Regulated secretion of GnRH

While both GnRH and GAP were found in the media of the GT-1 cell lines (Wetsel et al 1991), it was important to investigate the regulation of their secretion. In a static culture system, both K^+-induced depolarization and veratridine caused stimulation of GnRH release and the veratridine-stimulated release could be blocked by tetrodotoxin (Mellon et al 1990). Tetrodotoxin treatment also inhibited basal secretion, suggesting the existence of spontaneous action potentials which stimulated GnRH release. To develop a more physiological model of GnRH secretion that would closely mimic the *in vivo* situation, we put the GT-1 cells into a perifusion system. When the GT1-7 cells were cultured on Cytodex beads, loaded into 1 ml syringes and perifused with Krebs–Ringer bicarbonate glucose buffer, basal secretion of GnRH rapidly declined to reach a steady state within 60 minutes (Fig. 5C). When either 56 mM $[K^+]$ or 200 nM phorbol 12,13-dibutyrate (PDBu) were added to the media, release of GnRH was rapid. Removal of the stimulus resulted in a more rapid return to basal level after K^+-induced depolarization than after PDBu stimulation. Furthermore, noradrenaline and dopamine stimulate release via a positively coupled adenylate cyclase pathway (unpublished 1992), indicating the presence of neurotransmitter receptors.

Basal secretion was analysed using cells in individual perifusion chambers; GnRH release was found to be pulsatile (Wetsel et al 1992). Similar results were obtained when GT1-1 cells were cultured on coverslips and perifused in a Sykes Moore chamber. The interpulse frequency found with either method was similar to that observed in castrated male rats and mice. Interestingly, this mode of secretion may have a morphological foundation in that the GT1-7 cells establish numerous synapse-like connections among themselves, as well as showing biochemical evidence of coupling (Lucifer yellow dye transfer). The latter observation suggests that the GT-1 cells may be electrically coupled in culture. Regardless, our observations indicate that the GT1 cells possess an intrinsic ability to generate pulsatile GnRH secretory activity. Furthermore, the GAP:GnRH molar ratio rapidly declines from a value greater than 16 to a level of 2 within the first 60 minutes of perifusion (data not shown). This more normal ratio is maintained for the remainder of the experiment. These results indicate that perifusion may route some of the pro-GnRH from the constitutive pathway and/or influence processing of this precursor.

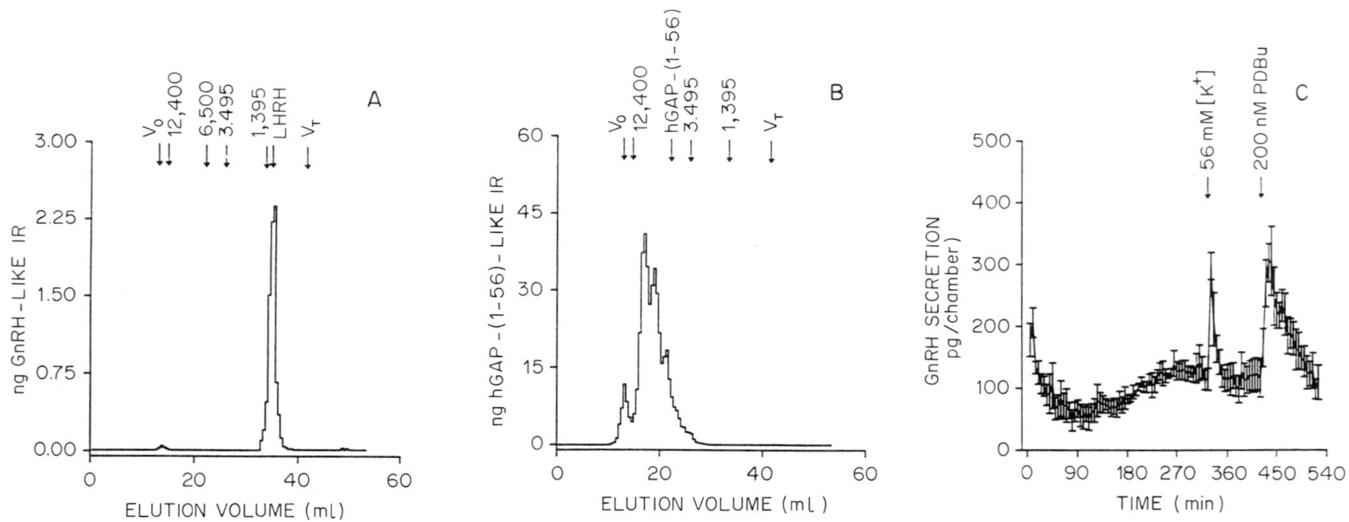

FIG. 5. Chromatographic analysis of GnRH and GnRH-associated peptide (GAP)-like immunoreactive materials and regulated secretion of GnRH. Cell lysates were fractionated by size and the fractions were screened with A772 anti-GnRH and MC-2 anti-GAP sera (Wetsel et al 1991). Molecular weight markers are shown at the top. V_0, void volume; V_t, total volume. (A) The immunoreactive profile for GnRH and (B) the GAP profile. (C) GT1-7 cells were grown on Cytodex beads, loaded into syringes and perifused for nine hours. Medium was collected at 5 minute intervals and GnRH measured from eight separate chambers by radioimmunoassay. Results are the mean $\pm$ SEM at each interval. The times of administration of 56 mM $[K^+]$ and 200 nM PDBu (phorbol 12,13-dibutyrate) are shown and the interval of administration was 15 minutes.

Discussion

We have produced tumours in transgenic mice and derived clonal GnRH-secreting cell lines which are immortal, differentiated hypothalamic neurons by targeting expression of an oncogene using the regulatory regions of the GnRH gene. Expression of the GnRH–T antigen hybrid gene produced uniform sterility in transgenic founder mice, indicating a specific effect of the transgene on reproductive function and/or development. A number of these mice developed brain tumours that were of non-hypothalamic origin, indicating that the transgene was sometimes also expressed ectopically. The development of immortal cell lines representing GnRH hypothalamic neurons required the recovery and successful culture of only one tumour (GT-1) caused by appropriately targeted oncogenesis.

The cells derived from the GT-4 and GT-1 tumours showed many different characteristics. The GT-4 cell cultures showed characteristics of epithelial monolayers and failed to express GnRH mRNA. These cells did express T antigen, indicating that they had been transformed by the ectopic expression of the transgene. However, this ectopic expression was not prevalent in other regions of the brain of this mouse or in other GnRH–Tag mice, since T antigen RNA was not detected by Northern blotting of total brain RNA (Mellon et al 1990 and data not shown). In contrast, the overall morphology of GT-1 cells is neuronal, with the spontaneous extension of neurites that end in growth cones or contact distant cells. During cell division, the GT-1 cells retract these extensions and divide as rounded cells, followed by re-extension of neurites and attachment to the dish. In the absence of cell division, the arbors of neurites are maintained and become more elaborate. Many specific characteristics of neurosecretory cells are maintained, such as the expression of specific neural and neuroendocrine markers, Golgi apparatus, rough endoplasmic reticulum, processing of the GnRH precursor to GnRH and GAP, secretory granules and regulated pulsatile secretion (Liposits et al 1991, Mellon et al 1990, Wetsel et al 1991). Thus, the GT-1 cell lines are an excellent cell culture model of the GnRH hypothalamic neuron *in situ* and provide an avenue for future molecular, physiological, cell biological and biochemical investigations into the characteristics of an isolated CNS neuron and the mechanisms involved in regulation of GnRH.

Acknowledgements

This research was supported by NIH grants HD 20377, DK 44838, and the Burns Foundation (P. L. M.) and NIH grant HD 08924 (R. I. W.). J. J. W. was a fellow of the American Cancer Society, California Division (#S-60-89) and D. B. W. a fellow of the N.I.H. (HD07299).

References

Baetge EE, Behringer RR, Messing A, Brinster RL, Palmiter RD 1988 Transgenic mice express the human phenylethanolamine N-methyl-transferase gene in adrenal medulla and retina. Proc Natl Acad Sci USA 85:3648–3652

Bond CT, Hayflick JS, Seeburg PH, Adelman JP 1989 The rat gonadotropin-releasing hormone: SH locus: structure and hypothalamic expression. Mol Endocrinol 3:1257–1262

Botteri FM, van der Putten H, Wong DF, Sauvage CA, Evans RM 1987 Unexpected thymic hyperplasia in transgenic mice harboring a neuronal promoter fused with simian virus 40 large T antigen. Mol Cell Biol 7:3178–3184

Brinster RL, Chen HY, Messing A, van Dyke T, Levine AJ, Palmiter RD 1984 Transgenic mice harboring SV40 T-antigen genes develop characteristic brain tumors. Cell 37:367–379

Cepko CL 1989 Immortalization of neural cells via retrovirus-mediated oncogene transduction. Annu Rev Neurosci 12:47–65

Choi Y, Lee I, Ross SR 1988 Requirement for the simian virus 40 small tumor antigen in tumorigenesis in transgenic mice. Mol Cell Biol 8:3382–3390

de Vitry F, Camier M, Czernichow P, Benda P, Cohen P, Tixier-Vidal A 1974 Establishment of a clone of mouse hypothalamic neurosecretory cells synthesizing neurophysin and vasopressin. Proc Natl Acad Sci USA 71:3575–3579

Eberwine JH, Roberts JL 1984 Glucocorticoid regulation of pro-opio-melanocortin gene transcription in the rat pituitary. J Biol Chem 259:2166–2170

Efrat S, Linde S, Kofod H et al 1988a Beta-cell lines derived from transgenic mice expressing a hybrid insulin gene–oncogene. Proc Natl Acad Sci USA 85:9037–9041

Efrat S, Teitelman G, Anwar M, Ruggiero D, Hanahan D 1988b Glucagon gene regulatory region directs oncoprotein expression to neurons and pancreatic α cells. Neuron 1:605–613

Frederiksen K, Jat PS, Valtz N, Levy D, McKay R 1988 Immortalization of precursor cells from the mammalian CNS. Neuron 1:439–448

Giotta GJ, Cohn M 1981 The expression of glial fibrillary acidic protein in a rat cerebellar cell line. J Cell Physiol 107:219–230

Giotta GJ, Heitzmann J, Cohn M 1980 Properties of two temperature-sensitive rous sarcoma virus transformed cerebellar cell lines. Brain Res 202:445–458

Hammang J, Baetge E, Behringer R, Brinster R, Palmiter R, Messing A 1990 Immortalized retinal neurons derived from SV40 T-antigen-induced tumors in transgenic mice. Neuron 4:775–782

Hanahan D 1985 Heritable formation of pancreatic β-cell tumours in transgenic mice expressing recombinant insulin/simian virus 40 oncogenes. Nature (Lond) 315:115–122

Hanahan D 1989 Transgenic mice as probes into complex systems. Science (Wash DC) 246:1265–1275

Imaki T, Nahon JL, Sawchenko PE, Vale W 1989 Widespread expression of corticotropin-releasing factor messenger RNA and immunoreactivity in the rat olfactory bulb. Brain Res 496:35–44.

Jenkins NA, Copeland NG 1989 Transgenic mice in cancer research. In: DeVita VT et al (eds) Important advances in oncology. Lippincott, Philadelphia, PA p 61–77

Lewis SA, Balcarek JM, Krek V, Shelanski M, Cowan NJ 1984 Sequence of a cDNA clone encoding mouse glial fibrillary acidic protein: structural conservation of intermediate filaments. Proc Natl Acad Sci USA 81:2743–2746

Liposits Z, Merchenthaler I, Wetsel WC et al 1991 Morphological characterization of immortalized hypothalamic neurons synthesizing luteinizing hormone-releasing hormone. Endocrinology 129:1575–1583

Mason AJ, Hayflick JS, Zoeller RT et al 1986a A deletion truncating the gonadotropin releasing hormone gene is responsible for hypogonadism in the *hpg* mouse. Science (Wash DC) 234:1366–1371
Mason AJ, Pitts SL, Nikolics K et al 1986b The hypogonadal mouse: reproductive functions restored by gene therapy. Science (Wash DC) 234:1372–1378
Mayo KE, Cerrelli GM, Rosenfeld MG, Evans RM 1985 Characterization of cDNA and genomic clones encoding the precursor to rat hypothalamic growth hormone-releasing factor. Nature (Lond) 314:464–467
McKay RDG 1989 The origins of cellular diversity in the mammalian central nervous system. Cell 58:815–821
Mellon PL, Windle JJ, Goldsmith P, Pedula C, Roberts J, Weiner RI 1990 Immortalization of hypothalamic GnRH neurons by genetically targeted tumorigenesis. Neuron 5:1–10
Montminy MR, Goodman RH, Horovitch SJ, Habener JF 1984 Primary structure of the gene encoding rat preprosomatostatin. Proc Natl Acad Sci USA 81:3337–3340
Ornitz DM, Hammer RE, Messing A, Palmiter RD, Brinster RL 1987 Pancreatic neoplasia induced by SV40 T-antigen expression in acinar cells of transgenic mice. Science (Wash DC) 238:188–193
Ronnett G, Hester L, Nye J, Connors K, Snyder S 1990 Human cortical neuronal cell line: establishment from a patient with unilateral megalencephaly. Science (Wash DC) 248:603–605
Rosenfeld MG, Crenshaw EB 3d, Lira SA et al 1988 Transgenic mice: applications to the study of the nervous system. Annu Rev Neurosci 11:353–372
Weiner R, Findell P, Kordon C 1988 Role of classic and peptide neuromediators in the neuroendocrine regulation of LH and prolactin. In: Knobil E, Neill J (eds) The physiology of reproduction. Raven Press, New York p 1235–1281
Wetsel W, Negro-Vilar A 1989 Testosterone selectively influences protein kinase-C-coupled secretion of proluteinizing hormone-releasing hormone-derived peptides. Endocrinology 125:538–547
Wetsel W, Culler M, Johnston C, Negro-Vilar A 1988 Processing of the luteinizing hormone-releasing hormone precursor in the preoptic area and hypothalamus of the rat. Mol Endocrinol 2:22–31
Wetsel WC, Mellon PL, Weiner RI, Negro-Vilar A 1991 Metabolism of pro-LHRH in immortalized hypothalamic neurons. Endocrinology 129:1584–1595
Wetsel WC, Valença MM, Merchenthaler I et al 1992 Intrinsic pulsatile secretory activity of immortalized LHRH secreting neurons. Proc Natl Acad Sci USA 89:4149–4153
Windle J, Weiner R, Mellon P 1990 Cell lines of the pituitary gonadotrope lineage derived by targeted oncogenesis in transgenic mice. Mol Endocrinol 4:597–603
Wool L, Theriault N, Vogeli G 1989 Vimentin cDNA clones covering the complete intermediate-filament protein are found in an EHS tumor cDNA library. Gene (Amst) 76:171–175

DISCUSSION

Lightman: What difference do you think it makes to the cell that it contains T antigen? Could that be changing those cells physiologically?

Mellon: It worries me less, the more we find that the cells' behaviour reflects what's known to be physiological *in vivo* behaviour. For instance, the cells secrete

GnRH in response to noradrenaline and dopamine, though we know they are not fully differentiated neurons in the sense defined by post mitosis. Morphologically, they carry cilia like other GnRH neurons. So for the kinds of investigations we are doing, I think these cells are accurate. I also think it is really important to take the knowledge that we gain and go back into the animal. In our gene expression studies that's rather straightforward, we just introduce the transgenes into mice again, and ask whether the GnRH gene enhancer that we have isolated can target expression only to the GnRH neurons on a heterologous promoter. Some of the other studies that people have done can be duplicated in slices once they have been worked out in the pure cultures.

Zoeller: If you applied a treatment that increased the amount of GnRH message, would you also increase the amount of T antigen and thereby drive an increase in cell division? If so, cellular levels of GnRH mRNA may not increase as much as the proportion of GnRH mRNA in an extract of the total RNA.

Mellon: I would think so. The question is whether T antigen is present in excess to begin with, so whether an increase in the amount of T antigen would change the rate of cell division. We haven't looked at the T antigen mRNA after various treatments. We know the level of T antigen mRNA goes down in response to phorbol esters, as does the level of GnRH mRNA.

Theodosis: From your images, these neurons, as many peptidergic neurons in culture, look quite dedifferentiated, even though they have granules and presumably secrete peptide. Have you looked for factors that facilitate their differentiation, such as different substrates, the presence or absence of glia, or the presence or absence of other neurons?

Mellon: We've tried different substrates: these cells like to sit on Matrigel, which is a complex medium; they are not so happy on polylysine. They do look a little more differentiated on the Matrigel but they are still dividing, so they are still bringing in their neurites. The way to make them look most differentiated is to take away serum; then they can't divide, so they don't pull in their neurites, the neurites get longer and the contacts get firmer.

We haven't done co-culturing with glia. I tried it once with a C6 glioma cell line, which did not work, the glioma cells just overgrew the GnRH cells. We have tried various growth factors. NGF seems to have very little effect. FGF has a morphological effect in that the neurites seem to get more branches, but these results are preliminary.

Clarke: The classical work done by Andrew & Dudek (1984) on non-identified neuroendocrine cells in the hypothalamus suggested that these cells had inherent phasic activity. From that it was suggested that all cells, like GnRH cells, might have their own inherent pulsatility and that it was only the projections to the cells that would regulate them in a coordinated manner. Do you have any electrical information about these cells? Do they have phasic electrical activity?

Mellon: Bertil Hille and Martha Bosma from the University of Washington, Seattle have done patch clamping and intracellular recording on these cells. The cells have spontaneous sodium spikes. The problem is that to demonstrate a half hour pulse, you really need to hold the patch about an hour and a half, and apparently that's very difficult, if not impossible. The cell electrophoreses itself and basically ceases activity much sooner than that. We are trying calcium imaging to investigate the phasic activity. That will help us answer the question of whether the clock is carried by the network of cells or the individual cells.

Clarke: Is there any intracellular recording?

Mellon: I think that patches may be easier to hold than intracellular recording.

Charlton: There is a simple explanation for this apparent clock. If a cell is a constitutive granule exocytoser and it is coupled with its neighbouring cells, then it exocytoses a set of granules at the periphery of the cell and it takes 45 minutes for the next set of granules to get there to be exocytosed. I don't necessarily think there must be a clock inside, it could be just something constitutive in exocytosing cells, as long as they are electrically coupled or coupled via gap junctions.

Pfaff: An apparent clock is a clock.

Plant: How many cells are coupled? Presumably there are millions of cells in the cultures; do you envisage that they are all talking to each other?

Mellon: I would say the majority of them are coupled. From the electron microscopy freeze fracture pictures there is no reason to doubt that they are all connected to each other. You would only need 50% or so to be in contact to see these kinds of spikes over background.

Plant: How does a spontaneous spike compare with a K^+-induced spike in the magnitude of its upstroke?

Mellon: I don't think the K^+ studies have been done in the perifusion system.

Plant: Presumably, in the transgenic animal these cells would be pulsing. Have you attempted to describe such activity using push–pull perfusion in the region of the brain bearing the tumour? This might shed light on their *in vivo* pattern of secretion.

Mellon: No. These animals are sterile and very rare, so we cannot generate enough to try this kind of experiment.

Pfaff: Could you tell us more about the GnRH promoter? Is there a cyclic AMP response element? Is there anything that looks like an AP-1 binding site? Is there an oestrogen or a glucocorticoid response element?

Mellon: We don't see the standard cAMP response element; we don't see any effect of forskolin on expression of either the transfected gene or the endogenous one. We do see down-regulation in response to phorbol ester. There is a canonical AP-1 site but it's possible to delete that region and still have full down-regulation.

When we transfect these cells with 3 kb of the 5′ flanking region of the rat GnRH gene, deplete the serum, then treat with oestrogen, we see no effect on transcription of the endogenous or the transfected GnRH gene. Binding studies show no oestrogen receptor. Oestrogen receptor mRNA cannot be detected by Northern blots or using PCR. If we co-transfect with oestrogen receptor expression vectors, we still see no regulation of the co-transfected GnRH reporter gene.

Pfaff: That's reassuring to us. It took us much work to show a disjunction between the oestrogen receptor and GnRH, thus defeating all of our previous hypotheses. Have you done any DNase footprinting experiments with the GnRH promoter?

Mellon: We have done some footprinting in the area of the promoter, the first 180 bp, and see a few footprints. We have also seen footprints in 300 bp that include the enhancer region. Footprinting often gives detectable signals, but the strength of the signals that we see in the enhancer region makes them very convincing.

Pfaff: Is there some obvious physical characteristic of the protected region in the enhancer? Is there a specific sequence—a preponderance of purines or pyrimidines?

Mellon: Not really; it is rather AT-rich. We have done a set of gel-shift competition experiments trying to determine how many of the different binding proteins are related to each other. There are some subclasses that may be related, so it may not be individual proteins binding to all the different sites. We are now purifying the binding proteins and screening for them by a number of different methods.

Young: Do these cells have GnRH receptors?

Mellon: They make so much GnRH that we could not see binding of exogenous labelled GnRH, probably because their receptors are blocked. The only indication that they might have GnRH receptors is in some preliminary results from Martha Bosman and Bertil Hille. They patch-clamped and were able to wash the patch enough, add GnRH and see some electrical changes. I believe these cells do have GnRH receptors but we haven't shown that.

Young: Do these cells also make eicosanoids of some sort as a transmitter?

Mellon: We haven't tested for that.

Theodosis: Have you looked for any other kinds of receptors on these cells?

Mellon: We have failed to detect oestrogen receptors. We have tested for glucocorticoid receptors and not found them, but we are not as confident about that. We have not tested for the presence of other specific receptors. We have found a large number of PCR products for seven-transmembrane domain proteins. My feeling is that these are novel receptors, but we don't know what they are.

There are adrenergic receptors and dopamine receptors; exactly which types is not yet determined. You can detect these receptor mRNAs by Northern blotting and you can detect them by secretion studies with different blockers and activators.

Gainer: Could the controversy over the different receptor subtypes be due to the use of different sublines of cells?

Mellon: I think everyone is using GT1–7. Some labs may be using GT1–3, which could account for the differing results.

Barraclough: Richard Weiner and colleagues at UCSF have found that noradrenaline releases GnRH from GT-1 cells and this effect is blocked by an α_2 receptor antagonist rather than an α_1-type antagonist, which one would normally expect to block GnRH release. *In vivo*, α_2 receptor antagonists do not block spontaneous GnRH release. Thus, these GT-1 cells seem to respond differently to normal GnRH cells.

Liposits: There is a subset of GnRH-synthesizing neurons that also make galanin. Have you looked for galanin mRNA expression?

Mellon: William Wetsel and Andres Negro-Vilar at Research Triangle Park have looked and not found it.

Gainer: If the GnRH gene is being expressed during the cell's migration from the nose to the brain, why is it not producing a tumour in the nose?

Mellon: The experiments of Douglas Hanahan at UCSF on the pancreas, which are the most well characterized, show that T antigen is not enough to cause a tumour, a second event is necessary. He has driven expression of the T antigen with the insulin promoter, so he is looking at the pancreatic islet cells. Most of the islets show a little hyperplasia, but only occasional islets develop into real tumours. He thinks that tumour formation requires angiogenesis, but it also may require a second oncogenic event.

From our analysis of the chromosomes it is hard to say whether these tumours are clonal. I would guess that they might be clonal and that's why you don't get them in most of the mice. T antigen expression only predisposes the cell to form a tumour. In our pituitary tumours we also get focal tumorigenesis through development. I think it has also been shown in the exocrine pancreas in transgenic animals.

Charlton: When Mason et al (1986) at Genentech replaced the GnRH gene in hypogonadal mice, they saw a weak *in situ* hybridization signal in the paraventricular nucleus of the thalamus.

Zoeller: They also found GnRH mRNA in liver and several other tissues of transgenic hypogonadal mice. Were you surprised that you didn't see expression of this gene in other tissues?

Mellon: I thought we would see expression in ovaries and placenta. We never did check placenta because we would have had to do Caesarian sections on the original transplanted animals to recover the placentas. We saw no expression of T antigen in the ovaries. We did find some non-hypothalamic brain tumours but we didn't find tumours of other tissues.

The mice carrying tumours outside the hypothalamus were sterile but we didn't look at those particular animals in more detail. We did look at the ones that didn't get any tumours and they had no projections of GnRH neurons to the median eminence.

Barraclough: Was the migration of the neurons normal in the brains of these animals?

Mellon: In the adult animals that we looked at, they were present in the right regions, but they didn't have extensions to the median eminence. Richard Weiner and Paul Goldsmith did additional microinjections of transgenes and killed the newborn pups or took the later fetuses and looked at migration of the T antigen/GnRH neurons. They were looking at a very small number of animals. Because we can't breed these animals and the efficiency of making primary animals is very low, it was a heroic experiment. They found some animals that were positive for the DNA. In those, the cells were migrating and as far as they could tell migration was normal.

Barraclough: Radovick et al (1991) found that in transgenic mice, expression of the SV40 T antigen driven by the promoter of the human GnRH gene had arrested migration during development. The cells formed tumours along the migratory pathway in the olfactory bulb rather than entering the hypothalamus.

Mellon: We did not find tumours from the nasal area, thus our results differed from theirs. They used a different piece of DNA: it is from the human gene cut at −1100. The enhancer in the rat gene is upstream of that. So if the human gene organization is similar, their construct wouldn't have had the analogous enhancer for the human gene. We don't know yet whether the human gene enhancer is upstream of −1100, but if it is anything like the rat gene, they would have not had the primary targeting elements.

Charlton: It is important to do a developmental study, because the GnRH cells in your animals might have arrived at the preoptic area too late to get whatever signals are essential for growth cones to follow their normal pathway to the median eminence.

Gainer: How do you interpret the fact that when you wash the cells you get processing of the GnRH precursor, and when you don't wash the cells, you don't?

Mellon: Negative feedback of GnRH on its own processing.

Gainer: But it was unprocessed, so it wasn't GnRH.

Mellon: Not really. The ratio of GnRH-associated peptide (GAP) and the precursor to properly processed GnRH goes from 16:1 to 2:1. We have two antibodies, one detects GAP and the precursor, the other detects GnRH. In static cultures that have been sitting around for a couple of days, we find 16-fold more signal with the first antibody than with the anti-GnRH antibody. After washing the cells, we detect twofold more with the first antibody than with the anti-GnRH antibody. The GnRH is still there, but there is less of it. In unwashed cultures there is probably more total GnRH than in the washed cultures, because in the latter you are washing it away rapidly.

Theodosis: Isn't the ratio of the precursor to mature GnRH also linked to the differentiation state of the cell?

Mellon: That's possible. I couldn't say whether the cells are more or less differentiated in the absence of GnRH or the precursor.

Gainer: It could be that there is degradation in the static circumstance and the thing that was more sensitive was the decapeptide not the GAP peptide. The antibodies against the decapeptide are very specific, the peptide has to be amidated, etc.

Mellon: The materials have been fractionated by HPLC; the precursor as opposed to the GAP is much more abundant in static culture. That suggests that GAP processing is increasing in the washed cultures.

Nagy: What is the effect of washing the cells on T antigen production?

Mellon: The T antigen is not secreted, it's in the nucleus. We haven't looked at the protein at all. We haven't compared the amounts of T antigen mRNA in washed cultures versus static cultures either.

Pfaff: You and Dr Negro-Vilar get some records in which the interpulse intervals are about 30–40 minutes. How frequently does that occur?

Mellon: These interpulse intervals are seen virtually every time. Richard Weiner has done the same experiment but in a very different way. He allows cells on coverslips to grow to confluency, then perfuses the coverslips in a different kind of apparatus. I understand that he also sees pulses at intervals of 30–40 minutes.

Poulain: What is the rate of sampling?

Mellon: I think Richard Weiner uses 2.5 minute intervals; William Wetsel and Andres Negro-Vilar use five minutes.

Pfaff: Could you tell us more details of the cell–cell contacts? You said they looked like synapses.

Liposits: We stained the immortalized GT1-7 cells with an antiserum raised against GAP. When we looked at these cells, we saw highly differentiated neurons (Fig. 1). They have short, dendrite-like and long, varicose, axon-like processes. The processes establish connections with each other that result in a neural network. Collateral processes arising from axons have also been shown to terminate on other GT1-7 cells (Liposits et al 1991). Ultrastructurally, the majority of the connections are tight junctions. There are also synapse-like formations between axonal processes and cell bodies. These connections are characterized by differentiated pre- and postsynaptic membranes and presynaptic densities.

We also cultured the cells on Cytodex beads (Fig. 1f) and put them into cylinders for studying hormone release from the cells (Wetsel et al 1992). Scanning electron microscopy revealed that the neurons covered the beads and seemed to bridge from one bead to another after a four hour incubation. Varicose axons have also been seen growing and climbing on the surface of other cells.

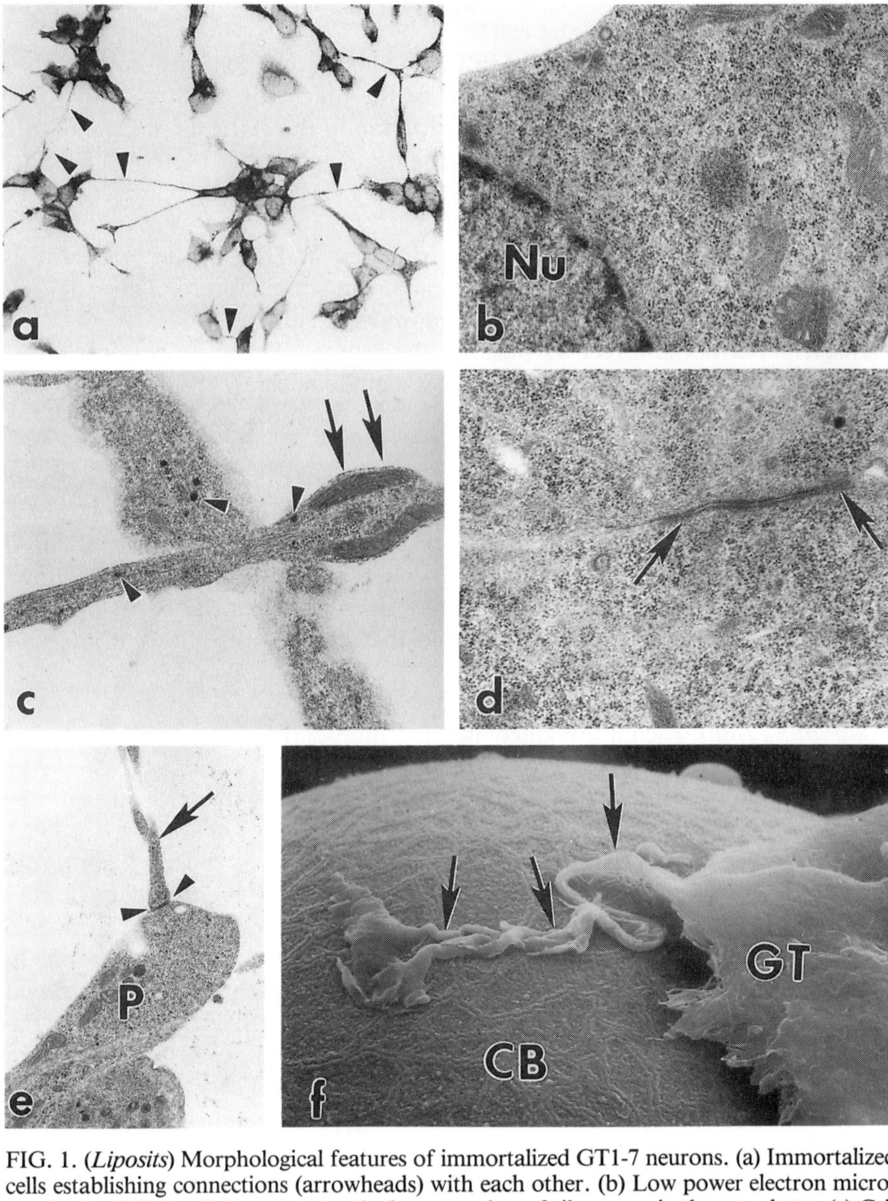

FIG. 1. (*Liposits*) Morphological features of immortalized GT1-7 neurons. (a) Immortalized cells establishing connections (arrowheads) with each other. (b) Low power electron micrograph of a GT1-7 cell demonstrates the large number of ribosomes in the cytoplasm. (c) Cell processes crossing each other. Arrowheads indicate secretory granules; arrows point to a varicosity. (d) Two GT1-7 cells coupled by a tight junction (arrows). (e) Cell process (arrow) terminating (arrowheads) on a perikaryon (P). (f) Scanning electron micrograph of a GT1-7 neuron (GT) growing on a Cytodex bead (CB). Arrows indicate the emerging axon. Nu, nucleus.

In addition, we have to consider the gap junctions as anatomical loci of communication. We injected Lucifer yellow into GT1-7 cells and found that the dye was transferred among the cells (Wetsel et al 1992). These connections might be responsible for the pulsatile secretion of GnRH from the immortalized neurons.

Palkovits: How uniform are those cells in terms of peptide content? Do they all contain GAP? Do they contain other peptides?

Liposits: We did immunocytochemical labelling and looked by both light and electron microscopy. We saw intense labelling for GAP and a less intense labelling for GnRH in all cells using the colloidal immunogold labelling technique with silver enhancement. The GAP is present over the ribosomes as well as in secretory vesicles. Double-labelling studies showed that GAP and GnRH are present in the same secretory vesicle (Liposits et al 1991). These are the only two peptides I have looked at so far.

Palkovits: Would you say that all synaptic contacts are GAP synapses?

Liposits: Yes, the contacts are immunopositive for GAP/GnRH.

Palkovits: Have you ever seen GAP-negative synapses there?

Liposits: No.

Barraclough: It's my impression that in the rat brain there are dendrodendritic-type contacts between GnRH neurons. The axons of these cells pass ventrally and medially through the anterior hypothalamus and terminate in the median eminence. To my knowledge, no one has described axons projecting rostrally to make contact with other GnRH neurons.

Liposits: I think the GnRH neurons in the rat are special: they are rather primitive fusiform cells with two processes. One of these is the varicose axon, which does contact other GnRH neurons (Liposits et al 1984).

Barraclough: It seems to me that the broad distribution of GnRH neurons, which are localized extending from the diagonal band of Broca to the medial preoptic area, would preclude that each cell can talk to each other GnRH cell. Hoffman et al (1990), in studies on expression of c-*fos* protein, showed that even when there are large releases of LH, there is not expression of c-*fos* protein in every GnRH cell: only GnRH cells in the preoptic area and anterior hypothalamus expressed c-*fos*, while more rostral GnRH neurons did not appear to be stimulated. These observations suggest that some cells are not active. If GnRH cells communicated with one another, as they seem to do on the Cytodex beads, you would expect activation of most or all of the GnRH neurons. This type of arrangement seen in culture does not exist in the rat brain.

Liposits: The signs of axosomatic communication were observed in the region of the diagonal band of Broca. This is the same pattern that we see among cultured immortalized cells.

Barraclough: But there are dendrodendritic synapses.

Liposits: I can't recall all the published data, but Silverman (1988) has shown axosomatic-type synapses between GnRH immunoreactive elements.

Palkovits: You couldn't find projections to the median eminence; does this mean those cells are not projecting neurons? GnRH cells could project to the hypothalamus, some fibres to the organum vasculosum laminae terminalis and some to other regions of the central nervous system.

Mellon: At the moment only Paul Goldsmith would know because I haven't seen the micrographs.

References

Andrew RD, Dudek 1984 Analysis of intracellularly recorded phasic bursting by mammalian neuroendocrine cells. J Neurophysiol 51:552–566

Hoffman GE, Lee WS, Attardi BN, Yann V, Fitzsimmons MD 1990 Luteinizing hormone-releasing hormone neurons express c-fos antigen after steroid activation. Endocrinology 126:1736–1741

Liposits Zs, Sétáló Gy, Flerkó B 1984 Application of the silver-gold intensified 3,3'-diaminobenzidine chromogen to the light and electron microscopic detection of the luteinizing hormone-releasing hormone system of the brain. Neuroscience 13:513–525

Liposits Zs, Merchenthaler I, Wetsel WC et al 1991 Morphological characteristics of immortalized hypothalamic neurons synthesizing luteinizing hormone-releasing hormone. Endocrinology 129:1575–1583

Mason AJ, Pitts SL, Nikolics K et al 1986 The hypogonadal mouse: reproductive functions restored by gene therapy. Science (Wash DC) 234:1372–1378

Radovick S, Wray S, Lee E et al 1991 Migratory arrest of gonadotropin-releasing hormone neurons in transgenic mice. Proc Natl Acad Sci USA 88:3402–3406

Silverman A-J 1988 The gonadotropin releasing hormone (GnRH) neuronal systems: immunocytochemistry. In: Knobil E, Neill DJ (eds) The physiology of reproduction. Raven Press, New York, p 1283–1304

Wetsel WC, Valenca MM, Merchenthaler M et al 1992 Intrinsic pulsatile secretory activity of immortalized LHRH secreting neurons. Proc Natl Acad Sci USA 89:4149–4153

Regulation of gene expression in the hypothalamus: hybridization histochemical studies

W. Scott Young III

Laboratory of Cell Biology, National Institute of Mental Health, Bethesda, Maryland 20892, USA

Abstract. Hybridization histochemistry has bridged molecular biology and neuroanatomy to provide nearly dynamic views of gene expression in the brain—perhaps especially in the hypothalamus. These snapshots of transcript levels with precise anatomical localization have revealed new insights into gene regulation in the hypothalamus under specific conditions. Magnocellular neurons in the paraventricular and supraoptic nuclei produce vasopressin and oxytocin. Transcript levels for these hormones are affected by hyperosmolality, as are those for many other neuropeptides. Patterns of gene expression in the magnocellular neurons in these nuclei during development and under different physiological conditions have been studied less extensively. The parvocellular neurons of the paraventricular nucleus produce corticotropin-releasing factor and thyrotropin-releasing hormone. Expression of the corticotropin-releasing factor gene is regulated by glucocorticoids. Physiological stresses, which activate the hypothalamo–pituitary–adrenal axis, also affect gene expression in the parvocellular paraventricular nucleus. Thyrotropin-releasing hormone is synthesized in a different set of parvocellular neurons in the paraventricular nucleus and in other neurons of the hypothalamus. Expression of the thyrotropin-releasing hormone gene is regulated by thyroid hormone. The suprachiasmatic nucleus contains neurons that produce vasopressin or vasoactive intestinal polypetide in a circadian rhythm. Future studies using combinations of classical neuroanatomical techniques, hybridization histochemistry and immunohistochemistry will further our understanding of hypothalamic responses to various stimuli.

1992 Functional anatomy of the neuroendocrine hypothalamus. Wiley, Chichester (Ciba Foundation Symposium 168) p 127–143

Hybridization histochemistry (*in situ* hybridization) is a powerful technique that can also provide a dynamic picture of gene expression. The hypothalamus, being a major site of convergence of diverse stimuli that arise from the internal and external milieus of the organism, has been a prime target of this technique. Investigators can manipulate the animal's environment and then observe changes in transcript levels in various regions of the hypothalamus. In this paper, I will focus on the paraventricular (PVN), supraoptic (SON) and suprachiasmatic (SCN) nuclei where many prototypical studies have been performed. The first two regions contain neurons that synthesize neuropeptides that are transported to and released from nerve endings of the median eminence portal circulation and posterior pituitary, whereas the SCN contains neurons that exhibit circadian rhythms. Because of space limitations, I will not be able to discuss every region of the hypothalamus (most notably, the arcuate and periventricular regions), comprehensive anatomical distribution, or all the relationships between changes in transcript levels and related physiological studies; I will also have to omit relevant biochemical studies that predate or confirm the hybridization histochemical studies. The gonadotropin-releasing hormone and other genes under gonadal steroid regulation will be discussed elsewhere in this volume by Drs Pfaff and Barraclough.

Paraventricular and supraoptic nuclei—magnocellular neurons

The PVN and SON are best known for producing vasopressin and oxytocin, which are involved in regulating salt and water balance, parturition and lactation (Leng & Bicknell 1986). The influence of hyperosmolality on gene expression in the PVN and SON has been studied extensively, in part because of the known increases in vasopressin and oxytocin release that it causes. As expected, rats that are dehydrated or given 2% saline chronically to drink show elevated levels of transcripts for vasopressin and oxytocin (Uhl et al 1985, McCabe et al 1986, Sherman et al 1986, Lightman & Young 1987b, Van Tol et al 1987). Between 4 and 10% of the neurons containing higher transcript levels project to the spinal cord or medullary vagal complex (Pretel & Piekut 1989). The onset of the vasopressin gene's responsiveness to hyperosmolality in the SON appears to develop between gestational Days 19 and 21 (E19 and 21), which is later than the onset of its expression (Reppert & Uhl 1988). Humans who are hypovolemic (have a reduced blood volume) before death also have raised levels of vasopressin in the SON (Rivkees et al 1989).

Numerous genes for other neuropeptides (Sawchenko et al, this volume) and proteins are also expressed in these nuclei and are similarly up-regulated in response to chronic hyperosmolality. These include carboxypeptidase H, corticotropin-releasing factor (CRF), cholecystokinin, dynorphin, galanin and tyrosine hydroxylase (Sherman et al 1986, 1988, Young 1986, Lightman & Young 1987b, Young et al 1987b, 1990a, Bondy et al 1989b, Meister et al 1990), as well as one for a ribosomal RNA (Kawata et al 1988). The coordinate control of these genes for neurohormone production undoubtedly leads to finer regulation of peptide release in the posterior pituitary (e.g. Falke 1988, Bondy & Gainer 1989, Bondy et al 1989a), and these peptides may regulate pituicyte function (e.g. degree of axon terminal development) as well.

Hyperosmolality also results in coexpression of vasopressin and oxytocin in 1–2% of the magnocellular neurons (Kiyama & Emson 1990). Levels of transcripts encoding several proteins that may be involved in regulating gene expression are also increased in response to hyperosmolality: c-*fos* (Sager et al 1988), that for protein kinase C-ζ (Young 1989) and those for an inhibitory and a stimulatory α-subunit of the GTP-binding protein family (Young et al 1987a). The amount of cAMP and the stimulatory α-subunit of the GTP-binding protein are also increased in the SON. Meister et al (1990) and Sherman et al (1988) have examined in detail the relationships between mRNA and peptide levels after salt-loading and support the general rule that more transcripts lead to more translation products. Interestingly, chronic ethanol ingestion by rats did not affect plasma vasopressin and reduced the number of vasopressin transcripts, despite increased plasma osmolality (Ishizawa et al 1990).

The Brattleboro rat has diabetes insipidus caused by a single base deletion in the second exon of the vasopressin gene (Schmale et al 1984). In this model of chronic hyperosmolality, mRNA levels for oxytocin, cholecystokinin, dynorphin, galanin and tyrosine hydroxylase are increased (Young et al 1986b, 1987b, Rökaeus et al 1988, Sherman et al 1988). Despite lower (compared to the parent strain, Long–Evans) levels of vasopressin transcripts, the Brattleboro rat does show increased levels of vasopressin transcripts in response to chronic intermittent salt-loading (Sherman et al 1988).

The effects of pregnancy and lactation on gene expression in the magnocellular neurons of the PVN and SON have been studied less extensively. Pregnant rats have higher levels of oxytocin transcripts (McCabe et al 1986); lactating rats have higher levels of oxytocin, dynorphin and vasopressin transcripts (Lightman & Young 1987b). Lactating mice have higher levels of transcripts for oxytocin (Young et al 1990b). Lactation leads to vasopressin expression in about 17% of the oxytocin neurons (Mezey & Kiss 1991), perhaps to compensate for fluid loss.

Developmental studies have also been rather limited. Vasopressin transcripts were detected at E16 and E18 in the rat SON and PVN, respectively; oxytocin transcripts were detected at E17 and E18, respectively (Reppert & Uhl 1987, Laurent et al 1989).

The effects of other physiological states or perturbations have been studied even more sporadically in the magnocellular neurons of the PVN and SON. Ovariectomy prevents the increase in oxytocin mRNA levels during sexual maturation of female rats (Miller et al 1989). Conversely, chronic treatment with oestradiol elevates oxytocin mRNA levels in the SON (and magnocellular neurons of the anterior commissural nucleus) of ovariectomized rats (Chung et al 1991). The use of various surgical lesions, especially deafferentations of the PVN and SON, should be a fruitful approach to understanding gene expression in these nuclei. There appears to be an input to the PVN and SON from the medial posterior hypothalamus that inhibits the expression of galanin without influencing that of the coexpressed vasopressin (Young et al 1990a,

Bachus SE, Palkovits M, Young WS, unpublished observations 1991). Hypophysectomy generally depresses vasopressin and oxytocin expression while increasing galanin and cholecystokinin expression (Villar et al 1990, Young et al 1990a). The significance of these changes, especially the increases, is unclear, but may relate to repair mechanisms.

Paraventricular nucleus—parvocellular neurons

Corticotropin-releasing factor and co-expressed genes

CRF and thyrotropin-releasing hormone (TRH) are each present in one of the two major separate populations of small cells in the PVN. CRF neurons express numerous other neuropeptides (Sawchenko et al, this volume) and these neurons have been the subject of many hybridization histochemical studies. As CRF is viewed as the primary neurohormone sitting atop the hypothalamo–pituitary–adrenal axis (Makara, this volume), initial studies examined the effect of adrenalectomy on CRF expression. As expected, the absence of corticosterone (and its negative feedback) leads to increased levels of transcripts encoding CRF (Young et al 1986a), as well as vasopressin (Young et al 1986b), in these parvocellular neurons. The latter observation agrees with immunohistochemical data showing vasopressin immunoreactivity in CRF neurons after adrenalectomy (Tramu et al 1983). Dexamethasone implants near one PVN unilaterally depress CRF expression (Kovács & Mezey 1987); however, parvocellular neurons with descending projections appear to respond in the opposite fashion to corticosterone (Swanson & Simmons 1989). The dorsal hippocampus also appears to relay glucocorticoid negative feedback to the parvocellular PVN (Herman et al 1989b), and afferents arising caudal to the PVN undoubtedly relay certain sensory-generated signals (for example Herman et al 1990 and intraperitoneal injections of hypertonic saline, as discussed below). Levels of transcripts for peptidyl-glycine α-amidating monooxygenase, which may be involved in amidating vasopressin and CRF, are also increased or decreased in parvocellular neurons after adrenalectomy or corticosterone treatment, respectively (Grino et al 1990). These treatments do not affect magnocellular vasopressin expression (Young et al 1986b, Grino et al 1990). The effect of adrenalectomy on vasopressin expression in the parvocellular PVN is augmented by lesions of the lateral septum (Baldino et al 1988).

Stresses, which activate the hypothalamo–pituitary–adrenal axis, were also examined for their effects on gene expression in the parvocellular neurons of the PVN. Initially, naloxone-precipitated morphine withdrawal and intraperitoneal hypertonic saline injections, but not drinking 2% saline (a larger osmotic load that actually depresses CRF mRNA levels in the parvocellular neurons, Young 1986), were shown to increase levels of transcripts encoding CRF, enkephalin and vasopressin (Lightman & Young 1987a, 1988). Further studies using the hypertonic saline injection model of acute stress showed that the CRF response is not appreciably affected by additional glucocorticoids, although the effect of adrenalectomy is additive (Lightman & Young 1986a). Levels of CRF, unlike enkephalin, transcripts are not sensitive to high doses of glucocorticoids (Harbuz et al 1990), unless these are given by local implantation (Harbuz & Lightman 1989a). Lesions of the ventral noradrenergic bundle or the PVN with 6-hydroxydopamine also did not affect basal or stress-induced levels of CRF or enkephalin transcripts, indicating an absence of major catecholamine influence (Harbuz et al 1991b). Chronic injections of hypertonic saline do not significantly increase CRF mRNA levels in the parvocellular PVN beyond that seen after a single injection, but do greatly increase enkephalin transcript levels in the magnocellular neurons of the PVN and SON (Young & Lightman 1992).

Lactation is a powerful inhibitor of this hypertonic saline stress response but does not affect the increase in CRF trancripts produced by adrenalectomy (Lightman & Young 1989b). Domperidone-induced hyperprolactinaemia does not affect the stress response either in male or non-lactating female rats, but the lactation-induced inhibition of the stress response is reversed by removing the litter for 48 hours (Lightman & Young 1989b). Increases in CRF and enkephalin transcripts caused by naloxone-induced morphine withdrawal (both agents delivered through intracerebroventricular infusions) are not blocked by urethane anaesthesia, even though there is loss of consciousness and inhibition of CRF release (Harbuz et al 1991a).

Other stresses have also been examined for their effects in the parvocellular PVN. Intraperitoneal injections of streptococcal cell wall preparations elicit arthritis in Lewis rats, but not in histocompatible Fisher rats which, unlike the Lewis rats, increase their levels of CRF and enkephalin transcripts after the injections (Sternberg et al 1989). Exposure to cold or to cold and isolation raises vasopressin mRNA levels in the parvocellular PVN (Zoeller et al 1990, Angulo et al 1991), but exposure to cold alone does not affect CRF or enkephalin transcript levels there (Harbuz & Lightman 1989b). Restraint and swim stresses each elevate CRF transcript levels (Harbuz & Lightman 1989b). Chronic electroconvulsive shock treatment increases CRF mRNA in the parvocellular PVN but increases vasopressin transcripts in only the magnocellular neurons of the SON (Herman et al 1989a). Chronic ethanol exposure (Rivier et al 1990) and intermittent foot-shocks (Imaki et al 1991) increase levels of CRF transcripts in the parvocellular PVN, whereas food restriction (Brady et al 1990) reduces them there. Intracerebroventricular colchicine and intraperitoneal reserpine each increase levels of transcripts for CRF, enkephalin, neurotensin and vasoactive intestinal polypeptide in parvocellular neurons of the PVN also (Ceccatelli et al 1991).

CRF transcripts in the parvocellular neurons are first detected on E17 in rats, then increase until E20. There is a subsequent decline in levels to a nadir on the first postnatal day (P1), followed by an increase to adult levels by P7

(Grino et al 1989). This period of decreased CRF mRNA levels corresponds to a period of reduced responsiveness to stress and is associated with an absence of responsiveness to adrenalectomy in the CRF neurons on P7; this response is restored by P14 (Grino et al 1989). The levels of CRF transcripts in the parvocellular PVN are also subject to periodic changes; there is a circadian cycle with a nadir at midnight (Watts & Swanson 1989) and an oestrus cycle with a sharp rise at the time of the ovulatory surge and a subsequent decline (Bohler et al 1990).

Thyrotropin-releasing hormone

TRH is present in parvocellular non-CRF neurons of the PVN as well as in other neurons of the hypothalamus (Segerson et al 1987a) and provides another example of regulation by negative feedback. Animals that have been chemically thyroidectomized have raised levels of TRH mRNA in the PVN, and this increase is reversed by treatment with injection of triiodothyronine (Koller et al 1987, Segerson et al 1987b). Triiodothyronine acts directly upon the PVN as shown by unilateral implants (Dyess et al 1988) and the regulation of the TRH mRNA levels is independent of thyroid-stimulating hormone, the pituitary (Zoeller et al 1988) and catecholamine input (Dyess et al 1988). In agreement with this site of feedback of triiodothyronine, transcripts encoding α_1- and β_1-thyroid hormone receptor genes are in the PVN (Bradley et al 1989, unpublished observations).

TRH mRNA is first detected in the hypothalamus on E14 and in the PVN on E16 and reaches an adult pattern at P21 (Burgunder & Taylor 1989). The onset of feedback regulation occurs between E20 and P7 (Taylor et al 1990). Cold exposure elevates TRH mRNA levels in the PVN; however, the effect is not seen at the circadian peak (at onset of darkness) in TRH mRNA levels (Zoeller et al 1990).

Suprachiasmatic nucleus

The SCN, a site of circadian rhythmicity, contains two segregated populations of neurons, one of which expresses vasopressin, the other vasoactive intestinal polypeptide (VIP). Levels of vasopressin (Uhl & Reppert 1986) and VIP transcripts in the SCN (Albers et al 1990) in rats are highest and lowest, respectively, in the morning. The rhythm first appears at E21 (Reppert & Uhl 1987) and is maintained in Brattleboro rats that do not translate their aberrant mRNA (Uhl & Reppert 1986). The levels of VIP mRNA in the SCN are increased by enucleation (Holtzman et al 1989) and lactation (Gozes et al 1989). Levels of transcripts encoding the immediate-early proteins c-*fos* and NGFI-A are not regulated during the light–dark cycle in rats or hamsters but an exposure to light during the dark phase dramatically raises the levels (Rusak et al 1990), suggesting a role for these proteins in resetting the rhythm.

Future directions

The melding of 'classical' neuroanatomical techniques with hybridization histochemistry will be important for our understanding of hypothalamic responses to various stimuli. In particular, using combinations of tract-tracing, simultaneous immunohistochemistry and/or double hybridization histochemistry (Young 1990), one should be able to study gene regulation within particular subsets of neurons. Selective lesions of afferent pathways should also provide insights into the nature of regulatory inputs. Ontogenetic studies will also become more common, especially with the use of the more sensitive ribonucleotide probes: it is likely that published dates of 'first' appearances of transcripts will be pushed back in many cases.

The use of primary cell cultures and immortalized neurons (see Mellon et al, this volume) will permit more controlled examinations of gene regulation. Hybridization histochemistry will be an important tool for studying these cultures. For example, Wray et al (1989) have shown that oestradiol reduces levels of transcripts encoding gonadotropin-releasing hormone in slice cultures.

Hybridization histochemistry will also play an increasingly important role in understanding gene regulation in transgenic animals, especially when antibodies cannot distinguish between endogenous and transgene translation products. The first neuropeptide transgene to be appropriately expressed in the hypothalamus was a 13.5 kb genomic fragment that contained the mouse gonadotropin-releasing hormone gene and it corrected the sterility caused by a deletion mutation in the hypogonadal mouse (Mason et al 1986). The vasopressin and oxytocin genes have also been studied in transgenic animals using hybridization histochemistry. Transgenic vasopressin expression in the PVN and SON is produced with a metallothionein–vasopressin fusion gene (Habener et al 1989), but not with vasopressin constructs alone. Rat oxytocin gene expression in mouse occurs in a tissue- and physiologically (lactation-)specific way with an oxytocin–vasopressin minilocus (Young et al 1990b). These mice translate and transport to the pituitary the rat oxytocin neurophysin coded for by the transgene. These types of studies establish minimal sequences needed for expression; subsequent alterations in the fragments will permit functional dissection of the genes. It is likely that transgenic animals with gain- or loss-of-function mutations will also be studied using hybridization histochemistry.

References

Albers HE, Stopa EG, Zoeller RT et al 1990 Day–night variation in preprovasoactive intestinal peptide/peptide histidine isoleucine mRNA within the rat suprachiasmatic nucleus. Mol Brain Res 7:85–89

Angulo JA, Ledoux M, McEwen BS 1991 Genomic effects of cold and isolation stress on magnocellular vasopressin mRNA-containing cells in the hypothalamus of the rat. J Neurochem 56:2033–2038

Baldino FJ, O'Kane TM, Fitzpatrick-McElligott S, Wolfson B 1988 Coordinate hormonal and synaptic regulation of vasopressin RNA. Science (Wash DC) 241:978–981

Barraclough CA 1992 Neural control of the synthesis and release of luteinizing hormone-releasing hormone. In: Functional anatomy of the neuroendocrine hypothalamus. Wiley, Chichester (Ciba Found Symp 168) p 233–251

Bohler HCLJ, Zoeller RT, King JC, Rubin BS, Weber R, Merriam GR 1990 Corticotropin-releasing hormone mRNA is elevated on the afternoon of proestrus in the parvocellular paraventricular nuclei of the female rat. Mol Brain Res 8:259–262

Bondy CA, Gainer H 1989 Corticotropin-releasing hormone stimulates neurohypophysial hormone release through an interaction with the intermediate lobe of the pituitary. J Neuroendocrinol 1:5–8

Bondy CA, Jensen RT, Brady LS, Gainer H 1989a Cholecystokinin evokes secretion of oxytocin and vasopressin from rat neural lobe independent of external calcium. Proc Natl Acad Sci USA 86:5198–5201

Bondy CA, Whitnall MH, Brady LS 1989b Regulation of carboxypeptidase H gene expression in magnocellular neurons: response to osmotic stimulation. Mol Endocrinol 3:2086–2092

Bradley DJ, Young WS, III, Weinberger C 1989 Differential expression of α and β thyroid hormone receptor genes in rat brain and pituitary. Proc Natl Acad Sci USA 86:7250–7254

Brady LS, Smith MA, Gold PW, Herkenham M 1990 Altered expression of hypothalamic neuropeptide mRNAs in food-restricted and food-deprived rats. Neuroendocrinology 52:441–447

Burgunder J-M, Taylor T 1989 Ontogeny of thyrotropin-releasing hormone gene expression in the rat diencephalon. Neuroendocrinology 49:631–640

Ceccatelli S, Cortés R, Hökfelt T 1991 Effect of reserpine and colchicine on neuropeptide mRNA levels in the rat hypothalamic paraventricular nucleus. Mol Brain Res 9:57–69

Chung SK, McCabe JT, Pfaff DW 1991 Estrogen influences on oxytocin mRNA expression in preoptic and anterior hypothalamic regions studied by in situ hybridization. J Comp Neurol 307:281–295

Dyess EM, Segerson TP, Liposits Z et al 1988 Triidothyronine exerts direct cell-specific regulation of thyrotropin-releasing hormone gene expression in the hypothalamic paraventricular nucleus. Endocrinology 123:2291–2297

Falke N 1988 Dynorphin (1-8) inhibits stimulated release of oxytocin but not vasopressin from isolated neurosecretory endings of the rat neurohypophysis. Neuropeptides 11:163–167

Gozes I, Avidor R, Biegon A, Baldino FJ 1989 Lactation elevates vasoactive intestinal peptide messenger ribonucleic acid in rat suprachiasmatic nucleus. Endocrinology 124:181–186

Grino M, Burgunder J-M, Eskay RL, Eiden LE 1989 Onset of glucocorticoid responsiveness of anterior pituitary corticotrophs during development is scheduled by corticotropin-releasing factor. Endocrinology 124:2686–2692

Grino M, Young WS, III, Burgunder J-M 1989 Ontogeny of expression of the corticotropin-releasing factor gene in the hypothalamic paraventricular nucleus and of the proopiomelanocortin gene in rat pituitary. Endocrinology 124:60–68

Grino M, Guillaume V, Bourdouresque F, Conte-Devolx B, Maltese J-Y, Oliver C 1990 Glucocorticoids regulate peptidyl-glycine α-amidating monooxygenase gene expression in the rat hypothalamic paraventricular nucleus. Mol Endocrinol 4:1613–1619

Habener JF, Cwikel BJ, Hermann H, Hammer RE, Palmiter RD, Brinster RL 1989 Metallothionein–vasopressin fusion gene expression in transgenic mice. J Biol Chem 264:18 844–18 852

Harbuz MS, Lightman SL 1989a Glucocorticoid inhibition of stress-induced changes in hypothalamic corticotrophin-releasing factor messenger RNA and proenkephalin A messenger RNA. Neuropeptides 14:17–20

Harbuz MS, Lightman SL 1989b Responses of hypothalamic and pituitary mRNA to physical and psychological stress in the rat. J Endocrinol 122:705–711

Harbuz MS, Nicholson SA, Gillham B, Lightman SL 1990 Stress responsiveness of hypothalamic corticotrophin-releasing factor and pituitary pro-opiomelanocortin mRNAs following high-dose glucocorticoid treatment and withdrawal in the rat. J Endocrinol 127:407–415

Harbuz M, Russell JA, Sumner BEH, Kawata M, Lightman SL 1991a Rapid changes in the content of proenkephalin A and corticotropin releasing hormone mRNAs in the paraventricular nucleus during morphine withdrawal in urethane-anaesthetized rats. Mol Brain Res 9:285–291

Harbuz MS, Chowdrey HS, Jessop DS, Biswas S, Lightman SL 1991b Role of catecholamines in mediating messenger RNA and hormonal responses to stress. Brain Res 551:52–57

Herman JP, Schäfer MK-H, Sladek CD et al 1989a Chronic electroconvulsive shock treatment elicits up-regulation of CRF and AVP mRNA in select populations of neuroendocrine neurons. Brain Res 501:235–246

Herman JP, Schäfer MK-H, Young EA et al 1989b Evidence for hippocampal regulation of neuroendocrine neurons of the hypothalamo–pituitary–adrenocortical axis. J Neurosci 9:3072–3082

Herman JP, Wiegand SJ, Watson SJ 1990 Regulation of basal corticotropin-releasing hormone and arginine vasopressin messenger ribonucleic acid expression in the paraventricular nucleus: effects of selective hypothalamic deafferentations. Endocrinology 127:2408–2417

Holtzman RL, Malch R, Gozes I 1989 Disruption of the optic pathway during development affects vasoactive intestinal peptide mRNA expression. New Biologist 1:215–221

Imaki T, Nahan J-L, Rivier C, Sawchenko PE, Vale W 1991 Differential regulation of corticotropin-releasing factor mRNA in rat brain regions by glucocorticoids and stress. J Neurosci 11:585–599

Ishizawa H, Dave JR, Liu L-I, Tabakoff B, Hoffman PL 1990 Hypothalamic vasopressin mRNA levels in mice are decreased after chronic ethanol ingestion. Eur J Pharmacol 189:119–127

Kawata M, McCabe JT, Harrington C, Chikaraishi D, Pfaff DW 1988 In situ hybridization analysis of osmotic stimulus-induced changes in ribosomal RNA in rat supraoptic nucleus. J Comp Neurol 270:528–536

Kiyama H, Emson PC 1990 Evidence for the co-expression of oxytocin and vasopressin messenger ribonucleic acids in magnocellular neurosecretory cells: simultaneous demonstration of two neurophysin messenger ribonucleic acids by hybridization histochemistry. J Neuroendocrinol 2:257–260

Koller KJ, Wolff RS, Warden MK, Zoeller RT 1987 Thyroid hormones regulate levels of thyrotropin-releasing mRNA in the paraventricular nucleus. Proc Natl Acad Sci USA 84:7329–7333

Kovács KJ, Mezey É 1987 Dexamethasone inhibits corticotropin-releasing factor gene expression in the rat paraventricular nucleus. Neuroendocrinology 46:365–368

Laurent FM, Hindelang C, Klein MJ, Stoekel ME, Felix JM 1989 Expression of the oxytocin and vasopressin genes in the rat hypothalamus during development: an in situ hybridization study. Develop Brain Res 46:145–154

Leng G, Bicknell RJ 1986 The neurohypophysis. In: Lightman SL, Everitt BJ (eds) Neuroendocrinology. Blackwell Scientific Publications, Boston p 177–206

Lightman SL, Young WS III 1987a Changes in hypothalamic preproenkephalin A mRNA following stress and opiate withdrawal. Nature (Lond) 328:643–645
Lightman SL, Young WS III 1987b Vasopressin, oxytocin, dynorphin, enkephalin, and corticotrophin releasing factor mRNA stimulation in the rat. J Physiol (Lond) 394:23–39
Lightman SL, Young WS III 1988 Corticotrophin-releasing factor, vasopressin and pro-opiomelanocortin mRNA responses to stress and opiates in the rat. J Physiol (Lond) 403:511–523
Lightman SL, Young WS III 1989a Influence of steroids on the hypothalamic CRF and enkephalin messenger ribonucleic acid responses to stress. Proc Natl Acad Sci USA 86:4306–4310
Lightman SL, Young WS III 1989b Lactation inhibits stress mediated secretion of corticosterone and oxytocin and hypothalamic accumulation of CRF and enkephalin mRNA. Endocrinology 124:2358–2364
Makara GB 1992 The relative importance of hypothalamic neurons containing corticotropin-releasing factor or vasopressin in the regulation of adrenocorticotropic hormone secretion. In: Functional anatomy of the neuroendocrine hypothalamous. Wiley, Chichester (Ciba Found Symp 168) p 43–53
Mason AJ, Hayflick JS, Zoeller RT et al 1986 A deletion truncating the gonadotropin-releasing hormone gene is responsible for hypogonadism in the *hpg* mouse. Science (Wash DC) 234:1366–1371
McCabe JT, Morrell JI, Pfaff DW 1986 Measurement of expression of the vasopressin and oxytocin genes in single neurons by *in situ* hybridization. In: Fink G, Harmar AJ, McKerns KW (eds) Neuroendocrine Molecular Biology. Plenum, New York p 219–229
Meister B, Cortés R, Villar MJ, Schalling M, Hökfelt T 1990 Peptides and transmitter enzymes in hypothalamic magnocellular neurons after administration of hyperosmotic stimuli: comparison between messenger RNA and peptide/protein levels. Cell Tissue Res 260:279–297
Mellon PL, Wetsel WC, Windle JJ et al 1992 Immortalized hypothalamic gonadotropin-releasing hormone neurons. In: Functional anatomy of the neuroendocrine hypothalamus. Wiley, Chichester (Ciba Found Symp 168) p 104–126
Mezey É, Kiss JZ 1991 Coexpression of vasopressin and oxytocin in hypothalamic supraoptic neurons of lactating rats. Endocrinology 129:1814–1820
Miller FD, Ozimek G, Milner RJ, Bloom FE 1989 Regulation of neuronal oxytocin mRNA by ovarian steroids in the mature and developing hypothalamus. Proc Natl Acad Sci USA 86:2468–2472
Pfaff DW, Brooks PJ, Funabashi T, Pfaus JG, Mobbs CV 1992 Gene memory in neuroendocrine and behavioural systems. In: Functional anatomy of the neuroendocrine hypothalamus. Wiley, Chichester (Ciba Found Symp 168) p 165–186
Pretel S, Piekut DT 1989 Mediation of changes in paraventricular vasopressin and oxytocin mRNA content to the medullary vagal complex and spinal cord of the rat. J Chem Neuroanat 2:327–334
Reppert SM, Uhl GR 1987 Vasopressin messenger ribonucleic acid in supraoptic and suprachiasmatic nuclei: appearance and circadian regulation during development. Endocrinology 120:2483–2487
Reppert SM, Uhl GR 1988 The vasopressin gene is expressed prior to regulation in the supraoptic nuclei of fetal rats. Brain Res 456:392–396
Rivier C, Imaki T, Vale W 1990 Prolonged exposure to alcohol: effect on CRF mRNA levels, and CRF- and stress-induced ACTH secretion in the rat. Brain Res 520:1–5
Rivkees SA, Chaar MR, Hanley DF, Maxwell M, Reppert SM, Uhl GR 1989 Localization and regulation of vasopressin mRNA in human neurons. Synapse 3:246–254

Rökaeus Å, Young WS III, Mezey É 1988 Galanin coexists with vasopressin in the normal rat hypothalamus and galanin's synthesis is increased in the Brattleboro (diabetes insipidus) rat. Neurosci Lett 90:45–50

Rusak B, Robertson HA, Wisden W, Hunt SP 1990 Light pulses that shift rhythms induce gene expression in the suprachiasmatic nucleus. Science (Wash DC) 248:1237–1240

Sager SM, Sharp FR, Curran T 1988 Expression of c-*fos* protein in brain: metabolic mapping at the cellular level. Science (Wash DC) 240:1328–1331

Sawchenko PE, Imaki T, Vale W 1992 Co-localization of neuroactive substances in the endocrine hypothalamus. In: Functional anatomy of the neuroendocrine hypothalamus. Wiley, Chichester (Ciba Found Symp 168) p 16–42

Schmale H, Ivell R, Briendl M, Darmer D, Richter D 1984 The mutant gene from diabetes insipidus (Brattleboro) rat is transcribed but the message is not efficiently translated. Eur Mol Biol Organ (EMBO) J 3:3289–3293

Segerson TP, Hoefler H, Childers H et al 1987a Localization of thyrotropin-releasing hormone prohormone messenger ribonucleic acid in rat brain by *in situ* hybridization. Endocrinology 121:98–107

Segerson TP, Kauer J, Wolfe HC et al 1987b Thyroid hormone regulates TRH biosynthesis in the paraventricular nucleus of the rat hypothalamus. Science (Wash DC) 238:78–80

Sherman TG, Civelli O, Douglass J, Herbert E, Burke S, Watson SJ 1986 Hypothalamic dynorphin and vasopressin mRNA expression in normal and Brattleboro rats. Fed Proc 45:2323–2327

Sherman TG, Day R, Civelli O et al 1988 Regulation of hypothalamic magnocellular neuropeptides and their mRNAs in the Brattleboro rat: coordinate responses to further osmotic challenge. J Neurosci 8:3785–3796

Sternberg EM, Young WS III, Bernardini R et al 1989 A central nervous system defect in corticotropin releasing hormone biosynthesis is associated with susceptibility to streptococcal cell wall arthritis in Lewis rats. Proc Natl Acad Sci USA 86:4771–4775

Swanson LW, Simmons DM 1989 Differential steroid hormone and neural influences on peptide mRNA levels in CRH cells of the paraventricular nucleus: a hybridization histochemical study in the rat. J Comp Neurol 285:413–435

Taylor T, Gyves P, Burgunder J-M 1990 Thyroid hormone regulation of TRH mRNA levels in rat paraventricular nucleus of the hypothalamus changes during ontogeny. Neuroendocrinology 52:262–267

Tramu G, Croix C, Pillez A 1983 Ability of CRF immunoreactive neurons of the paraventricular nucleus to produce a vasopressin-like material. Neuroendocrinology 37:467–469

Uhl GR, Reppert SM 1986 Suprachiasmatic nucleus vasopressin messenger RNA: circadian variation in normal and Brattleboro rats. Science (Wash DC) 232:390–393

Uhl GR, Zingg HH, Habener JF 1985 Vasopressin mRNA in situ hybridization: localization and regulation studied with oligonucleotide probes in normal and Brattleboro rat hypothalamus. Proc Natl Acad Sci USA 82:5555–5559

Van Tol HHM, Voorhuis DTAM, Burbach JPH 1987 Oxytocin gene expression in discrete hypothalamic magnocellular cell groups is stimulated by prolonged salt loading. Endocrinology 120:71–76

Villar MJ, Meister B, Cortés R, Schalling M, Morris M, Hökfelt T 1990 Neuropeptide gene expression in hypothalamic magnocellular neurons of normal and hypophysectomized rats: a combined immunohistochemical and *in situ* hybridization study. Neuroscience 36:181–199

Watts AG, Swanson LW 1989 Diurnal variations in the content of preprocorticotropin-releasing hormone messenger ribonucleic acids in the hypothalamic paraventricular nucleus of rats of both sexes as measured by *in situ* hybridization. Endocrinology 125:1734–1738

Wray S, Zoeller RT, Gainer H 1989 Differential effects of estrogen on luteinizing hormone-releasing hormone gene expression in slice explant cultures prepared from specific rat forebrain regions. Mol Endocrinol 3:1197–1206

Young WS III 1986 Corticotropin-releasing factor mRNA in the hypothalamus is affected differently by drinking saline and by dehydration. FEBS (Fed Eur Biochem Soc) Lett 208:158–162

Young WS III 1989 Levels of transcripts encoding a member of the protein kinase C family in the paraventricular and supraoptic nuclei are increased by hyperosmolality. J Neuroendocrinol 1:1–4

Young WS III 1990 In situ hybridization histochemistry. In: Björklund A, Hökfelt T, Wouterlood FG, van den Pol A (eds) Methods for the analysis of neuronal microcircuits and synaptic interactions. Elsevier, New York p 481–512

Young WS III, Lightman SL 1992 Chronic stress elevates enkephalin expression in the paraventricular and supraoptic nuclei and modifies opioid receptor binding in the median eminence and posterior pituitary of the rat. In: Kvetnansky R, McCarty R, Axelrod J (eds) Stress: neuroendocrine and molecular approaches. Gordon and Breach Science, New York p 221–234

Young WS III, Mezey E, Siegel RE 1986a Quantitative in situ hybridization histochemistry reveals increased levels of corticotropin-releasing factor mRNA after adrenalectomy in rats. Neurosci Lett 70:198–203

Young WS III, Mezey E, Siegel RE 1986b Vasopressin and oxytocin mRNAs in adrenalectomized and Brattleboro rats: analysis by quantitative in situ hybridization histochemistry. Mol Brain Res 1:231–241

Young WS III, Shepard EA, Burch RM 1987a Plasma hyperosmolality increases G-protein and cAMP synthesis in the paraventricular and supraoptic nuclei. Mol Endocrinol 1:884–888

Young WS III, Warden M, Mezey E 1987b Tyrosine hydroxylase mRNA is increased by hyperosmotic stimuli in the paraventricular and supraoptic nuclei. Neuroendocrinology 46:439–444

Young WS III, Horvath S, Palkovits M 1990a The influences of hyperosmolality and synaptic inputs on galanin and vasopressin expression in the hypothalamus. Nueroscience 39: 115–125

Young WS III, Reynolds K, Shepard EA, Gainer H, Castel M 1990b Cell-specific expression of the rat oxytocin gene in transgenic mice. J Neuroendocrinol 2:917–925

Zoeller RT, Wolff RS, Koller KJ 1988 Thyroid hormone regulation of messenger ribonucleic acid encoding thyrotropin (TSH)-releasing hormone is independent of the pituitary gland and TSH. Mol Endocrinol 2:248–252

Zoeller RT, Kabeer N, Albers HE 1990 Cold exposure elevates cellular levels of messenger ribonucleic acid encoding thyrotropin-releasing hormone in paraventricular nucleus despite elevated levels of thyroid hormones. Endocrinology 127:2955–2962

DISCUSSION

Pfaff: The oxytocin–vasopressin minilocus that was expressed in the transgenic mice contained only 360 bases of oxytocin promoter (Young et al 1990). Are you surprised that it works when you have so little there?

Young: I was expecting that the vasopressin portion would work and that the oxytocin construct wouldn't. We have never been able to get vasopressin expression with any of the constructs. Nor have we ever been able to get the oxytocin portion alone incorporated into transgenic mice. We have examined over 100 mice both after birth and at embryonic Days 13 and 14 to see if resorbing embryos had incorporation. We haven't pursued that any further, for example by injecting stem cells with the oxytocin portion and then injecting those cells into blastocysts. My original reason for trying the minilocus was that the two exons are so similar.

Robinson: You said that a proportion of the cells were double labelled with both oxytocin and vasopressin messages and this proportion increased after dehydrating the animals. Do you ever see expression of the oxytocin transgene in those double labelled cells?

Young: That was in the rat. I wouldn't be surprised if 1–2% of our oxytocin transgene cells also express mouse vasopressin. I haven't noticed it happen specifically.

Robinson: My impression is that there is a relatively high percentage of cells double labelled for oxytocin and vasopressin in the rat compared to results from immunocytochemical double labelling for oxytocin and vasopressin. Could it be that one of those messages is not translated?

Young: Dr Éva Mezey has looked at this in a lot of detail (Mezey & Kiss 1991). She has used all the combinations of immunostaining and *in situ* hybridization and gets similar data: in lactating animals approximately 17% of the oxytocin cells are making vasopressin, although at lower levels than presumably cells that make only vasopressin. Her hypothesis is that the oxytocin cells have turned on expression of vasopressin to help with fluid balance during lactation.

Pfaff: The relationships between the oxytocin and vasopressin genes suggest that they may have arisen by gene duplication events. How similar are the promoters between the two?

Gainer: I would say that upstream of the first 30 bases they are very divergent. If you compare the consensus sequences of the gene promoter regions from the four species that have been cloned, it is very hard to see any similarity between the vasopressin and oxytocin promoters beyond the modified Goldberg–Hogness box. At first glance that would suggest that if you took the promoters of each together with some construct to make a reporter gene and made transgenic mice, it would work fine, but it didn't. Murphy, Carter and colleagues had up to 13 kb of upstream sequence in one case (Pardy et al 1990, Carter et al 1989). It has never worked except with this minilocus construct and then only for oxytocin.

Liposits: Jirikowski et al (1990) reported the presence of oxytocin mRNA in axon terminals; what is your view of this?

Young: We haven't done those experiments but there is evidence for segregation of different transcripts among different parts of cells, dendrites versus axons, even different dendritic fields. That is another whole area of regulation of translation—where the RNA is.

Sawchenko: Is there any evidence that oxytocin or vasopressin mRNA in axons and terminals in the posterior pituitary can be translated?

Gainer: No. The original theory of where vasopressin and oxytocin were synthesized was that this occurred in the posterior pituitary. Howard Sach's original experiments disproved that. We repeated those experiments when we had antibodies that could precipitate prohormones and could never see any translation. Furthermore, there are no ribosomes in those nerve terminals.

Halász: Have you looked at the expression of any peptides other than TRH after hypothalamic intervention?

Young: There is a depression of vasopressin synthesis in the supraoptic nucleus after retrochiasmatic lesion that may be due to injury. There is an increase in galanin synthesis that may result from a combination of interrupting inhibitory input and stimulating some response to the injury. Galanin expression also increases after sectioning the pituitary stalk. If one looks further back in the hypothalamus with caudal hemisections, where we are not damaging the axons, there is not much of an effect on vasopressin, whereas the effect on galanin expression is still seen. So we have dissociated an injury component from release from an inhibition.

Gainer: Do you know anything about the input controlling galanin expression in the paraventricular nucleus?

Young: We have looked at the cells in the region of the posterior arcuate premamillary area that project to the paraventricular nucleus. We don't know the route and we don't know which cells they are. We made progressively finer lesions, but the effects have been less than with the posterior hypothalamic hemisection, perhaps because it may be a more diffuse system.

Palkovits: There is not a simple vasopressin–galanin interaction, and such an interaction does not occur only in the supraoptic nucleus: it seems to be a more general phenomenon. There is an excellent model in the lower brainstem, in the inferior olive, namely the climbing fibres. They cross over 100% at the inferior olive level. It means the output from the inferior olive is crossed, the input to it is ipsilateral. If you transect the axon, you change the pattern of synthesis of certain peptides in cells on the contralateral side. If you transect the input to the cells, you get the same effect on the ipsilateral side. It seems that the axon and the input to the cells are both important for peptide synthesis.

We found such alterations in several systems, for example acetylcholine versus galanin in the preganglionic neurons. When you transect any of the cranial nerves that have climbing fibres, there is depletion of acetylcholine staining in cells and an immediate increase in galanin staining. CRF versus aspartate in the climbing fibres is another example: the amount of aspartate in the inferior olive declines after a lesion in the cerebellum while the amount of CRF there increases. Calcitonin gene-related peptide is involved in other systems. Alterations in glutamate versus enkephalin can be observed after manipulations in the optic system. If you look at this phenomenon as a function of time, the amount of the peptide increases for seven days, then falls to the normal level.

Young: It is almost as if, if you know the right stimulus, you can cause one of these cells to make any peptide you want.

Gainer: As long as they are programmed to be that promiscuous.

Hatton: Scott, how far back did you make these cuts that still increase the galanin?

Young: We went caudal to the mamillary bodies and there was no effect.

Hatton: But cutting just anterior to the mamillary has? There is a big histaminergic input from the histamine-containing neurons of the posterior hypothalamus. They have effects that are apparently differential on the oxytocin and vasopressin cells, but they seem to synapse on both types.

Palkovits: That is not the case. We have destroyed separately the so-called prelateral mamillary nucleus (magnocellular premamillary area), which contains histaminergic cells, and the lesion had no effect on galanin mRNA levels in the supraoptic nucleus.

Hatton: Both the medial and the lateral parts?

Palkovits: Yes.

Zoeller: I have always been a little concerned about the use of autoradiography standards for film and emulsion in the context of *in situ* hybridization. For example, I think it is important to characterize the range over which there is a linear response of each film and each batch of emulsion to different intensities of radioactivity. In this way, you can ensure that your signals are within the film's ability to detect differences (i.e. the signal is not saturated). However, using autoradiography standards to calculate the numbers of probe molecules hybridized in the tissue is more complicated. I have tried the simple experiment of following *in situ* hybridization for vasopressin mRNA on rat brain sections containing the paraventricular and supraoptic nuclei with film autoradiography or scintillation counting of tissues scraped from the sections and dissolved in liquid scintillant. When we compare the d.p.m. values calculated from the film with those provided by a β-counter, we see a positive correlation but the calculated values are always higher. Perhaps we are not using the proper denominator in the construction of the standard curve, i.e. d.p.m./unit area versus film density.

Young: We have done similar studies with enkephalin so that a larger portion of the brain tissue can be labelled. We get a linear relationship between the amount of radioactivity that we measure when we scrape off the tissue and that which we measure by *in situ* hybridization. There is also a correlation with the thickness of the sections, so at least up until 30 μm in thickness the probes seem to penetrate well into the tissue. The line doesn't go through the origin, but through the axis at about 2 μm, so the bottom 2 μm may be inaccessible to probe or may be destroyed.

Pfaff: I also espouse the use of *in situ* hybridization as a quantitative technique, but there are a few people who say that no histochemical technique can be quantitative. What do you think are the most serious limitations of *in situ* hybridization as a quantitative technique?

Young: One question concerns its sensitivity. What percentage of the message are you detecting? It is difficult to answer because how do we know how much is there in the first place? Any manipulation you make in the animal to extract the RNA, could destroy mRNA. That is the most serious limitation. Theoretically, the sensitivity is one copy per cell, if the probe is hot enough and the background is low enough. But that doesn't mean that the one copy per cell will survive the processing for *in situ* hybridization.

Pfaff: What about problems guaranteeing that one has achieved saturating probe concentrations and about limitations on the linear range?

Young: There's some controversy in the literature about whether or not you can saturate the RNA by increasing the levels of probe. I think that most people agree that to saturate RNA, you have to use lots of probe and then it becomes very expensive to do these experiments. Practically, one uses certain amounts, say 10^6 counts per 50 μl, and that seems to work fine. If you keep the conditions fairly constant, you can see relative changes. Looking at relative changes is the strongest application of this technique.

Palkovits: What percentage would you accept as a real change?

Pfaff: That depends on the inherent biological variability. We have measured an induction with *in situ* hybridization as 3.1-fold and with slot blots as 3.15-fold. That seemed close enough.

Sawchenko: A major limitation of *in situ* methods as quantitative tools derives from the fact that they are histological techniques. We don't know how consistently we perfuse animals; we don't know how consistently tissue is frozen and how consistently the probes penetrate. I think much of the variability arises when we try to compare the relative numbers we get from *in situ* hybridizations among animals and with determinations of tissue concentrations of RNA.

Pfaff: When a message is very rare because only a very small number of cells express that message, *in situ* looks great, and no Northern blot or even PCR will detect it. Whereas if a message is rare because large numbers of cells express it but with extremely low abundance per cell, then *in situ* looks terrible.

Young: I agree. The use of riboprobes reduces that occurrence tremendously; riboprobes are so much more sensitive than oligonucleotides.

Zoeller: On the problem of saturation—since we are using ^{35}S-labelled probes, it might be that probes begin to form disulphide bonds between each other within cells that have a high abundance of target mRNA. If so, the signal may be amplified within specific cells, unless you use a high concentration of an antioxidant like dithiothreitol (DTT). We have found that in the presence of little or no DTT, the signal over the paraventricular and supraoptic nuclei continues to increase with increasing concentrations of ^{35}S-vasopressin probe.

But if we use a high level of DTT, we see saturation of the probe at a reasonable concentration. You wouldn't pick that up as background because it is manifested only in cells that have vasopressin message. It looks like signal but it may be amplified because of the interaction of the sulphur groups in the probe molecules.

References

Carter DA, Ang HL, Murphy D 1989 Expression pattern of neuropeptide genes in transgenic mice. Soc Neurosci Abstr 15:1125

Jirikowski GF, Sanna PP, Bloom FE 1990 Localization of oxytocin mRNA in axons of the hypothalamo-hypophyseal tract. Soc Neurosci Abstr 16:390

Mezey É, Kiss JZ 1991 Coexpression of vasopressin and oxytocin in hypothalamic supraoptic neurons of lactating rats. Endocrinology 129:1814–1820

Pardy K, Ang HL, Carter D, Murphy D 1990 Bovine specificity of vasopressin and oxytocin gene expression in transgenic mice. Neuroendocrinology Abstr 52(suppl):157

Young WS III, Reynolds K, Shepard EA, Gainer H, Castel M 1990 Cell-specific expression of the rat oxytocin gene in transgenic mice. J Neuroendocrinol 2:917–925

Feedback regulation of thyrotropin-releasing hormone gene expression by thyroid hormone in the hypothalamic paraventricular nucleus

Ronald M. Lechan and Ildikó Kakucska

Department of Medicine, Division of Endocrinology, New England Medical Center, 750 Washington Street, Boston, MA 02111, USA

Abstract. Hypothyroidism caused by chemical or surgical thyroidectomy or hypophysectomy causes a substantial increase in the content of thyrotropin-releasing hormone (TRH) mRNA and proTRH exclusively in cells of the medial and periventricular paravocellular subdivisions of the hypothalamic paraventricular nucleus (PVN). This response may be important to raise the anterior pituitary thyrostat to promote increased secretion of thyroid-stimulating hormone (TSH) and to induce the secretion of a more biologically active TSH. The increase in TRH mRNA can be obliterated by stereotaxic implants of hormonally active L-triiodothyronine (T3) placed into the anterior hypothalamus but not by implants of the hormonally inactive 3,5′-diiodo-L-thyronine (T2); we therefore suggested that T3 has a direct action on TRH-containing cells of the PVN. Ablation of brainstem catecholaminergic projection fields to the PVN (known to stimulate TRH secretion) has no effect on TRH mRNA expression; β1 thyroid hormone receptor mRNA is present in extracts of the PVN. Euthyroid levels of serum T3 in hypothyroid animals achieved via intraperitoneally implanted osmotic minipumps are not associated with a return of PVN levels of TRH mRNA to normal unless circulating T3 levels are raised into the hyperthyroid range (1.7 times normal). This requirement is similar to that needed to normalize nuclear thyroid hormone receptor levels in the anterior pituitary of hypothyroid animals, suggesting that in addition to circulating T3 monodeiodination of T4 to T3 within the brain must also contribute to feedback inhibition of TRH mRNA. As Type II deiodinase activity is absent or very low in the PVN and does not rise with hypothyroidism, we propose that an alternative source for T4 monodeiodination exists within the central nervous system.

1992 Functional anatomy of the neuroendocrine hypothalamus. Wiley, Chichester (Ciba Foundation Symposium 168) p 144–164

Even before thyrotropin-releasing hormone (TRH) was characterized as a tripeptide amide (Burgus et al 1970, Nair et al 1970) and shown to be present in high concentrations in the brain (Jackson & Reichlin 1974), studies by

Greer (1952) and by other workers (Bogdanove & Halmi 1953, Florsheim 1958, Reichlin 1960, Halász et al 1967) had identified a locus within the hypothalamus (the 'thyrotropic area') that is critical for the control of thyroid function. This region is now recognized to be the paraventricular nucleus (PVN), a symmetrical, winged-like nucleus at the dorsal margin of the third ventricle that contains hypophysiotropic neurons of the TRH–tuberoinfundibular system (Fig. 1A). As opposed to oxytocin and vasopressin, which originate primarily in large or 'magnocellular' neurons of the PVN, TRH immunoreactive neurons are found in small to medium sized or 'parvocellular' neurons in the PVN contained within medial and periventricular parvocellular subdivisions (Lechan & Jackson 1982). These cells project to the median eminence mainly through the lateral retrochiasmatic area (Palkovits et al 1982) after extending laterally from the PVN, dorsal to the fornix and arching medially toward the base of the hypothalamus, although some axons also descend along the wall of the third

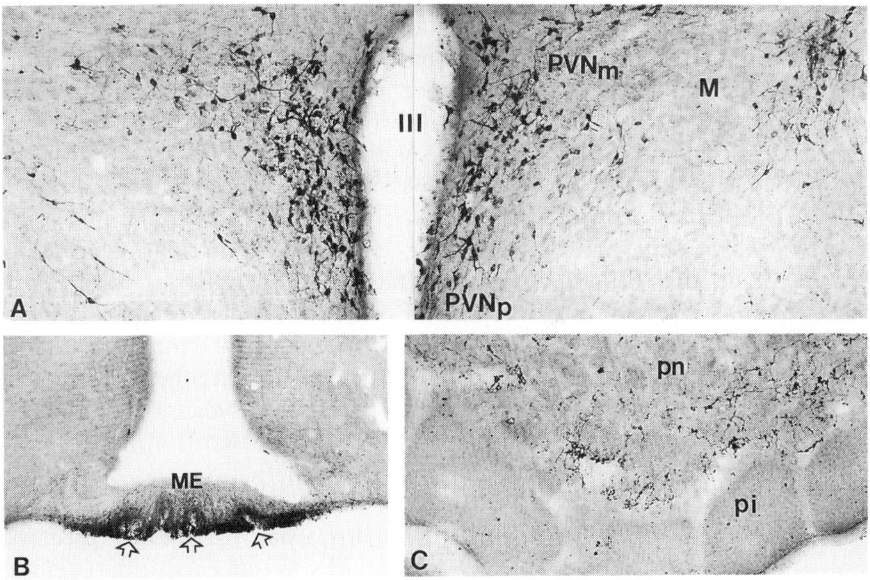

FIG. 1. Coronal sections through the (A) paraventricular nucleus, (B) median eminence and (C) posterior pituitary, demonstrating the origin of TRH-containing neurons of the hypothalamic tuberoinfundibular tract and the location of their axon terminals in the external zone of the median eminence (ME) and posterior pituitary (pn). Note tendency for TRH immunoreactive perikarya to cluster in medial and periventricular parvocellular subdivisions of the PVN (PVNm, PVNp), whereas the magnocellular division (M) is relatively devoid of immunoreactivity. Arrow heads in (B) denote TRH-containing axon terminals in close contact with portal capillaries. III, third ventricle; pi, intermediate lobe. From Lechan & Jackson (1982), with permission.

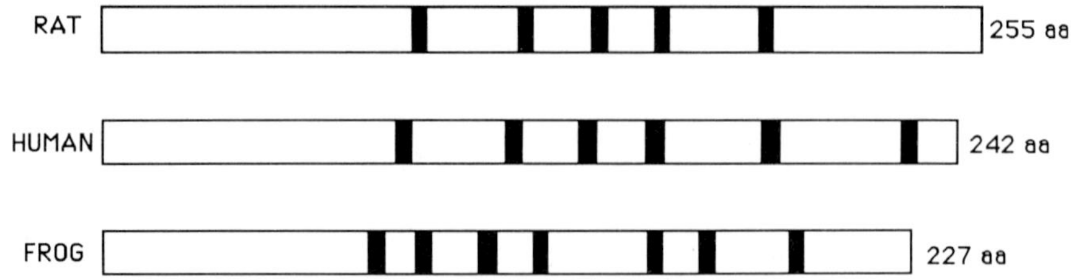

FIG. 2. Schematic representation of the deduced sequences of rat, human and frog preproTRH. Multiple copies of a progenitor sequence for TRH (Gln-His-Pro-Gly) (black box) occur in each species.

ventricle in the subependymal neuropil. Axon terminals in the median eminence can be identified along its entire rostrocaudal extent in the external zone in close apposition to fenestrated capillaries of the portal system (Fig. 1B), into which the tripeptide can be secreted for regulation of anterior pituitary secretion of thyroid-stimulating hormone (TSH). Some axons extend further distally along the stalk to terminate in the posterior pituitary (Fig. 1C), although the physiological significance of this pathway is unknown. TRH-containing perikarya in the anterior parvocellular subdivision of the PVN also project to the median eminence (Ishikawa et al 1988), but appear to be under a different set of physiological controls (see later).

In addition to the paraventricular nucleus, TRH-containing neurons are present in many other regions of the hypothalamus as well as in extrahypothalamic regions of the central nervous system (Lechan & Jackson 1982). These populations have no known projections to the median eminence and, therefore, it is presumed that they do not subserve a direct hypophysiotropic function. Collections of TRH neurons in the lateral hypothalamus, for example, project to the septal nuclei where they may induce analeptic effects (Kalivas & Horita 1980), whereas TRH neurons in the preoptic area may be involved in thermoregulation (Hori et al 1988).

TRH arises from the post-translational processing of a larger precursor molecule, as demonstrated by the isolation of cDNAs that encode the complete sequences of rat (Lechan et al 1986), human (Yamada et al 1990) and frog (Kuchler et al 1990) preproTRH. All contain a repetitive progenitor sequence for TRH, Gln-His-Pro-Gly (Fig. 2); each sequence is flanked by cleavage sites. Glycine is necessary for C-terminal amidation of the proline residue in mature TRH (Bradbury et al 1982). Further post-translational modification to TRH occurs by conversion of the N-terminal glutamyl residue into pyroglutamyl (Fischer & Speiss 1987). The presence of multiple copies of TRH in the precursor may be important for magnification of the response to signals that trigger TRH secretion. In addition to TRH, other peptides derived from the TRH prohormone are also secreted from the median eminence (Bruhn et al 1991, Bulant et al 1990). One of these peptides, preproTRH(160–169), which separates the third and fourth progenitor sequences of TRH, has been reported by Boulant et al (1990) to bind to specific receptors on anterior pituitary tissue and may potentiate the action of TRH; thus the non-TRH components of proTRH may have physiological functions. Evidence for differential processing of proTRH to peptides other than TRH, including C-terminally extended forms of TRH, in certain regions of the central nervous system has raised the possibility that region-specific processing of proTRH gives rise to a number of biologically active peptides (Wu & Jackson 1988, Cockle & Smyth 1987, Bulant et al 1990).

Antisera to proTRH and probes for hybridization to proTRH mRNA have provided powerful, new approaches to study the physiology of TRH neuronal systems in the central nervous system. One area of classical interest that has lent itself to this type of analysis has been the role of thyroid hormone in feedback regulation of hypophysiotropic neurons in the hypothalamic

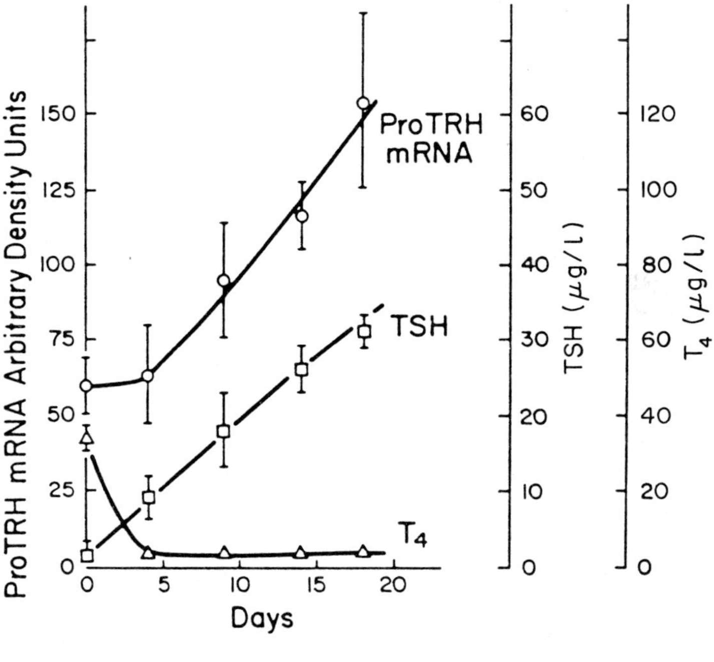

FIG. 3. The effect of thioamide-induced hypothyroidism on the amount of proTRH mRNA extracted from the hypothalamic paraventricular nucleus at four day intervals over 18 days of study. Equal amounts of RNA were separated by electrophoresis and the amount that hybridized to a cRNA probe for proTRH mRNA was determined by densitometry.

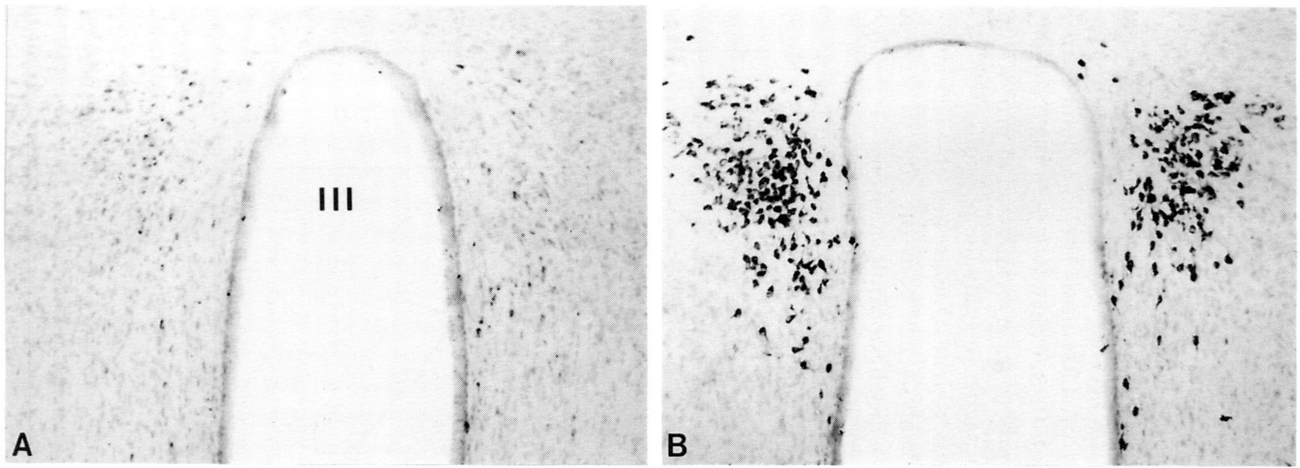

FIG. 4. Immunocytochemical staining for proTRH in the paraventricular nucleus of (A) a control, euthyroid animal and (B) a hypothyroid animal. Note increased intensity of immunostaining in the hypothyroid animal and tendency for reaction product to fill the cytoplasm. III, third ventricle. From Segerson et al (1987), with permission.

TRH–tuberoinfundibular system. Although feedback control by thyroid hormone on the secretion of TSH from anterior pituitary thyrotrophs has been well established, the effect of thyroid hormone on TRH has been controversial. The same alterations in thyroid state have been reported to increase, decrease or not change the amount of TRH in the hypothalamus (Bassiri & Utiger 1974, Montoya et al 1975, Utsumi et al 1977, Roti et al 1978). At least part of the explanation for these discrepant results relates to measurement of TRH extracted from the entire hypothalamus rather than the PVN alone: only TRH present in the PVN has hypophysiotropic function therefore inclusion of TRH from regions of the hypothalamus not involved in the regulation of anterior pituitary function in samples for radioimmunoassay could confound the results. In addition, since the content of TRH in the hypothalamus could be independently altered by changes in the rate of synthesis, transport and secretion, radioimmunoassay of the total TRH content could lead to inconclusive results.

If one assumes that neuronal content of a peptide mRNA reflects the biosynthetic activity of the neuron more closely than does its peptide content, a more direct approach to determine the effect of hormonal signals on a discrete population of cells would be to quantitate peptide mRNA. Accordingly, studies by Segerson et al (1987) used Northern blot hybridization to show that hypothyroidism induced by surgical thyroidectomy or by the oral administration of thioamines causes an increase in the amount of proTRH mRNA extracted from a micropunch of the PVN. The increase was estimated to be from approximately 469 attomoles in euthyroid animals to 914 attomoles in the hypothyroid animals. This response follows the fall in serum T4 (L-thyroxine) to undetectable levels and parallels the gradual rise in serum TSH (Fig. 3). Changes in the amount of proTRH mRNA in the PVN are not mediated by TSH itself, as proTRH mRNA also increases in animals made hypothyroid by hypophysectomy in which TSH is undetectable (Lechan & Segerson 1989). Conversely, the administration of pharmacological doses of T4 inhibits the hypothyroid-induced rise in proTRH mRNA, while supraphysiological levels cause an even further decline to approximately 60% of euthyroid values, establishing an inverse relationship between circulating thyroid hormone and paraventricular nucleus proTRH mRNA concentrations (Lechan & Segerson 1989). Similar findings have been reported by Koller et al (1987) and Zoeller et al (1988).

Immunocytochemical studies using an antiserum directed against proTRH also show a dramatic increase in the staining characteristics of paraventricular cells in hypothyroid animals (Fig. 4A,B). We have therefore proposed that thyroid hormone regulates both transcription (and/or stability) of proTRH mRNA and translation of the TRH prohormone in the PVN (Segerson et al 1987). Post-translational processing of proTRH may then allow for increased secretion of TRH into the portal capillary system for conveyance to the anterior pituitary gland and thereby contribute to the hypothyroid-induced elevation

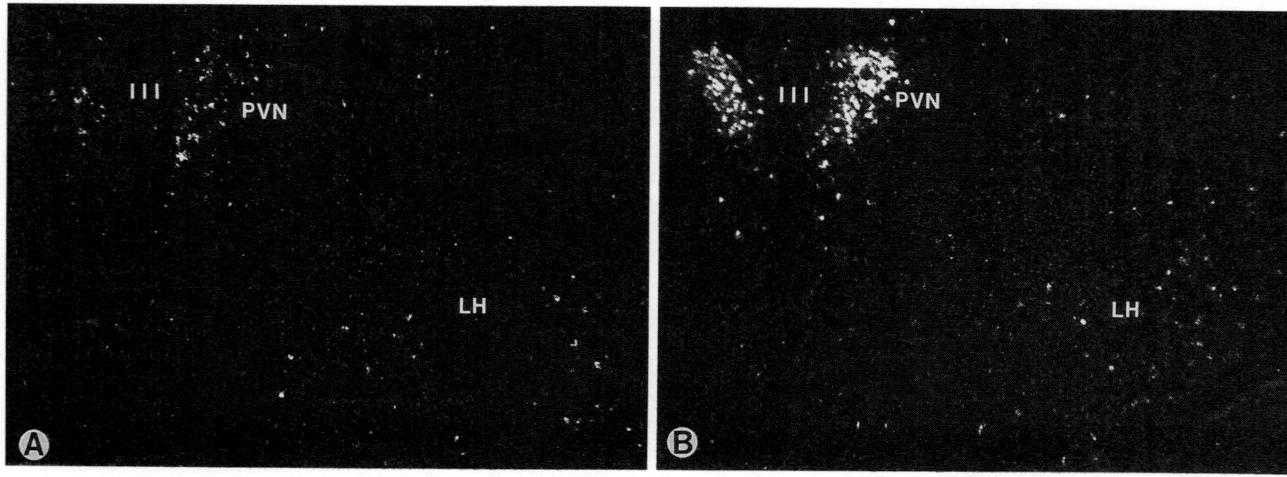

FIG. 5. *In situ* hybridization of proTRH mRNA in the hypothalamic paraventricular nucleus (PVN) and lateral hypothalamus (LH) in (A) a euthyroid and (B) a hypothyroid animal. Note marked increase in hybridization signal in the PVN of hypothyroid animal but the absence of any change in the LH. III, third ventricle. From Segerson et al (1987), with permission.

in TSH. Simultaneous co-secretion of other products of proTRH may potentiate this action (Bulant et al 1990).

Increased secretion of TRH into the portal plexus may explain the decreased immunostaining for TRH seen in the median eminence in hypothyroid animals (Mori & Yamada 1987). This would be consistent with observations by Bruhn et al (1991) that hypothyroidism enhances the release of TRH and proTRH fragments from explants of the median eminence following a depolarizing stimulus. Studies by Martin et al (1970) and several other groups (Ishikawa et al 1984, Murakami et al 1988) showed that in the absence of TRH after destruction of the PVN, anterior pituitary TSH secretion is more sensitive to feedback inhibition by T4. Kaplan et al (1986) found that conversely, chronic, intrathecal infusion of TRH causes elevation of TSH and free and total thyroid hormones. It may be concluded that alterations in the concentration of TRH in the portal blood result in readjustment of the set point for feedback regulation by thyroid hormone. TRH may also influence the biological activity of TSH, as suggested by its ability to alter the glycosylation of the secreted hormone (Taylor & Weintraub 1989).

As demonstrated by *in situ* hybridization histochemistry, the influence of thyroid hormone on proTRH mRNA is limited exclusively to medial and periventricular parvocellular neurons of the PVN (Fig. 5). Adjacent TRH neurons in the anterior parvocellular subdivision, as well as other TRH neurons in the hypothalamus, show no change in their concentration of TRH mRNA. This highly selective effect of thyroid hormone feedback regulation on medial and periventricular parvocellular neurons of the PVN but not other TRH-containing groups in the hypothalamus had raised the possibility that these effects are mediated by the rich afferent input to the PVN. This nucleus receives inputs from several regions of the brain, including the diencephalon, telencephalon and brainstem (Sawchenko & Swanson 1983). Studies by Liposits et al (1987) have provided morphological evidence for the presence of numerous catecholaminergic axons in synaptic association with TRH neurons in the paraventricular nucleus. Antisera to the catecholamine-synthesizing enzymes, dopamine β-hydroxylase (E.C.1.14.17.1) and phenylethanolamine *N*-methyltransferase (E.C.2.1.1.28) can be used as markers for noradrenaline- and adrenaline-containing axons, respectively. These show that the paraventricular nucleus receives a dense innervation in medial and periventricular subdivisions, which originates mainly from the medullary noradrenergic and adrenergic cell groups. Similar morphological observations have been made by Shioda et al (1986). These results are consistent with those of numerous physiological studies indicating that noradrenaline and adrenaline stimulate the secretion of hypophysiotropic TRH (see Morely 1981 for review). Toni et al (1990) have also shown a dense innervation of axon terminals containing neuropeptide Y on both perikarya and dendrites of TRH-containing cells in the paraventricular nucleus. Some of these axons may also contain catecholamines and originate

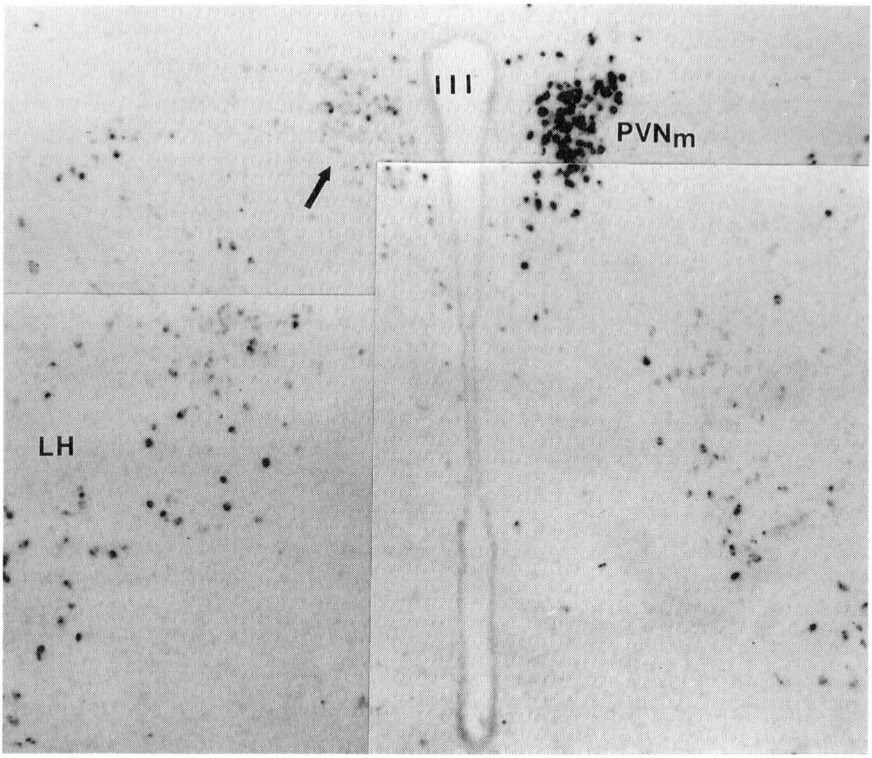

FIG. 6. Bright field illumination of a coronal section through the hypothalamus showing a reduction in signal intensity of autoradiographic grains representing TRH mRNA on one side of the paraventricular nucleus (PVNm) (arrow) after implantation of crystalline triiodothyronine into the anterior hypothalamus on that side. No reduction in TRH mRNA is noted on the opposite side or in adjacent cells in the lateral hypothalamus (LH). III, third ventricle. From Dyess et al (1988), with permission.

from the lower brainstem catecholaminergic cell groups (Sawchenko et al 1985).

Studies by Dyess et al (1988), however, indicate that central regulation of proTRH mRNA in the PVN is exerted directly on these neurons. Stereotaxic placement of crystalline triiodothyronine (T3) but not the relatively inactive diiodothyronine (T2) into hypothyroid animals adjacent to one side of the PVN causes marked suppression of the content of proTRH mRNA and proTRH on that side (Fig. 6). These findings confirm those of Yamada & Greer (1959) that microinjection of T4 into the 'thyrotropic area' reduces thyroid uptake of radioiodine and of Belchetz et al (1978) that bilateral injection of T3 into hypothyroid monkeys reduces plasma TSH concentration. By using a more proximal endpoint for the response (i.e. proTRH mRNA content), Dyess et al (1988) avoid

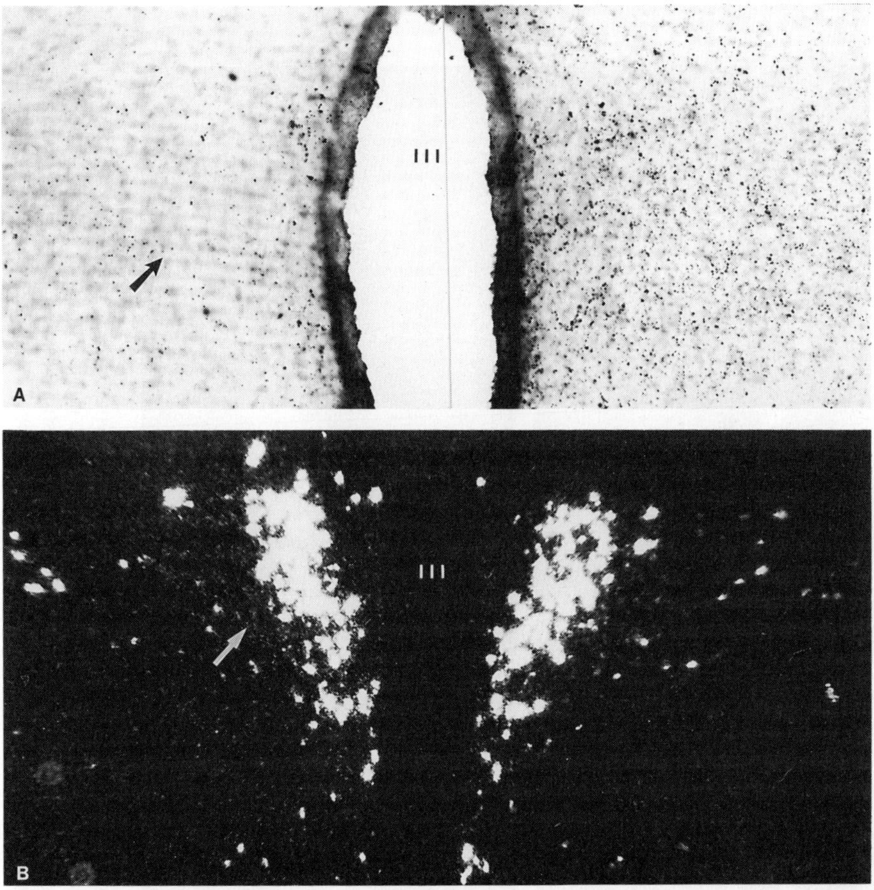

FIG. 7. Adjacent sections through the paraventricular nucleus in a hypothyroid rat stereotaxically injected with the neurotoxin, 6-hydroxydopamine, on one side of the nucleus to deplete that side of its innervation by catecholaminergic fibres. A marked diminution in immunostaining is seen for dopamine β-hydroxylase, a marker of catecholamine-containing neurons, on the side of the injection (arrow) in (A). Despite catecholamine depletion, hybridization to proTRH mRNA on the treated side (arrow) is unchanged (B). III, third ventricle. From Dyess et al (1988), with permission.

the concern that implanted thyroid hormone has its effect by diffusing to the pituitary or stimulating somatostatin secretion into the portal capillary system. In contrast, ablation of the afferent input to the PVN from ascending fibres from the lower brainstem by lesions at the rostral level of the mesencephalon or stereotaxic injection of the neurotoxin, 6-hydroxydopamine, adjacent to the PVN, causes no change in the extent or pattern of hybridization to proTRH mRNA in hypothyroid animals (Fig. 7). This excludes the importance of

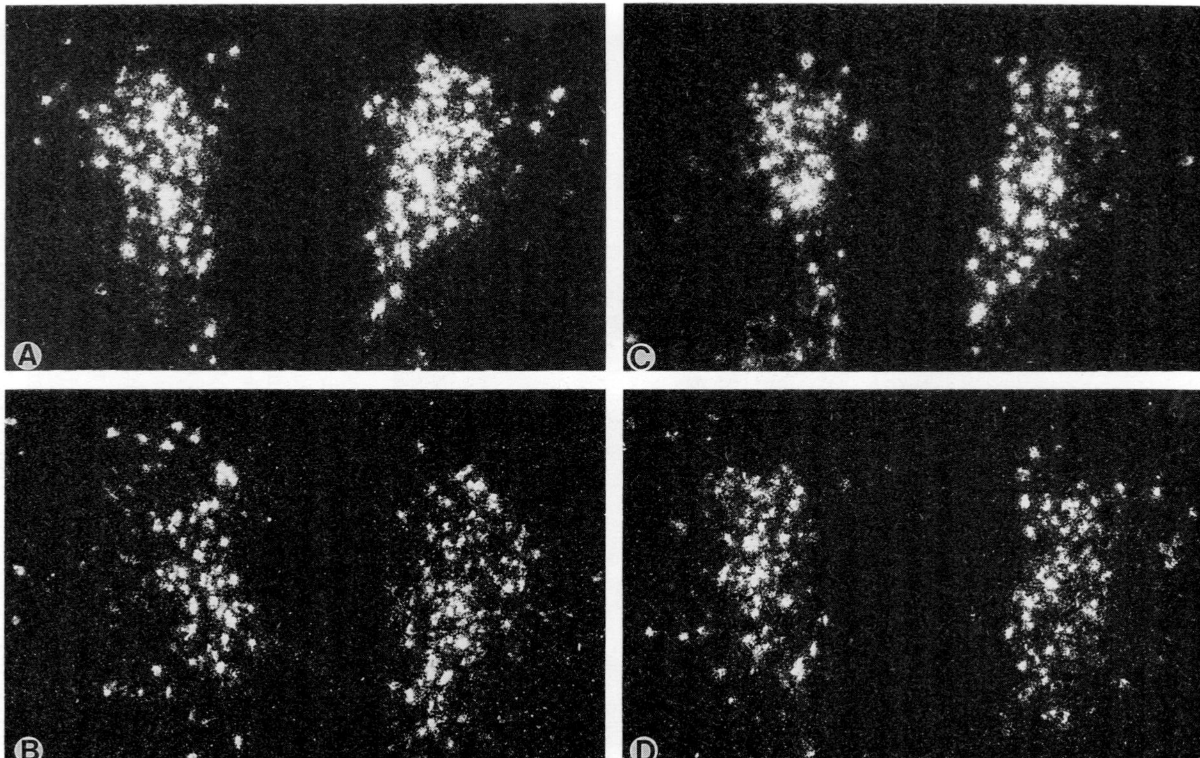

FIG. 8. *In situ* hybridization of proTRH mRNA in the paraventricular nucleus. (A) Hypothyroid and (B) euthyroid animals. (C, D) Hypothyroid animals receiving triiodothyronine (T3) by constant infusion from osmotic minipumps. (C) 0.5 μg/100 g body weight per day T3 causes only an intermediate reduction in proTRH mRNA despite return of plasma T3 levels to normal. (D) 0.75 μg/100 g body weight per day T3 raises plasma T3 levels to 1.7 times normal and fully suppresses proTRH mRNA expression to normal. III, third ventricle. From Kakucska et al (1992).

catecholamines and possibly neuropeptide Y in the thyroid hormone-induced changes in proTRH mRNA in the PVN but does not dismiss the probability that these substances have an important regulatory role in the secretion of TRH by acting at a post-transcriptional level.

The PVN was shown in early work by Ford & Gross (1958) to concentrate radiolabelled thyroid hormone preferentially, it has a rich vascular supply (Ambach & Palkovits 1974) and contains high levels of the β1 thyroid hormone receptor mRNA (Bradley et al 1989). Therefore, it may be hypothesized that feedback regulation of TRH gene expression in the PVN is mediated by circulating levels of thyroid hormone, acting directly on PVN neurons. The PVN, however, contains little if any Type II 5′-deiodinase (Riskind et al 1987), an enzyme which catalyses the conversion of T4 to biologically active T3. Nor is there an increase in enzyme activity during hypothyroidism, unlike that observed in the anterior pituitary and cerebral cortex (M. M. Kaplan et al, personal communication). Thus, whereas T3 in the anterior pituitary and cerebral cortex is derived primarily from the intracellular monodeiodination of T4 (Silva & Larsen 1977), the PVN may respond only to levels of T3 in the circulation. The infusion of constant, graded doses of T3 to hypothyroid animals from intraperitoneally implanted osmotic minipumps (Kakucska et al 1992), however, shows that raising plasma T3 concentration into the normal or slightly hyperthyroid range does not reduce proTRH mRNA levels to normal (Fig. 8A–C). Only doses that raise plasma T3 into the hyperthyroid range (approximately 1.7 times normal) suppress proTRH mRNA expression in the PVN to normal (Fig. 8C,D). This response is similar to that observed for feedback regulation of thyroid hormone on anterior pituitary TSH secretion in the absence of T4 (Connors & Hedge 1980) and suggests that the PVN requires both T3 and T4 for feedback inhibition of TRH gene expression. Because the PVN has little or no deiodinase activity, these observations raise the possibility that T4 is converted to T3 within the brain at a site other than the PVN before exerting its effect. Potential loci include neurons distant from the PVN with the subsequent delivery of T3 to the PVN by axonal transport (Dratman & Crutchfield 1978) and the choroid plexus. The latter accumulates systemically injected T4 for delivery to the cerebrospinal fluid by expressing the T4-binding protein, transthyretin, in extraordinary amounts (Schreiber et al 1990). Further insight into the mechanism by which T4 contributes to feedback regulation of TRH biosynthesis in the PVN may add to the understanding of the complexity of the diverse effects of thyroid hormone on the central nervous system.

Acknowledgements

The authors wish to thank Yanping Qi for expert technical assistance. The work was supported by NIH grant DK 37021.

References

Ambach B, Palkovits M 1974 Blood supply of the rat hypothalamus II. Nucleus paraventricularis. Acta Morphologica Acad Sci Hung 22:311–320

Bassiri RN, Utiger RD 1974 Thyrotropin releasing hormone in the hypothalamus of the rat. Endocrinology 94:188–197

Belchetz PE, Gredley G, Bird D, Himsworth RL 1978 Regulation of thyrotrophin secretion by negative feedback of tri-iodothyronine on the hypothalamus. J Endocrinol 76:439–448

Bogdanove EM, Halmi NS 1953 Effects of hypothalamic lesions and subsequent propylthiouracil treatment on pituitary structure and function in the rat. Endocrinology 53:274–292

Bradbury AF, Finnie MDA, Smyth DG 1982 Mechanism of C-terminal amide formation by pituitary enzymes. Nature (Lond) 298:686–688

Bradley DJ, Young WS III, Weinberger C 1989 Differential expression of alpha and beta thyroid hormone receptor genes in rat brain and pituitary. Proc Natl Acad Sci USA 86:7250–7254

Bruhn TO, Taplin JH, Jackson IMD 1991 Hypothyroidism reduces content and increases in vitro release of pro-thyrotropin-releasing hormone peptides from the median eminence. Neuroendocrinology 53:511–515

Bulant M, Roussell J-P, Astier H, Nicholas P, Vaudry H 1990 Processing of thyrotropin-releasing hormone prohormone (pro-TRH) generates a biologically active peptide, prepro-TRH-(160–169), which regulates TRH-induced thyrotropin secretion. Proc Natl Acad Sci USA 87:4439–4443

Burgus R, Dunn TF, DeSiderio D, Ward DW, Vale W, Guillemin R 1970 Characterization of ovine hypothalamic hypophysiotropic TSH-releasing factor. Nature (Lond) 226: 321–325

Cockle SM, Smyth DG 1987 Specific processing of the thyrotropin-releasing prohormone in rat brain and spinal cord. Eur J Biochem 165:693–698

Connors JM, Hedge GA 1980 Feedback effectiveness of periodic versus constant triiodothyronine replacement. Endocrinology 106:911–917

Dratman MB, Crutchfield FL 1978 Synaptosomal [^{125}I]triiodothyronine after intravenous [^{125}I]thyroxine. Am J Physiol 4:E638–647

Dyess EM, Segerson TP, Lechan RM et al 1988 Triiodothyronine exerts direct cell-specific regulation of thyrotropin-releasing hormone gene expression in the hypothalamic paraventricular nucleus. Endocrinology 123:2291–2297

Fischer WH, Speiss J 1987 Identification of a mammalian glutaminyl cyclase converting glutaminyl into pyroglutamyl peptides. Proc Natl Acad Sci USA 84:3628–3632

Florsheim WH 1958 The effect of anterior hypothalamic lesions on thyroid function and goitre development in the rat. Endocrinology 62:783–789

Ford DH, Gross J 1958 The localization of I^{131}-labeled triiodothyronine and thyroxine in the pituitary and brain of the male guinea pig. Endocrinology 63:549–560

Greer MA 1952 Role of the hypothalamus in control of thyroid function. J Clin Endocrinol 12:1259–1268

Halász B, Florsheim WH, Corcorran NL, Gorski RA 1967 Thyrotrophic hormone secretion in rats after partial or total interruption of neural afferents to the medial basal hypothalamus. Endocrinology 80:1075–1082

Hori T, Yamasaki M, Asami T, Koga H, Kiyohara T 1988 Responses of anterior hypothalamic-preoptic thermosensitive neurons to thyrotropin releasing hormone and cyclo(His-Pro). Neuropharmacology 27:895–901

Ishikawa K, Kekegawa T, Suzuki M 1984 Role of the hypothalamic paraventricular

nucleus in the secretion of thyrotropin under adrenergic and cold-stimulated conditions in the rat. Endocrinology 114:352–358
Ishikawa K, Taniguchi Y, Inoue K, Kurosumi K, Suzuki M 1988 Immunocytochemical delineation of the thyrotropic area: origin of thyrotropin-releasing hormone in the median eminence. Neuroendocrinology 47:384–388
Jackson IMD, Reichlin S 1974 Thyrotropin-releasing hormone (TRH); distribution in hypothalamic and extrahypothalamic brain tissues of mammalian and submammalian chordates. Endocrinology 95:854–862
Kakucska I, Rand W, Lechan RM 1992 Thyrotropin-releasing hormone gene expression in the hypothalamic paraventricular nucleus is dependent upon feedback regulation by both triiodothyronine and thyroxine. Endocrinology 130:2845–2850
Kalivas PW, Horita A 1980 Thyrotropin-releasing hormone: neurogenesis of actions in the pentobarbital narcotized rat. J Pharmacol Exp Ther 212:203–210
Kaplan MM, Taft JA, Reichlin S, Munsat TL 1986 Sustained rises in serum thyrotropin, thyroxine, and triiodothyronine during long term, continuous thyrotropin-releasing hormone treatment in patients with amytrophic lateral sclerosis. J Clin Endocrinol & Metab 63:808–814
Koller KJ, Wolff RS, Warden MK, Zoeller RT 1987 Thyroid hormones regulate levels of thyrotropin-releasing-hormone mRNA in the paraventricular nucleus. Proc Natl Acad Sci USA 84:7329–7333
Kuchler K, Richter K, Trnovsky J, Egger R, Kreil G 1990 Two precursors of thyrotropin-releasing hormone from skin of *Xenopus laevis*. J Biol Chem 265:11731–11733
Lechan RM, Jackson IMD 1982 Immunohistochemical localization of thyrotropin-releasing hormone in the rat hypothalamus and pituitary. Endocrinology 111:55–65
Lechan RM, Segerson TP 1989 Pro-TRH gene expression and precursor peptides in rat brain: observations by hybridization analysis and immunocytochemistry. Ann NY Acad Sci 553:29–59
Lechan RM, Wu P, Jackson IMD et al 1986 Thyrotropin-releasing hormone precursor: characterization in rat brain. Science (Wash DC) 231:159–161
Liposits Zs, Paull WK, Wu P, Jackson IMD, Lechan RM 1987 Hypophysiotrophic thyrotropin releasing hormone (TRH) synthesizing neurons: ultrastructure, adrenergic innervation and putative transmitter action. Histochemistry 88:1–10
Martin JB, Boshans R, Reichlin S 1970 Feedback regulation of TSH secretion in rats with hypothalamic lesions. Endocrinology 87:1032–1040
Montoya E, Seibel MJ, Wilber JF 1975 Thyrotropin releasing hormone secretory physiologic studies by radioimmunoassay and affinity chromatography. Endocrinology 96:1413–1418
Mori M, Yamada M 1987 Thyroid hormones regulate the amount of thyrotrophin-releasing hormone in the hypothalamic median eminence of the rat. J Endocrinol 114:443–448
Morley JE 1981 Neuroendocrine control of thyrotropin secretion. Endocr Rev 2:396–436
Murakami M, Tanaka K, Greer MA, Mori M 1988 Anterior pituitary type II thyroxine 5′-deiodinase activity is not affected by lesions of the hypothalamic paraventricular nucleus which profoundly depress pituitary thyrotropin secretion. Endocrinology 123:1676–1681
Nair RMG, Barrett JF, Bowers CY, Schally AV 1970 Structure of porcine thyrotropin releasing hormone. Biochemistry 9:1103–1106
Palkovits M, Eskay RL, Brownstein MJ 1982 The course of thyrotropin releasing hormone fibers to the median eminence in rats. Endocrinology 110:1526–1528
Reichlin S 1960 Thyroid function, body temperature regulation and growth in rats with hypothalamic lesions. Endocrinology 66:340–354
Riskind PN, Kolodny JM, Larsen PR 1987 The regional hypothalamic distribution of type II 5′-monodeiodinase in euthyroid and hypothyroid rats. Brain Res 420:194–198

Roti E, Christianson D, Harris AR, Braverman LE, Vagenakis AG 1978 "Short" loop feedback regulation of hypothalamic and brain thyrotropin-releasing hormone content in the rat and dwarf mouse. Endocrinology 103:1662–1667

Sawchenko PE, Swanson LW 1983 The organization of forebrain afferents to the paraventricular and supraoptic nuclei of the rat. J Comp Neurol 218:121–144

Sawchenko PE, Swanson LW, Grzanna R, Howe PRC, Bloom SR, Polak JM 1985 Co-localization of neuropeptide Y immunoreactivity in brainstem catecholaminergic neurons that project to the paraventricular nucleus of the hypothalamus. J Comp Neurol 241:138–153

Schreiber G, Aldred AR, Jaworowski A, Nilsson C, Achen MG, Segal MB 1990 Thyroxine transport from blood to brain via transthyretin synthesis in choroid plexus. Am J Physiol 258:R338–345

Segerson TP, Kauer J, Lechan RM et al 1987 Thyroid hormone regulates TRH biosynthesis in the paraventricular nucleus of the rat hypothalamus. Science (Wash DC) 238:78–80

Shioda S, Nakai Y, Sato A, Sunayama S, Shimoda Y 1986 Electron-microscopic cytochemistry of the catecholaminergic innervation of TRH neurons in the rat hypothalamus. Cell Tissue Res 245:247–252

Silva JE, Larsen PR 1977 Pituitary nuclear 3,5,3′-triiodothyronine and thyrotropin secretion: an explanation for the effect of thyroxine. Science (Wash DC) 198:617–620

Taylor T, Weintraub BD 1989 Altered thyrotropin (TSH) carbohydrate structures in hypothalamic hypothyroidism created by paraventricular nuclear lesions are corrected by in vivo TSH-releasing hormone administration. Endocrinology 125:2198–2203

Toni R, Jackson IMD, Lechan RM 1990 Neuropeptide Y-immunoreactive innervation of thyrotropin-releasing hormone-synthesizing neurons in the rat hypothalamic paraventricular nucleus. Endocrinology 126:2444–2453

Utsumi M, Makimura H, Tateiwa M, Sakoda M, Baba S 1977 Effects of thyroxine and cold exposure on hypothalamic levels in rats with various pituitary-thyroid states. Endocrinol Jpn 24:537–543

Wu P, Jackson IMD 1988 Post-translational processing of thyrotropin releasing hormone precursor peptides in rat brain: identification of three novel peptides derived from proTRH. Brain Res 456:22–28

Yamada M, Greer MA 1959 Studies on the mechanism of hypothalamic control of thyrotropin secretion: effect of thyroxine injection into the hypothalamus or the pituitary on thyroid hormone release. Endocrinology 64:559–566

Yamada M, Radovick S, Wondisford FE, Nakayama Y, Weintraub BD, Wilber JF 1990 Cloning and structure of human genomic DNA and hypothalamic cDNA encoding human preprothyrotropin-releasing hormone. Mol Endocrinol 4:551–556

Zoeller RT, Wolff RS, Koller KJ 1988 Thyroid hormone regulation of messenger ribonucleic acid encoding thyrotropin (TSH)-releasing hormone is independent of the pituitary gland and TSH. Mol Endocrinol 2:248–252

DISCUSSION

Lightman: Is it possible that you might simply need T4 in a permissive role? May not a low dose of T4 actually be permissive to the effect of T3 at the paraventricular nucleus?

Lechan: In general, T4 is thought to act more as a 'prohormone' in the CNS, causing its biological effects through monodeiodination to T3, which binds with approximately ten times higher affinity to the nuclear thyroid hormone receptor than does T4. Obregon et al (1980), for example, have shown that if one blocks the conversion of T4 to T3 completely with iopanoic acid, T4 has few inhibitory feedback effects on anterior pituitary TSH secretion in hypothyroid rats. As the paraventricular nucleus appears to have little if any deiodinase activity, the mechanism for intracellular monodeiodination of T4 is likely deficient or absent. Nevertheless, Abend et al (1991) have evidence to suggest that T4 may directly decrease pituitary deiodinase activity without increasing nuclear T3 content. The mechanism for this action remains unknown.

Zoeller: Since T3 is probably down-regulating TRH message by decreasing transcription, the rate of decline in the amount of TRH message after treatment with T3 might be related to turnover rate. How long did you perfuse T3 in these experiments?

Lechan: The animals were chronically infused for seven days.

Plant: You measured total levels of T3. What about changes in free levels? These may not be as high as they would be in a euthyroid animal.

Lechan: Free T3 levels were not measured but, as an index of T3 activity, we did measure plasma TSH in these animals. The fall in proTRH mRNA in the paraventricular nucleus appeared to parallel the fall in TSH. In other words, reduction of proTRH mRNA in the paraventricular nucleus and plasma TSH to the euthyroid range in hypothyroid animals both required supranormal amounts of T3. This would indicate that the paraventricular nucleus is not more sensitive than the anterior pituitary to the feedback effects of T3.

Palkovits: What do you think about brain-born TSH? It has been reported that TSH is present in the PVN of hypophysectomized animals, or that during the early prenatal period, neuronal cells in developing animals can synthesize TSH before pituitary development.

Lechan: The CNS contains a vast number of peptides, including several hormones originally described in the anterior pituitary that likely subserve functions independent of a hypophysiotropic action. I do not believe it is known what the significance of this TSH-like material is in the rat hypothalamus.

Palkovits: You mentioned the catecholamine input to TRH cells. Which catecholamine is it? There is quite a lot of dopamine in that part of the PVN.

Lechan: Studies performed in collaboration with Dr Liposits et al (1987) and by Shioda et al (1986) have shown that both adrenaline and noradrenaline are present in axon terminals that make contacts with TRH neurons in the paraventricular nucleus. I am not aware of any study that has specifically looked at whether dopamine-containing axon terminals also make synaptic contacts with these neurons. Dopamine may exert an important neuroregulatory influence on the secretion of hypophysiotropic TRH, but as indicated by Andersson & Eneroth (1987), this probably occurs on TRH axons in the median eminence.

Palkovits: You mentioned also neuropeptide Y (NPY). Your study supposes that the NPY comes from the lower brainstem.

Lechan: The distribution and characteristics of the synapses of NPY-containing axon terminals on TRH cell bodies in the paraventricular nucleus are very similar to those of catecholamine-containing axon terminals. I cannot exclude, and feel that it is even likely, that some of these axons might also originate from NPY-producing neurons in the arcuate nucleus which have known projection fields to the paraventricular nucleus.

Palkovits: We have done some double staining studies on animals with brainstem transections, and on rats with monosodium glutamate-induced lesions of the arcuate nucleus, to look at the NPY input to the PVN—not specifically to the TRH cells, but to cells containing corticotropin-releasing factor and other peptides. We found cells containing both NPY and catecholamines from the lower brainstem, they constitute about one third of the NPY input to the PVN. Another third of the NPY fibres come from the arcuate nucleus, and one third are local, especially in the ventromedial parvicellular portion of the PVN.

The extraparaventricular TRH cells, are they interneurons or projecting neurons?

Lechan: I believe that a good majority of those cells are projecting neurons.

Palkovits: Projecting to where?

Lechan: There are many populations of TRH-synthesizing neurons in the hypothalamus outside the paraventricular nucleus as well as in extrahypothalamic regions of the brain. These neurons do not project to the median eminence. Some TRH-containing neurons in the lateral hypothalamus, perifornical group and lateral preoptic area, for example, project to the lateral septum. Perhaps the best characterized trajectory of extrahypothalamic TRH neuronal populations is that originating from cells in the medullary raphe. These neurons send axons in the lateral funiculus and central grey of the spinal cord to innervate sympathetic, preganglionic neurons in the intermediolateral column and α motoneurons in the ventral horn of lamina VII.

Sawchenko: Ron, you showed some years ago that cells in the rostral part of the PVN project to the median eminence. In a separate study, you showed that there are TRH cells in this rostral region that seem to be relatively inert to perturbations in thyroid hormone status. Do you know whether those rostral TRH cells themselves project to the median eminence? If they do not, what are they doing?

Lechan: As you know, the parvocellular portion of the paraventricular nucleus is divided into a number of subdivisions including anterior, medial, dorsal, lateral and periventricular parts, on the basis of cytoarchitectonic studies by Swanson & Kuypers (1980). TRH-containing neurons are primarily located in anterior, medial and periventricular subdivision neurons. We have previously shown that parvocellular neurons in all three locations can be retrogradely labelled after injection of the marker, wheat germ agglutinin, directly into the median eminence. Because only the medial and periventricular parvocellular TRH-producing neurons show increases in proTRH mRNA content during hypothyroidism, we have proposed that TRH-containing neurons in the paraventricular nucleus are functionally diverse. The function of anterior parvocellular TRH neurons remains unknown.

Zoeller: Have you looked at the effect of T3 on TRH message in horizontal sections through the paraventricular nucleus? I was looking specifically for subpopulations of TRH neurons that did not respond to T3, but I never saw them. I saw regulation of TRH mRNA within the rostral area as well as in the caudal part of the PVN. When you say rostral, how far rostral do you mean?

Lechan: Rostral to the suprachiasmatic nucleus, where the paraventricular nucleus becomes more diffuse.

Zoeller: Yes. I also see no relation within that area.

Clarke: How much of the proTRH(160–169) peptide is in the median eminence relative to how much TRH you might see?

Lechan: In the rat, there is one molecule of proTRH(160– 169) for every five molecules of a progenitor sequence for TRH, Gln-His-Pro-Gly. If proTRH were completely processed in the median eminence and the two peptides were equally stable, one would expect a ratio of proTRH(160–169):TRH of 1:5.

Lightman: There have been several studies of portal blood measurements in hypothyroid animals where people had enormous difficulty showing any change in TRH concentrations in portal blood despite high levels of pituitary TSH-β mRNA and plasma TSH. Do you think there may be processing differences of proTRH that are important in the regulation of pituitary TSH responses?

Lechan: Rondeel et al (1988) have shown that TRH concentrations in portal plasma sampled from hyperthyroid rats fall to approximately 50% of those in euthyroid controls but that there is no change in hypothyroid animals. In view of our observations, as well as the results of other studies mentioned previously that support the concept of increased secretion of TRH during hypothyroidism, the findings by Rondeel et al are somewhat surprising. One must keep in mind that there are potential difficulties imposed by the methods required to collect portal blood: the animals are under constant stress because of the need for continuous anaesthesia, which might affect TRH levels; also, transection of the pituitary stalk for collection of portal blood could lead to disruption and leakage of TRH from axons *en route* to the posterior pituitary, causing artificially elevated TRH values in the control population. The latter might explain the unusually high concentrations of TRH measured in the portal blood of euthyroid animals in Rondeel's paper of 2.1 ng/h, which is nearly an order of magnitude greater than the total content of TRH in the median eminence.

Lightman: Is that a problem with the antisera?

Lechan: It is possible.

Smith: Cockle et al (1989) described a series of TRH-like peptides which have been substituted at the histidine position. Have you any evidence that these peptides exist in the CNS or in the hypothalamus, in particular?

Lechan: We have not studied the anatomical distribution of these peptides, although I understand that immunoreactive material has been found in the hippocampus. These peptides are likely to be unrelated to TRH and are, in fact, derived from a separate gene.

Pfaff: When TRH axons project to other parts of the brain, is the TRH acting as a neurotransmitter or a neuromodulator? Does its action there depend upon the classical TRH receptor?

Lechan: TRH is widely distributed in the CNS and many of its functions are unknown. In addition to its hypophysiotropic function, there is evidence for TRH acting both as a neurotransmitter and as a neuromodulator. For example, in the spinal cord, TRH has been reported to have both direct and indirect effects on α motoneurons. Presumably the actions of TRH are all mediated by the same TRH receptor. This is supported by a fairly good 1:1 association between the location of TRH and the distribution of the TRH receptor (Manaker et al 1985).

Palkovits: Early electrophysiological studies suggested that most of the TRH transmitter action is inhibitory.

Lechan: I believe most of the evidence suggests that TRH has excitatory actions in the CNS. TRH increases the excitability of α motoneurons, stimulates autonomic activity including colonic activity, heart rate and respiration, and induces arousal.

Lightman: Your lab and Neil Blake et al (1991) have shown that food deprivation can also cause a marked reduction in TRH mRNA levels. How do you think this might be working and at what site might it be having its effect?

Lechan: In collaboration with Dr Charles Emerson at the University of Massachusetts Medical Center, we have been looking at food deprivation, largely as a model for non-thyroidal illness. This syndrome occurs both in humans and in animals during starvation, infection or chronic disease states. It is characterized by low plasma thyroid hormone levels and normal or low plasma TSH levels.

We have used starvation in a rat model to determine whether TRH contributes to the inappropriate levels of TSH in this syndrome. We have observed a substantial fall in TRH mRNA in the paraventricular nucleus in the food-deprived animals, indicating that a reduction in the biosynthesis of hypophysiotropic TRH does occur. This response is exactly the opposite of what one might expect on the basis of the data I presented today, showing that thyroid hormone has negative feedback effects on TRH-containing paraventricular neurons. This suggests that in non-thyroidal illness there must be overriding signals in the brain that are activated and capable of preventing the normal physiological response of hypophysiotropic TRH neurons to falling plasma levels of thyroid hormone. Intuitively, this might be important for the survival of the animal by reducing nitrogen loss and thereby preserving muscle mass. What the mechanisms might be and how they modulate TRH neurons in the paraventricular nucleus are actively being explored in my laboratory.

Young: Do you know anything about the thyroid hormone response element in the TRH gene and what receptors may be involved in regulating the expression of the TRH gene?

Lechan: Dr Stephanie Lee in our division, who has characterized the TRH gene, has been studying this question using a plasmid expression vector containing the promoter region of the TRH gene co-transfected into GH4 cells with the mouse c-*erbA* gene that encodes the T3 receptor. As she has been able to demonstrate a dose-dependent decrease in TRH gene expression in the presence of T3, it should now be possible to identify the thyroid hormone inhibitory element in the TRH gene.

Young: Could you tell us which receptor mediates the effect?

Lechan: Preliminary studies by Dr Lee indicate that both the $\alpha 1$ and $\beta 1$ thyroid hormone receptors have equal effects in her transfection system.

Sawchenko: In contrast to magnocellular neurosecretory neurons and parvocellular CRF neurons, relatively few, if any, co-localized substances have been found in neuroendocrine TRH or somatostatin cells. Do you ascribe any functional import to that?

Lechan: It is true that in the paraventricular nucleus, TRH is not known to be co-localized with another peptide in most animal species, except in the bullfrog, where it coexists with mesotocin. Ultrastructurally, however, these neurons do contain small, clear vesicles that could indicate the coexistence of a separate neurotransmitter. In addition, the TRH prohormone, itself, could give rise to other biologically active peptides that might modulate the activity of TRH. This has already been proposed for proTRH(160–169) by Bulant et al (1990). In other areas of the CNS, TRH is well recognized to co-localize with other substances. Johansson et al (1981), for example, have shown that TRH coexists with serotonin and substance P in neurons of the medullary raphe.

References

Abend SL, Fang SL, Alex S, Braverman LE, Leonard JL 1991 Rapid alteration in circulating free thyroxine modulates pituitary type II 5′-deiodinase and basal thyrotropin secretion in the rat. J Clin Invest 88:898–903

Andersson K, Eneroth P 1987 Thyroidectomy and central catecholamine neurons of the male rat: evidence for the existence of an inhibitory dopaminergic mechanism in the external layer of the median eminence and for a facilitatory noradrenergic mechanism in the paraventricular hypothalamic nucleus regulating TSH secretion. Neuroendocrinology 45:14–27

Blake NG, Eckland DJA, Foster OJF, Lightman SL 1991 Inhibition of hypothalamic TRH mRNA during food deprivation. Endocrinology 129:2714–2718

Bulant M, Roussell J-P, Astier H, Nicholas P, Vaudry H 1990 Processing of thyrotropin-releasing hormone prohormone (pro-TRH) generates a biologically active peptide, prepro-TRH-(160-169), which regulates TRH-induced thyrotropin secretion. Proc Natl Acad Sci USA 87:4439–4443

Cockle SM, Aitken A, Beg F, Smyth DG 1989 A novel peptide, pyroglutamylglutamylproline amide, in the rabbit prostate complex, structurally related to thyrotrophin-releasing hormone. J Biol Chem 264:7788–7791

Johansson O, Hökfelt T, Pernow B et al 1981 Immunohistochemical support for three putative transmitters in one neuron: coexistence of 5-hydroxytryptamine, substance P- and thyrotropin releasing hormone-like immunoreactivity in medullary neurons projecting to the spinal cord. Neuroscience 6:1857–1881

Liposits Zs, Paull WK, Wu P, Jackson IMD, Lechan RM 1987 Hypophysiotropic thyrotropin releasing hormone (TRH) synthesizing neurons: ultrastructure, adrenergic innervation and putative transmitter action. Histochemistry 88:1–10

Manaker S, Winokur A, Rostene WH, Rainbow TC 1985 Autoradiographic localization of thyrotropin-releasing hormone and catecholamines in the rat central nervous system. J Neurosci 5:167–174

Obregon MJ, Pascual A, Mallol J, Morreale de Escobar G, Escobar del Ray F 1980 Evidence against a major role of L-thyroxine at the pituitary level: studies in rats treated with iopanoic acid (telepaque). Endocrinology 106:1827–1836

Rondeel JMM, deGreef WJ, van der Schoof P, Karels B, Klootwijk W, Visser TJ 1988 Effect of thyroid status and paraventricular area lesions on the release of thyrotropin-releasing hormone and catecholamines into hypophysial portal blood. Endocrinology 123:523–527

Shioda S, Nakai Y, Sato A, Sunayama S, Shimoda Y 1986 Electron-microscopic cytochemistry of the catecholaminergic innervation of TRH neurons in the rat hypothalamus. Cell Tissue Res 245:247–252

Swanson LW, Kuypers HGJM 1980 The paraventricular nucleus of the hypothalamus: cytoarchitectonic subdivisions and organization of projections to the pituitary, dorsal vagal complex, and spinal cord as demonstrated by retrograde fluorescence double-labeling methods. J Comp Neurol 194:555–570

Gene memory in neuroendocrine and behavioural systems

D. W. Pfaff, P. J. Brooks, T. Funabashi, J. G. Pfaus and C. V. Mobbs

Laboratory of Neurobiology and Behavior, The Rockefeller University, 1230 York Avenue, NY 10021-6399, New York, USA

Abstract. Several examples of sex steroid hormone actions on rat brain and behaviour show that initial hormone exposures may be followed by enduring neuronal alterations, apparent long after the hormone itself has disappeared. Precedents from non-neuronal systems led to the concept of 'gene memory'. We are studying genomic structural alterations in rat hypothalamic neurons to account for these effects. The preproenkephalin gene is turned on by oestradiol in rat brain neurons in a tissue-specific and genetic sex-specific manner. Levels of preproenkephalin mRNA in the ventromedial hypothalamus correlate tightly with oestradiol-dependent reproductive behaviour. Our results indicate a tissue-specific pattern of DNA methylation in the enkephalin promoter. Putative binding sites for several transcription factors have been described in the preproenkephalin gene promoter; a role for some of these factors in regulating expression of the gene has been demonstrated.

1992 Functional anatomy of the neuroendocrine hypothalamus. Wiley, Chichester (Ciba Foundation Symposium 168) p 165–186

Concepts from reproductive neuroendocrine systems

Some neuroendocrine phenomena clearly work by rapid synaptic signalling causing rapid alterations in action potentials, leading to a neuroendocrine output. Other neuroendocrine mechanisms, which we have worked on, depend in part on steroid hormones working through nuclear receptors, effecting direct changes in levels of specific nerve cell messenger RNAs whose products govern behaviour.

A third type of mechanistic possibility is potentially more subtle and long lasting. Influences on nerve cells—be they developmental, hormonal or experiential—might cause changes in genomic structure such that subsequent transcriptional responses to normal controlling signals are altered for some time. These alterations might persist long after the initial cause has disappeared and might be due to changes in DNA itself or to protein–DNA interactions. Previous work on other cell types (see below) named this type of alteration 'gene memory'.

Some areas in the neuroendocrinology of reproduction, which we have worked on, yielded results which are reminiscent of the 'gene memory' reports from non-neural cells. Such results have helped to stimulate new studies of genomic structure, with an emphasis on the transcriptional control of preproenkephalin in rat forebrain neurons.

Developmental effects: neonatal hormone manipulations

A generation of neuroendocrine work has shown that a single exposure to androgenic or oestrogenic hormone at a specific time in brain development (in the rat, during neonatal life) can determine neural control of gonadotrophs and reproductive behaviour during the organism's entire adult life. Following the pioneering work of Carl Pfeiffer, it was demonstrated by Phoenix et al (1959) that exposure of the guinea pig brain to testosterone at a critical period changed the animal's reproductive neuroendocrine controls from female-typical to male-typical, permanently. These observations were quickly extended to encompass copulatory behaviours (Barraclough & Gorski 1962). Sexually differentiated non-copulatory behaviours follow the same developmental pattern: in the rat, exposure to testosterone just after birth dramatically changes these behaviours from a feminine to a masculine pattern (Pfaff & Zigmond 1971). For all of these phenomena, therefore, a single epoch of hormone exposure alters subsequent neuronal performance for the lifetime of the animal.

Experiments with the adult brain

Not all the actions of steroid sex hormones on adult brain and behaviour are acute ones. For example, ovariectomized female rats which have received periodic injections of oestradiol benzoate are more sensitive to subsequent doses of oestradiol benzoate many days later than animals which have received injections of only the oil vehicle (Whalen & Nakayama 1965, Gerall & Dunlap 1973, Beach & Orndorff 1974). In two ways, these observations suggested long-lasting effects of steroid hormones in the brain: on the one hand there appeared to be a long-lasting decline in hormone sensitivity following ovariectomy. On the other hand, a single treatment with oestrogen appeared to have a long-lasting effect on subsequent hormone sensitivity. However, the method of treatment with oestradiol benzoate permitted the possibility that some of the hormone persisted in the circulation for several days. The purpose of our work (Parsons et al 1979) was to prove that the long-lasting reversal of the decline in hormone sensitivity following ovariectomy could endure even in the absence of oestrogen and to test one set of hypotheses about intervening mechanisms. Ovariectomized female rats were pretreated with subcutaneous silastic capsules of oestradiol or (as a control) cholesterol. After removal of the capsules, we waited seven days before behavioural testing (Parsons et al 1979). Lordosis behaviour testing with a second oestrogen challenge showed significantly better reproductive behaviour performance, whether tested with sexually active males or by artificial manual stimulation, in animals which had received the oestradiol pretreatment compared to the animals pretreated with cholesterol. We showed by radioimmunoassay that circulating oestrogens had declined to the levels in control ovariectomized rats within 12 hours of removal of the capsules, therefore, the long-lasting facilitation of subsequent hormone sensitivity by the oestradiol pretreatment could not have been due simply to oestrogen persisting in the blood. One set of hypotheses about the mechanisms for this long-term change centred on the possibility that oestrogen receptors or progestin receptors in the brain or pituitary are altered by oestradiol pretreatment. However, we were not able to obtain evidence, either in the hypothalamus or in the pituitary, that oestrogen receptors or progesterone receptors were present in different quantities according to oestradiol versus cholesterol pre-treatment. Thus, it appeared that long-term potentiation by oestradiol of subsequent responsivity to oestradiol did not result from a change in oestrogen or progestin receptor dynamics in the hypothalamus (Parsons et al 1979). Besides, oestrogen actually *down*-regulates the mRNA for its own receptor in the hypothalamus (Lauber et al 1991a,b), which would not be consistent with heightened behavioural sensitivity. Other mechanisms for the long-lasting behavioural effect must be sought.

Even on a shorter time scale, there is evidence that brief exposures to oestradiol can facilitate later responsiveness of hypothalamic neurons to the hormone, after the steroid from the first pulse of administration has essentially disappeared. Starting from the work of Harris & Gorski (1978), which provided evidence of discontinuous requirements for oestrogen in the stimulation of cell division in the rat uterus, we initiated a two-pulse paradigm for the analysis of hypothalamic and behavioural mechanisms (Parsons et al 1981). Short and discontinuous periods of oestradiol treatment were sufficient to induce progesterone-facilitated sexual receptivity in female rats (Parsons et al 1982a). Lordosis responses comparable to those after six or 24 hours of continuous oestrogen treatment were seen in ovariectomized female rats which received only two one-hour pulses of oestradiol, provided that the second treatment began not fewer than four or more than 13 hours after the first. Thus, the animals' response to the second pulse was behaviourally detectable, if the first pulse had preceded it by the stated amount. Since these two treatments presented essentially discontinuous exposure (less than 3.6% nuclear oestrogen receptor capacity between the two peaks), it can be concluded that the first pulse promoted the response to the second, even after the hormone from the first pulse had essentially disappeared.

A parallel set of measurements was made for progesterone receptor occupation in the hypothalamus and preoptic area, because of the close temporal relationship between the presence of oestrogen-inducible progestin receptors and the course of oestrogen stimulation of mating behaviour (Parsons et al 1980). Treatments that caused a behavioural response were associated with threshold levels of progestin receptors in the medial basal hypothalamus and preoptic area

(25–35% of maximal induction), whereas behaviourally insufficient treatments were not (Parsons et al 1982a). By this measure as well, therefore, the first oestradiol pulse potentiated the brain's response to the second. Furthermore, new protein synthesis is required for at least some steps in this train of events. If the protein synthesis inhibitor anisomycin was administered 15 minutes before either the first oestradiol pulse or the second, or between the two pulses, a high frequency of lordosis behaviour was not seen (Parsons et al 1982b). In studies with different doses of anisomycin, inhibition of protein synthesis by approximately 60% in the medial basal hypothalamus and preoptic area for the duration of either pulse of oestradiol significantly decreased sexual receptivity at 24 hours after onset of hormone treatment (Parsons et al 1982b). These results are consistent with the notion that when an earlier hormone treatment facilitates response to a later treatment, protein synthesis is required.

With respect to *mechanisms*, various possibilities exist for the maintenance of long-lasting changes in behaviour, as promoted by effects of steroid hormones on the adult brain. In this paper, we highlight genomic structure modifications but further possibilities include synaptic changes and other synthetic changes (reviewed by Cohen & Pfaff 1992). A productive approach is to take a specific transcriptional system, ask which transacting factors are responsible for constitutive and regulated expression, and then ask which of them are changed during long-term alterations in hormone sensitivity. While some of the best current opportunities are offered by study of control over the preproenkephalin gene, oestrogen effects on oxytocin expression (Chung et al 1991a,b), effects of lactation on oxytocin mRNA (Brooks et al 1990) and effects of oestrogen on progesterone receptor mRNA (Romano et al 1989a, Lauber et al 1991c) also will be worth exploring. Differences in DNA methylation are expected not to explain acute hormone effects but to help determine tissue specificities regarding levels of constitutive expression. Alterations in hypersensitivity to DNase or dimethyl sulphate will reveal modifications in DNA–protein relations which are expected to cause acute changes in transcription rate and long-term changes in hormone sensitivity. The proteins involved in controlling preproenkephalin transcription are being analysed using gel-shift techniques with the oestrogen response element (ERE) and progesterone (glucocorticoid) response element (PRE, GRE), as well as the cyclic AMP response element, the AP-1 site and, notably, the sequence responsible for the binding of NF-κB and related proteins. In this series of experiments, well-established molecular techniques can be used to probe *genomic structure* for insights into not only acute hormone effects on neuronal gene expression, but also the long-lasting changes in hormone sensitivity.

Precedents from other systems

During the last two decades, many studies outside the nervous system gave evidence of long-lasting hormonal effects consistent with the notion of 'gene memories'. This phrase metaphorically denotes the possibility that hormones, such as oestrogen (Tam et al 1986, Burch & Evans 1986) and glucocorticoids (Zaret & Yamamoto 1984), can cause changes in chromatin state such that the efficiency of transcription is altered well after the initial hormonal challenge has disappeared. This type of effect is widespread, and is not exclusive of other sources of neuroendocrine plasticity such as synaptic changes (reviewed in Cohen & Pfaff 1992).

One of the most prominent examples of 'gene memory' in a hormone-stimulated system came from work with a human hepatoma cell line (Tam et al 1986) in which a brief oestrogen treatment altered responses to subsequent hormonal treatment. Enhanced apolipoprotein synthesis, as well as induced type II oestrogen-binding sites, could be propagated in these cells for several generations in the absence of further hormone treatment. This type of effect is not restricted to an individual cell line. A classical oestrogen-dependent system, that for the expression of vitellogenin, a protein in egg yolk, also demonstrated more rapid responses and higher levels of expression, indicating heightened hormone sensitivity, following a restricted exposure to oestrogen (Deeley et al 1977, Jost et al 1978, Searle & Tata 1981, Baker & Shapiro 1977, Burch & Evans 1986, Hache et al 1987). It has been argued that the increased sensitivity of the transcriptional system for the vitellogenin gene correlates with the appearance of type II oestrogen receptors, a controversial concept (Hache et al 1987). Nevertheless, it is clear even from a few references in the restricted domain of oestrogenic action that the notion of 'gene memory' in hormone-stimulated neural tissues finds ample precedent in other cell types.

Does prolactin expression by cells in the anterior pituitary qualify as another example of 'gene memory'? Prolactin synthesis and secretion are extraordinarily sensitive to oestrogen treatment (Yen & Jaffe 1978, p 160 ff.). The opposite manipulation, castration, markedly reduces prolactin cell size and, in turn, even brief replacement treatment with oestrogen induces cell hypertrophy and several signs of active protein synthesis (Tougard & Tixier-Vidal 1988, p 1318 ff.). The molecular basis for the oestrogen sensitivity of the rat prolactin gene lies in the DNA sequence of the 5′-flanking region: a region responsible for conferring responsivity to oestrogen contains a sequence with similarity to the consensus oestrogen response element (Maurer & Notides 1987). Deletion of this upstream region also damages cooperative responses with cyclic AMP (Day & Maurer 1989). Similarly, gel mobility shift, nuclease protection and gene transfer experiments indicate that an element responsible for tissue-specific expression is near, but separate from, the sequence which interacts with the oestrogen receptor (Kim et al 1988). Long-term treatment with oestrogens can cause pituitary adenomas composed of prolactin-secreting cells which can persist after the initial oestrogenic challenge and which might, therefore, qualify as an example of 'oestrogen memory'. In some strains of rats, prolactin secretion does not return to normal despite cessation of oestrogen treatment. However, it is not clear whether this enduring prolactin secretion results from changes in the chromatin upstream of the prolactin gene start site, or from the simple presence of additional cells resulting from the initial, prolonged oestrogen exposure. The safest paradigms for investigations of 'gene memory' in brain will be those in which studies of chemical changes in DNA itself (for example, methylation) and explorations of protein–DNA interactions (for example, by DNase hypersensitivity, footprinting, gel mobility shift) can be carried out with relatively small numbers of cells from a heterogeneous population.

Explorations with the gene for the neuropeptide, enkephalin

Preproenkephalin (Fig. 1) is a multivalent 31 kDa protein precursor that contains four copies of the pentapeptide Met^5-enkephalin (ME), and one copy each of the octapeptide ME-Arg^6-Gly^7-Leu^8 and the pentapeptide Leu^5-enkephalin (LE). Larger peptides with variable opioid activity are also cleaved from the proenkephalin precursor, including peptide E, peptide F, BAM-12P, BAM-20P and BAM-22P (Hook 1991). Both ME and LE have a higher affinity for δ opioid receptors than μ receptors and are almost inactive at κ receptors (Paterson et al 1983). However, both the octapeptide and heptapeptide extensions of ME are equally active at μ and δ receptors (Magnan et al 1982).

Differential regulation of female sexual behaviour by opioids: a facilitatory role for enkephalins

Part of the impetus to explore the steroid regulation of proenkephalin transcription and synthesis in brain comes from behavioural studies indicating that opioid peptides can inhibit or facilitate female sexual behaviours depending upon both the *receptor subtype* stimulated and the *site of action in the brain* (reviewed in Pfaus & Gorzalka 1987a, Wiesner & Moss 1989). For example, Pfaus & Gorzalka (1987b) demonstrated that infusions of low doses (20 or 200 ng) of β-endorphin or the μ receptor agonist morphiceptin into the lateral ventricles inhibited lordosis behaviour whereas high doses of these agonists (2 μg) produced a slight facilitation. Infusions of all doses of the relatively selective δ agonist DSTLT facilitated lordosis. The highest dose of the κ agonist dynorphin(1–9) (2 μg) produced a trend toward facilitation 30 minutes after infusion. In contrast, Wiesner & Moss (1989) reported that infusions of β-endorphin into the third ventricle produced a dose-dependent inhibition of lordosis behaviour. Infusions of β-endorphin into the ventromedial hypothalamus or mesencephalic reticular formation also inhibited lordosis, whereas infusions into the mesencephalic central grey, dorsomedial hypothalamus, amygdala or striatum were without effect. Sirinathsinghji (1986) found that infusions of β-endorphin into the medial preoptic area of oestrogen-primed females inhibited lordosis. Although Sirinathsinghji (1984) reported no effect of ME on lordosis behaviour

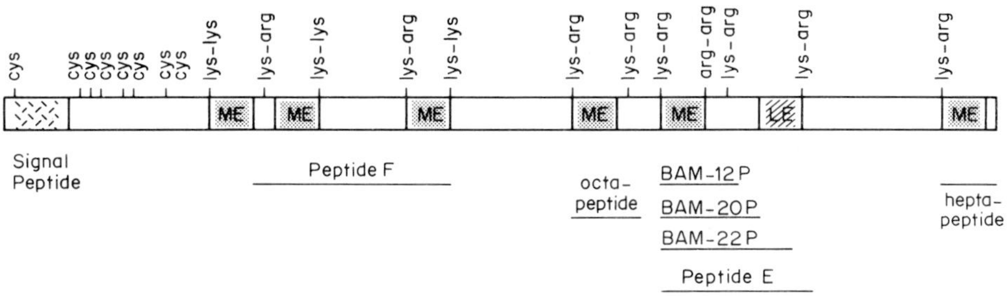

FIG. 1. Structure of rat preproenkephalin showing dibasic processing sites (lys-lys, lys-arg, arg-arg) and peptide products. ME, Met^5-enkephalin; LE, Leu^5-enkephalin. From Hook (1990).

following infusion into the mesencephalic central grey, infusions of ME in conjunction with kelotorphan, an inhibitor of enkephalin-degrading enzymes, into either the lateral ventricle or the mesencephalic central grey inhibited lordosis (Bednar et al 1987). Interestingly, β-endorphin possesses equal affinity for μ and δ receptors (Paterson et al 1983). Because high doses of morphiceptin can also act at δ receptors (Chang et al 1981), Pfaus & Gorzalka (1987b) suggested that μ receptors may inhibit, whereas δ receptors may facilitate, lordosis. Moreover, the inhibitory effects of both morphiceptin and β-endorphin may occur through an interaction with μ_1 sites, as the selective μ_1 antagonist, naloxazone, reversed the inhibitory effect of these peptides (Pfaus et al 1986, Wiesner & Moss 1986).

The availability of more selective opioid receptor agonists allows questions of receptor selectivity to be answered definitively. We recently examined the effects of more selective μ, δ and κ receptor agonists on the sexual behaviour of female rats (Pfaus & Pfaff 1992). Sexually experienced ovariectomized rats were implanted with guide cannulae aimed at the right lateral ventricle. One group of rats was treated with oestradiol benzoate (10 μg) 48 hours and progesterone (250 μg) four hours before testing, whereas the other was treated with oestradiol benzoate alone. Rats were infused with different doses of the selective μ receptor agonist DAMGO, the selective δ agonist DPDPE, or the selective κ receptor agonist U50-488H. The females were placed with a sexually vigorous male in a bilevel chamber (Pfaus et al 1990) for three tests of sexual behaviour, beginning 15, 30, and 60 minutes after infusion. As shown in Fig. 2, DAMGO reduced lordosis quotients and magnitudes significantly in rats treated with oestradiol benzoate and progesterone, but not in rats treated with oestradiol benzoate alone. In contrast DPDPE and U50-488H increased lordosis quotients and magnitudes significantly in both steroid-treated groups (Fig. 2). Measures of proceptivity, rejection responses and locomotor activity were not affected significantly by any drug treatment. These results were in good agreement with

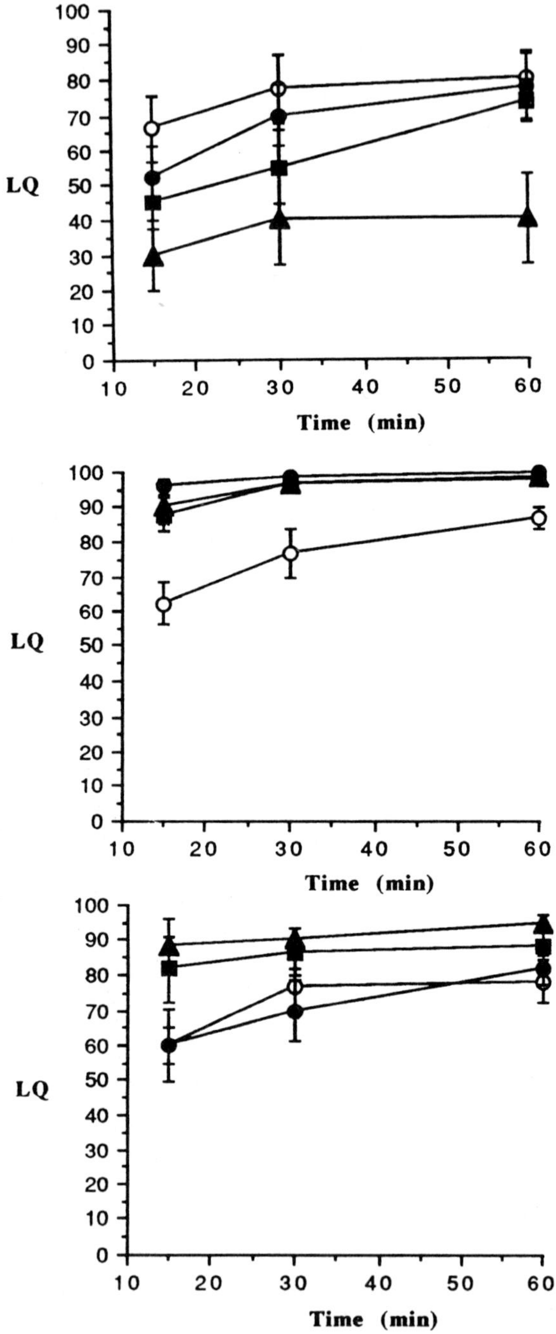
LQ
100
90
80
70
60
50
40
30
20
10
0
10
20
30
40
50
60
Time (min)
LQ
Time (min)
LQ
Time (min)

FIG. 2.

the findings of Pfaus & Gorzalka (1987b), and suggest that the facilitation of lordosis by δ and ϰ receptor agonists is independent of progesterone treatment, whereas the inhibitory effect of μ receptor agonists may require the presence of progesterone. Although it is not yet known at what forebrain sites δ or ϰ agonists facilitate sexual behaviour, the results of these studies lend strong support to the idea that enkephalins could facilitate female sexual behaviour by an interaction with δ opioid receptors.

Oestrogen can increase preproenkephalin mRNA levels

Consistent with the potential facilitatory actions of enkephalin through δ receptors on female reproductive behaviour, oestradiol has a robust positive effect on the levels of preproenkephalin mRNA. We measured oestrogen effects on preproenkephalin gene expression in hypothalamic and other rat forebrain neurons using both slot blot hybridization and *in situ* hybridization techniques (Romano et al 1988). Slot blot hybridization analysis indicated that the levels of enkephalin mRNA in the ventromedial hypothalamus of ovariectomized rats increase 3.1-fold after two weeks of oestrogen treatment. *In situ* hybridization revealed that the oestrogen-inducible enkephalinergic neurons are located in the ventrolateral aspect of the ventromedial nucleus, a subnucleus known to contain many oestrogen-concentrating neurons. The increase in mRNA levels was due to both a 63% increase in the number of detectable enkephalin-expressing neurons and a twofold increase in the level of enkephalin mRNA per expressing neuron: this yielded a 3.3-fold overall induction. Methodologically, the excellent agreement between the two techniques was expected, since previous work with haloperidol challenge of the preproenkephalin gene in the caudate similarly yielded good agreement between different types of assays (Romano et al 1987). The oestrogen results indicate a strong regulation of preproenkephalin transcription.

Tissue specificity is a major issue in preproenkephalin expression, both with comparisons among organs and comparisons among different brain regions. Previous *in situ* hybridization evidence demonstrated specific patterns of high or low constitutive expression across different brain regions (Harlan et al 1987). In the oestrogen experiments, as well, specificity was indicated in the comparisons among brain regions, as to whether there was hormonal regulation (Romano et al 1988). Besides the work with RNA from hypothalamic dissections,

---

FIG. 2. Dose-dependent effects of selective opioid receptor agonists on the lordosis behaviour of oestrogen- and progesterone-primed ovariectomized rats after infusions into the lateral ventricle. *Top*: Inhibition of lordosis quotients (LQs) produced by the selective μ receptor agonist DAMGO. *Middle*: Facilitation of LQs produced by the selective δ agonist DPDPE. *Bottom*: Facilitation of LQs produced by the selective ϰ receptor agonist U50-488H. ○, control; ●, 20 ng; ■, 200 ng; ▲, 2000 ng. From Pfaus & Pfaff (1992).

total RNA was isolated from the medial amygdala (another nucleus which contains both oestrogen receptors and enkephalin mRNA) and the caudate–putamen, which contains enkephalin mRNA but not oestrogen receptors. In neither of these tissues did oestrogen significantly alter enkephalin mRNA levels, even using the same rats in which significant hypothalamic effects had been detected. Therefore, the oestrogen regulation of enkephalin gene expression is tissue specific.

In another series of experiments, we examined the temporal features of the hormone induction by determining both the latency of the oestrogen-induced increase in enkephalin mRNA levels and the rate at which message levels decline following removal of oestrogen (Romano et al 1989b). The latency was relatively short: preproenkephalin mRNA levels were increased twofold within one hour of oestrogen treatment. After oestrogen removal, the levels of enkephalin mRNA declined rapidly. Progesterone treatment attenuated this decline, thus prolonging the oestrogen-induced rise in enkephalin mRNA levels. Together, the short latency of the oestrogen-induced elevation and rapid rate of decay clearly permit enkephalin gene expression to react quickly enough to hormone levels to follow the dynamic changes of the rat oestrous cycle. In this respect, and in consideration of the positive effect of progesterone, enkephalin gene expression clearly could be a major factor in promoting female reproductive behaviour.

The rapidity of the oestrogen effect on preproenkephalin mRNA, the presence of a putative oestrogen response element in the promoter region of the gene, and the presence of oestrogen receptors in enkephalinergic cells of the ventromedial hypothalamus are all consistent with a direct effect of oestradiol on the preproenkephalin gene mediated via the classical nuclear oestrogen receptor. However, given the large pool of cytoplasmic mRNA in cells, studies using exon-derived probes may not always accurately represent the magnitude and time course of steroid-regulated changes in preproenkephalin gene expression. For this reason, we have used intronic probes to study the regulation of preproenkephalin heteronuclear RNA (hnRNA) levels in response to various steroid treatments. For these studies, the polymerase chain reaction (PCR) was used to amplify a region of intron A of the preproenkephalin gene. This DNA fragment was used as a template for the synthesis of single-stranded, internally labelled DNA probes using a variant of the asymmetric PCR procedure (Gyllenstein & Ehrlich 1989). These probes are sensitive enough to be used for *in situ* hybridization experiments to detect preproenkephalin hnRNA in individual ventromedial hypothalamic neurons (PJ Brooks et al, unpublished 1991) (Fig. 3). Such studies should allow a better understanding of the exact temporal relationships between changes in ovarian steroid levels, preproenkephalin gene expression and reproductive behaviour.

Correlations of enkephalin gene expression with reproductive behaviour were also tested with a dose–response experiment in which all rats were assayed both for lordosis and for preproenkephalin mRNA (Lauber et al 1990). In the

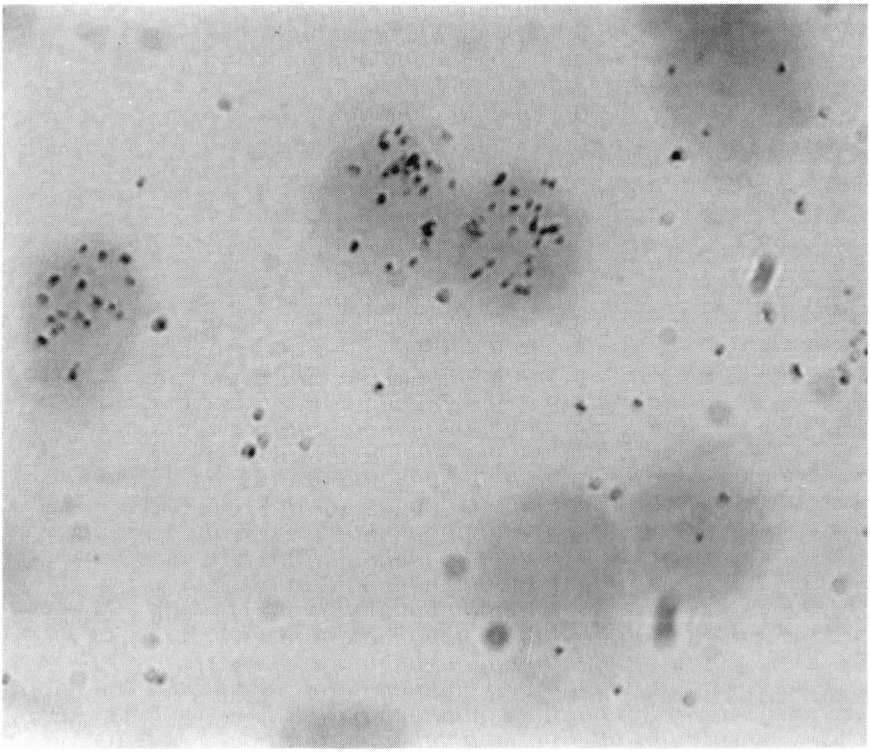

FIG. 3. Detection of preproenkephalin heteronuclear RNA in the ventromedial hypothalamus by *in situ* hybridization. This section through the ventromedial hypothalamus of an oestrogen-treated rat was hybridized with a ^{3}H-labelled single-stranded DNA probe to intron A of the rat preproenkephalin gene. The probe was labelled using a modification of the asymmetric PCR technique for synthesizing single-stranded DNA. Three labelled cell nuclei expressing the gene are shown. Exposure time: six weeks.

ventromedial hypothalamus, oestradiol treatment led to an impressively regular monotonic dose-dependent increase in enkephalin mRNA, while in the preoptic area only a small effect was observed at the highest oestradiol doses. Furthermore, when actual reproductive behaviour scores were plotted against preproenkephalin mRNA levels in the ventromedial hypothalamic nucleus, there was a clear positive correlation (Lauber et al 1990). This orderly relationship between mRNA and behaviour held whether the behaviour was measured in standard mating tests or with a lordosis reflex assay.

To determine whether there are sex differences in the hormonal regulation of enkephalin gene expression, we examined the effects of oestrogen and

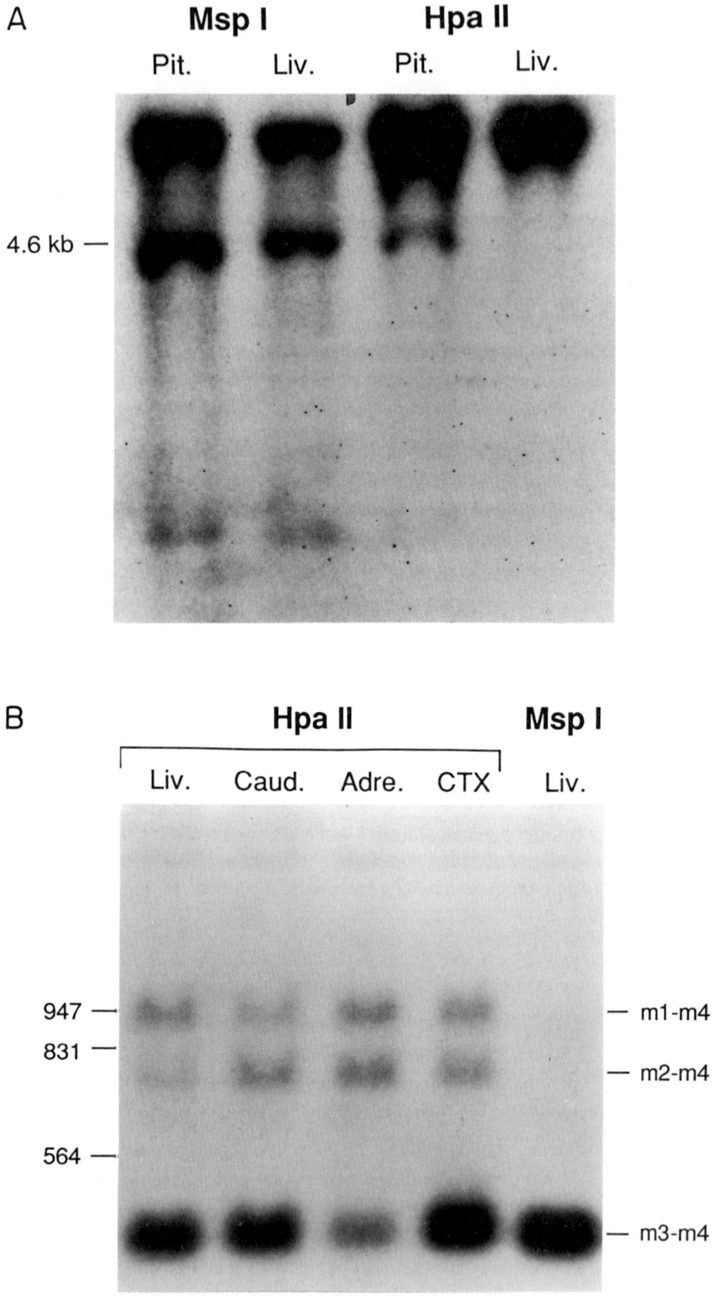
A
Msp I
Hpa II
Pit.
Liv.
Pit.
Liv.
4.6 kb
B
Hpa II
Msp I
Liv.
Caud.
Adre.
CTX
Liv.
947
831
564
m1-m4
m2-m4
m3-m4

FIG. 4.

testosterone on enkephalin mRNA levels in male as well as female rats (Romano et al 1990). Slot blot hybridization analysis of RNA isolated from the ventromedial hypothalamus indicated that oestrogen treatment significantly increased enkephalin mRNA levels of ovariectomized females, but had no effect on enkephalin mRNA levels of castrated males. Testosterone treatment of gonadomized males had no effect on enkephalin gene expression.

Overall, the regularity of the oestrogen effect on preproenkephalin mRNA in specific brain regions and the tight correlations of this molecular result with reproductive behaviour recommend the subject for detailed analysis of transcriptional controls. In subsequent work we will emphasize both direct mechanisms of hormone actions and the possibility of long-term hormone effects consistent with the concept of 'gene memory'.

Factors involved in the control of preproenkephalin gene expression

DNA methylation. In vertebrates this occurs primarily at C residues in CpG dinucleotides (Razin & Cedar 1977) and is proposed to be an important regulatory factor in the control of gene expression. In general, high levels of DNA methylation are found in transcriptionally silent DNA, while transcriptionally active DNA is typically unmethylated (Strobl 1990). In several cases, the methylation pattern of individual genes has been found to be cell specific. For example, the prolactin gene is hypermethylated in the liver, where it is not expressed, as opposed to the pituitary, where it is highly expressed by lactotrophs (Durrin et al 1984) (Fig. 4). In the case of prolactin, the differential methylation is seen at CpG residues throughout the coding region and promoter region of the gene. In other instances, unmethylated stretches of DNA in the promoter regions of specific genes can determine the 5′ end of transcriptionally active DNA (Strobl 1990).

FIG. 4. Tissue-specific methylation of the prolactin and preproenkephalin genes. (A) Genomic DNA from liver (liv.) or pituitary (pit.) was digested with *Msp*I or *Hpa*II as indicated and hybridized with a prolactin cDNA probe. The 4.6 kb band is present in *Hpa*II digests of pituitary DNA, but not liver DNA, indicating complete methylation of two CCGG sites within the transcribed region of the prolactin gene in liver. See Durrin et al (1984) for detailed discussion. Data from T. Funabashi & P. Brooks, unpublished observations 1991. (B) Genomic DNA from caudate nucleus (Caud.), liver (Liv), adrenal (Adre.) and cerebral cortex (CTX.) was digested with *Hpa*II or *Msp*I and hybridized with a single-stranded DNA probe complementary to a region of the preproenkephalin gene promoter. The relative intensities of bands within each lane after *Hpa*II digestion reflect the degree of methylation at four CCGG sites (m1–m4) in the promoter. Differences in the band pattern between lanes reflect tissue-specific differences in the methylation of the promoter. Band m2–m4 is consistently less intense than band m1–m4 in liver DNA compared to caudate DNA, indicating hypermethylation of site m2 in liver relative to caudate.

There are several possible mechanisms whereby DNA methylation could influence gene expression. Methylated DNA has been reported to produce DNase-insensitive structures (Keshet et al 1986) and may affect promoter function (Cedar 1984). In some, though not all, cases, methylation of specific DNA residues can prevent the binding of various stimulatory transcription factors, thereby inhibiting gene expression (Saluz et al 1988, also see Boyes & Bird 1991). In general, the DNA methylation pattern of a specific gene is fairly stable and would seem unlikely to account for acutely regulated behavioural changes. However, there are some notable examples of fairly rapid changes in DNA methylation in the endocrine literature. Saluz et al (1986) have shown that oestrogen stimulation of vitellogenin synthesis in the avian liver is accompanied by demethylation of specific C residues in the promoter region of the gene. This demethylation allows transcription factors to bind to these regions.

Of obvious relevance to neuroendocrinology is the regulation of methylation of the prolactin gene. Oestrogen stimulates transcription of the prolactin gene in the pituitary (see above), and long-term treatment of ovariectomized rats results in a demethylation of additional C residues in the pituitary (Durrin et al 1984). During lactation, prolactin synthesis increases dramatically and this is accompanied by demethylation of additional C residues within the coding region of the prolactin gene in the pituitary (Kumar & Biswas 1988).

As an initial step towards understanding the regulation of preproenkephalin gene expression, we have begun to examine the methylation pattern of this gene in different rat tissues. For these studies, genomic DNA is isolated from various organs and digested with the methylation-sensitive restriction enzyme *Hpa*II, which recognizes the sequence CCGG. Unlike its isoschizomer *Msp*I, *Hpa*II will not cut the DNA if the internal C residue is methylated. After digestion, the resulting Southern blots are hybridized with probes specific to the preproenkephalin promoter. Differences in hybridization patterns following *Hpa*II cleavage reflect differences in the degree of methylation at C residues around sequences recognized by the probe.

Our studies so far have yielded two major results. First, the preproenkephalin gene shows some degree of methylation in all tissues examined. This result is not so surprising, because even in tissues which do express the preproenkephalin gene, such cells do not account for all the DNA in the samples. The cellular heterogeneity of tissues, especially the brain, as opposed to cultured cells represents a significant technical challenge. Histochemical approaches will be useful in this regard. Second, we have detected significant differences in the methylation-dependent hybridization patterns between different tissues (Fig. 4b). Such differences reflect tissue-specific methylation of the preproenkephalin gene. Their relationship to differential gene expression remains to be studied in various brain regions. Future studies will explore possible long-term effects of oestrogen on methylation of the preproenkephalin gene in the ventromedial hypothalamus.

Transcription factors. Dynamic changes in gene expression are controlled by the binding to specific DNA sequences of particular combinations of transcription factors, which, via protein–protein interactions, stimulate RNA polymerase II-dependent transcription (Mitchell & Tjian 1989). Putative binding sites for several transcription factors upstream of the preproenkephalin gene start site have been described: these factors include AP-1 (Sonnenberg et al 1989), the cAMP response element binding protein (Comb et al 1986), NFκB, factors that bind to BETA sequences (Korner et al 1989) and oestrogen receptors (Romano et al 1989b). A direct role for some of these factors in regulating preproenkephalin gene expression has been demonstrated. Of particular relevance to the regulation of female sexual behaviour is the presence of putative EREs 365 to 379 bases upstream of the transcription start site (Romano et al 1989b). One way to determine whether this sequence actually functions as an ERE is remove the promoter region of the gene and attach it to the coding sequence of a reporter gene such as β-galactosidase in a specially designed neurotropic vector (Kaplitt et al 1991a). The construct could then be transfected into cultured cells which express the oestrogen receptor or into the brain (Kaplitt et al 1991b) and assayed for oestrogen activation of the reporter gene.

Regulation of preproenkephalin gene expression via the AP-1 site may also be important in regulating the activity of ventromedial hypothalamic neurons during copulation (JG Pfaus et al, unpublished 1991).

Future directions: genomic structure. Of particular importance will be experiments designed to assess the *in vivo* chromatin state of regulated genes under different behaviourally relevant conditions. One method for doing this is the DNase hypersensitivity assay, which takes advantage of the observation that transcriptionally active regions of chromatin exist in a conformation that is particularly sensitive to the actions of nucleases such as DNaseI or S1 (Garel & Axel 1976, Weintraub & Groudine 1976). Nuclei are treated with increasing concentrations of nuclease, then digested with restriction enzymes and analysed by Southern blotting. Additional bands appearing on the blot after treatment with low concentrations of nuclease indicate regions where the DNA is particularly sensitive to cleavage by the enzyme and may represent sites where transcription factors are bound.

For neuroendocrine purposes perhaps the most promising approaches to studying promoter function are the '*in vivo* footprinting' techniques (Cartwright & Kelly 1991). Unlike traditional footprinting techniques, which identify *potential* sites of transcription factor binding under non-physiological conditions, *in vivo* footprinting allows the visualization of transcription factor-binding sites in living cells. Because of the heterogeneous nature of cell populations in the nervous system, supplementing these techniques with well-designed histochemical approaches will be valuable.

References

Baker HJ, Shapiro DJ 1977 Kinetics of oestrogen induction of *Xenopus laevis* vitellogenin messenger RNA as measured by hybridization to complementary DNA. J Biol Chem 252:8428–8434

Barraclough CA, Gorski RA 1962 Studies on mating behaviour in the androgen-sterilized female rat in relation to the hypothalamic regulation of sexual behaviour. J Endocrinol 25:175–182

Beach FA, Orndorff RK 1974 Variation in the responsiveness of female rats to ovarian hormones as a function of preceding hormonal deprivation. Horm Behav 5: 202–205

Bednar I, Forsberg G, Sodersten P 1987 Inhibition of sexual behaviour in female rats by intracerebral injections of Met-enkephalin in combination with an inhibitor of enkephalin degrading enzymes. Neurosci Lett 79:341–345

Boyes J, Bird A 1991 DNA methylation inhibits transcription indirectly via a methyl CpG binding protein. Cell 64:1123–1134

Brooks PJ, Lund PK, Stumpf WE, Pedersen CA 1990 Oxytocin messenger ribonucleic acid levels in the medial preoptic area are increased during lactation. J Neuroendocrinol 2:621–626

Burch JBE, Evans MI 1986 Chromatin structural transitions and the phenomenon of vitellogenin gene memory in chickens. Mol Cell Biol 6:1886–1893

Cartwright IL, Kelly SE 1991 Probing the nature of chromosomal DNA–protein contacts by in vivo footprinting. BioTechniques 11:188–203

Cedar H 1984 DNA methylation and gene expression. In: Razin A, Cedar H, Riggs AD (eds) DNA methylation: biochemistry and biological significance. Springer-Verlag, New York, p 147–164

Chang K-J, Killian A, Hazum E, Cuatrecasas P, Chang J-K 1981 Morphiceptin (NH_2-Tyr-Pro-Phe-Pro-$CONH_2$): a potent and specific agonist for morphine (μ) receptors. Science (Wash DC) 212:75–77

Chung SK, McCabe JT, Pfaff DW 1991a Oestrogen influences on oxytocin mRNA expression in preoptic and anterior hypothalamic regions studied by in situ hybridization. J Comp Neurol 307:281–295

Chung SK, Haldar J, Pfaff DW 1991b Combination of in situ hybridization with retrograde marker to study hormone effect on oxytocin mRNA-expressing neurons which project to spinal cord. Soc Neurosci Abstr 17:715 (abstr 284.12)

Cohen RS, Pfaff DW 1992 Mechanisms of long term neuronal changes following oestrogen treatment. Prog Brain Res, in press

Comb M, Birnberg NC, Seasholtz A, Herbert E, Goodman HM 1986 A cyclic AMP and phorbol ester inducible DNA element. Nature (Lond) 323:353–356

Day RN, Maurer RA 1989 The distal enhancer region of the rat prolactin gene contains elements conferring response to multiple hormones. Mol Endocrinol 3:3–9

Deeley RG, Udell DS, Burns ATH, Gordon JI, Goldberger RF 1977 Kinetics of avian vitellogenin messenger RNA induction. J Biol Chem 252:7913–7915

Durrin LK, Weber JL, Gorski J 1984 Chromatin structure, transcription, and methylation of the prolactin gene domain in pituitary tumours of Fisher 344 rats. J Biol Chem 259:7086–7093

Garel A, Axel R 1976 Selective digestion of transcriptionally active ovalbumin genes from oviduct nuclei. Proc Natl Acad Sci USA 73:3966–3970

Gerall AA, Dunlap JL 1973 The effect of experience and hormones on the initial receptivity in female and male rats. Physiol & Behav 10:851–854

Gyllenstein U, Ehrlich HA 1989 Generation of single-stranded DNA by the polymerase chain reaction and its application to direct sequencing of the HLA-DQA locus. Proc Natl Acad Sci USA 85:7652–7656

Hache RJG, Tam S-P, Cochrane A, Nesheim M, Deeley RG 1987 Long-term effects of oestrogen on avian liver: oestrogen-inducible switch in expression of nuclear, hormone-binding proteins. Mol Cell Biol 7:3538–3547

Harlan RE, Shivers BD, Romano GJ, Howells RD, Pfaff DW 1987 Localization of preproenkephalin mRNA in the rat brain and spinal cord by in situ hybridization. J Comp Neurol 258:159–184

Harris J, Gorski J 1978 Evidence of a discontinuous requirement for oestrogen in stimulation of deoxyribonucleic acid synthesis in the immature rat uterus. Endocrinology 103:240–248

Hook VT 1990 Biosynthesis of endogenous opiate peptides by proteolytic precursor processing mechanisms. In: Watson RR (ed) Biochemistry and physiology of substance abuse. CRC Press, Baton Rouge vol 3:367–377

Jost J-P, Ohno T, Panyim S, Schuerch AR 1978 Appearance of vitellogenin synthesis in chicken liver following primary and secondary stimulation by 17β-oestradiol. Eur J Biochem 84:355–361

Kaplitt MG, Rabkin SD, Pfaff DW 1991a Molecular alterations in nerve cells: direct manipulation and physiological mediation. Curr Top Neuroendocrinol, in press

Kaplitt MG, Pfaus JG, Kleopoulos SP, Hanlon BP, Rabkin SD, Pfaff DW 1991b Expression of a foreign gene in adult mammalian brain following in vivo transfer via a herpes simplex virus type 1 defective viral vector. Mol & Cell Neurosci 2:320–330

Keshet I, Lieman-Hurwitz J, Cedar H 1986 DNA methylation affects the formation of active chromatin. Cell 44:535–543

Kim KE, Day RN, Maurer RA 1988 Functional analysis of the interaction of a tissue-specific factor with an upstream enhancer element of the rat prolactin gene. Mol Endocrinol 2:1374–1381

Korner M, Rattner A, Mauxion F, Sen R, Citri Y 1989 A brain specific transcription factor. Neuron 3:563–572

Kumar V, Biswas DK 1988 Dynamic state of site specific DNA methylation concurrent to altered prolactin and growth hormone gene expression in the pituitary gland of pregnant and lactating rats. J Biol Chem 263:12645–12652

Lauber AH, Romano GJ, Mobbs CV, Howells RD, Pfaff DW 1990 Oestradiol induction of proenkephalin messenger RNA in hypothalamus: dose-response and relation to reproductive behaviour in the female rat. Mol Brain Res 8:47–54

Lauber AH, Romano GJ, Mobbs CV, Pfaff DW 1991a Oestradiol regulation of oestrogen receptor messenger ribonucleic acid in rat mediobasal hypothalamus: an in situ hybridization study. J Neuroendocrinol 2:605–611

Lauber AH, Mobbs CV, Muramatsu M, Pfaff DW 1991b Estrogen receptor messenger RNA expression in rat hypothalamus as a function of genetic sex and estrogen dose. Endocrinology 129:3180–3186

Lauber AH, Romano GJ, Pfaff DW 1991c Sex difference in oestradiol regulation of progestin receptor mRNA in rat mediobasal hypothalamus as demonstrated by in situ hybridization. Neuroendocrinology 53:608–613

Magnan J, Paterson SJ, Kosterlitz HW 1982 The interaction of [Met^5] enkephalin and [Leu^5] enkephalin sequences, extended at the C-terminus, with the μ-,δ-, and κ-binding sites in the guinea-pig brain. Life Sci 31:1359–1361

Maurer RA, Notides AC 1987 Identification of an oestrogen-responsive element from the 5′-flanking region of the rat prolactin gene. Mol Cell Biol 7:4247–4254

Mitchell PJ, Tjian R 1989 Transcriptional regulation in mammalian cells by sequence-specific DNA-binding proteins. Science (Wash DC) 245:371–378
Parsons B, MacLusky NJ, Krieger MS, McEwen BS, Pfaff DW 1979 The effects of long-term oestrogen exposure on the induction of sexual behaviour and measurements of brain oestrogen and progestin receptors in the female rat. Horm Behav 13:301–313
Parsons B, MacLusky NJ, Krey L, Pfaff DW, McEwen BS 1980 The temporal relationship between oestrogen-inducible progestin receptors in the female rat brain and the time course of oestrogen activation of mating behaviour. Endocrinology 107:774–779
Parsons B, Rainbow TC, Pfaff DW, McEwen BS 1981 Oestradiol, sexual receptivity and cytosol progestin receptors in rat hypothalamus. Nature (Lond) 292:58–59
Parsons B, McEwen BS, Pfaff DW 1982a A discontinuous schedule of oestradiol treatment is sufficient to activate progesterone-facilitated feminine sexual behaviour and to increase cytosol receptors for progestins in the hypothalamus of the rat. Endocrinology 110:613–619
Parsons B, Rainbow TC, Pfaff DW, McEwen BS 1982b Hypothalamic protein synthesis essential for the activation of the lordosis reflex in the female rat. Endocrinology 110:620–624
Paterson SJ, Robson LE, Kosterlitz HW 1983 Classification of opioid receptors. Br Med J 39:31–36
Pfaff DW, Zigmond RE 1971 Neonatal androgen effects on sexual and nonsexual behaviour of adult rats tested under various hormone regimes. Neuroendocrinology 7:129–145
Pfaus JG, Gorzalka BB 1987a Opioids and sexual behaviour. Neurosci Biobehav Rev 11:1–34
Pfaus JG, Gorzalka BB 1987b Selective activation of opioid receptors differentially affects lordosis behaviour in female rats. Peptides (NY) 8:309–317
Pfaus JG, Pfaff DW 1992 Dual effect of selective opioid receptor agonists on the sexual behaviour of female rats: differential modulation by progesterone. Submitted
Pfaus JG, Pendleton N, Gorzalka BB 1986 Dual effect of morphiceptin on lordosis behaviour: possible mediation by different opioid receptor subtypes. Pharmacol Biochem Behav 24:1461–1464
Pfaus JG, Mendelson SD, Phillips AG 1990 A correlational and factor analysis of anticipatory and consummatory measures of sexual behaviour in the male rat. Psychoneuroendocrinology 15:329–340
Phoenix C, Goy R, Gerall A, Young WC 1959 Organizing action of prenatally administered testosterone propionate on the tissues mediating mating behaviour in the female guinea pig. Endocrinology 65:369–382
Razin A, Cedar H 1977 Distribution of 5-methyl cytosine in chromatin. Proc Natl Acad Sci USA 74:2725–2728
Romano GJ, Shivers BD, Harlan RE, Howells RD, Pfaff DW 1987 Haloperidol increases proenkephalin mRNA levels in the caudate-putamen of the rat: a quantitative study at the cellular level using in situ hybridization. Mol Brain Res 2:33–41
Romano GJ, Harlan RE, Shivers BD, Howells RD, Pfaff DW 1988 Oestrogen increases proenkephalin messenger ribonucleic acid levels in the ventromedial hypothalamus of the rat. Mol Endocrinol 2:1320–1328
Romano GJ, Krust A, Pfaff DW 1989a Expression and oestrogen regulation of progesterone receptor mRNA in neurons of the mediobasal hypothalamus: an in situ hybridization study. Mol Endocrinol 3:1295–1300
Romano GJ, Mobbs CV, Howells RD, Pfaff DW 1989b Oestrogen regulation of proenkephalin gene expression in the ventromedial hypothalamus of the rat: temporal qualities and synergism with progesterone. Mol Brain Res 5:51–58

Romano GJ, Mobbs CV, Lauber A, Howells RD, Pfaff DW 1990 Differential regulation of proenkephalin gene expression by oestrogen in the ventromedial hypothalamus of male and female rats: implications for the molecular basis of a sexually differentiated behaviour. Brain Res 536:63–68

Saluz HP, Jiricny J, Jost JP 1986 Genomic sequencing reveals a positive correlation between the kinetics of strand specific DNA demethylation of the overlapping oestrogen/glucocorticoid-receptor binding sites and the rate of avian vitellogenin mRNA synthesis. Proc Natl Acad Sci USA 83:7167–7171

Saluz HP, Feavers IM, Jiricny J, Jost JP 1988 Genomic sequencing and in vivo footprinting of an expression specific DNAse I hypersensitive site of avian vitellogenin promoter reveal a demethylation of an mCpG and a change in specific interactions of proteins with DNA. Proc Natl Acad Sci USA 85:6697–6700

Searle PF, Tata JR 1981 Vitellogenin gene expression in male *Xenopus* hepatocytes during primary and secondary stimulation with oestrogen in cell cultures. Cell 23:741–746

Sirinathsinghji DJS 1984 Modulation of lordosis behaviour of female rats by naloxone, β-endorphin and its antiserum in the mesencephalic central gray: possible mediation via GnRH. Neuroendocrinology 39:222–230

Sirinathsinghji DJS 1986 Regulation of lordosis behaviour in the female rat by corticotropin-releasing factor, β-endorphin/corticotropin, and luteinizing hormone-releasing hormone neuronal systems in the medial preoptic area. Brain Res 375:49–56

Sonnenberg JL, Rauscher FJ III, Morgan JI, Curran T 1989 Regulation of proenkephalin by Fos and Jun. Science (Wash DC) 246:1622–1625

Strobl JS 1990 A role for DNA methylation in vertebrate gene expression? Mol Endocrinol 4:181–185

Tam S-P, Hache RJG, Deeley RG 1986 Oestrogen memory effect in human hepatocytes during repeated cell division with hormone. Science (Wash DC) 234:1234–1237

Tougard C, Tixier-vidal 1988 Lactotropes and gonadotropes. In: Knobil E, Neill J (eds) The physiology of reproduction. Raven Press, New York, vol 1:1305–1333

Weintraub H, Groudine M 1976 Transcriptionally active and inactive conformations of chromosomal subunits. Science (Wash DC) 93:848–858

Whalen RE, Nakayama K 1965 Induction of estrous behaviour: facilitation by repeated hormone treatment. J Endocrinol 33:525–526

Wiesner JB, Moss RI 1986 Behavioural specificity of β-endorphin suppression of sexual behaviour: differential receptor antagonism. Pharmacol Biochem Behav 24:1235–1239

Wiesner JB, Moss RL 1989 A psychopharmacological characterization of the opioid suppression of sexual behaviour in the female rat. In: Dyer RG, Bicknell RJ (eds) Brain opioid systems in reproduction. Oxford University Press, Oxford, p 187–202

Yen SSC, Jaffe RB (eds) Reproductive endocrinology: physiology, pathophysiology and clinical management. Saunders, Philadelphia

Zaret KS, Yamamoto KR 1984 Reversible and persistent changes in chromatin structure accompany activation of a glucocorticoid-dependent enhancer element. Cell 38:29–38

DISCUSSION

Zoeller: E. F. LaGamma has shown that when he stresses animals, in the caudate multiple start sites are used for the up-regulation of enkephalin mRNA expression (Weisinger et al 1990). Do you think there is a lot of regulatory flexibility with respect to oestrogen induction of enkephalin, etc?

Pfaff: I do believe there is variation in the use of start sites. We have discussed the virtues of *in situ* hybridization (p 141–143); the ability to detect the use of an individual start site depends firstly on which probe you are using. Also, in an individual cell group, there can be neurons using not only the start sites that LaGamma talked about but also a start site in intron A. Such cells may be surrounded by neurons that are not using that start site (PJ Brooks et al, unpublished 1991).

Another question is whether the use of different start sites is a biologically meaningful and regulated event.

Mellon: When you are looking at DNA methylation in a tissue like the liver or the pituitary, you are looking at many different kinds of cells at once. In the caudate, what percentage of the nuclei from which you have extracted the DNA are glia as opposed to neurons expressing enkephalin?

Pfaff: I don't know the exact percentages of neurons and glia, especially in the caudate. And although enkephalin is wonderful to study in the caudate, not all neurons there express enkephalin, so as well as the presence of different cell types, we also have to consider expressing neurons versus non-expressing neurons.

Gainer: Does the spleen have a lot of preproenkephalin mRNA?

Pfaff: Not in the parenchyma but in lymphocytes.

Gainer: And their content is substantial, like the caudate and the adrenal?

Pfaff: Yes. Lymphocytes have been used for analysis of the enkephalin gene promoter. There is a transcription factor called NFκB which we are studying with gel shift experiments using the preproenkephalin promoter. This factor was initially identified in lymphocytes.

Gainer: You mentioned transfection of a construct containing the preproenkephalin gene promoter into cultured cells. What type of cells were these?

Pfaff: They were 293 cells, which are kidney cells. The vector was expressed, perhaps surprisingly. I can reason backwards and say that the transcription factors which are required to operate the rat preproenkephalin promoter are present in those cells, but why that should be the case biologically, I don't know.

Young: The kidney cell expresses a lot of enkephalin during development; the expression is shut off in the adult. Are the 293 cells of fetal origin?

Mellon: The 293 cell is transformed by adenovirus E1A, so one might expect them to be a bit dedifferentiated.

Pfaff: Our main aim is to put this enkephalin promoter construct directly into the brain to identify areas that do express or that do not express the reporter gene, and then to find out why.

Meanwhile, there are advantages to using the cell line. If we want to do gel shift studies, for example, and we want nuclear extracts from a homogeneous cell line to study the preproenkephalin promoter, and if we know the promoter works in a particular cell line, we can use that as our first source of nuclear extract rather than trying to get enough brain cells. On the one hand such studies have all the advantages of work with cell lines; on the other hand I would be the last to say that results in a cell line are going to substitute for studies in differentiated brain cells.

Zoeller: Have you thought about injecting the putative oestrogen receptor-binding sequence from the proenkephalin promoter to compete out endogenous oestrogen receptor?

Pfaff: We certainly have thought of that. We can do two things with respect to interfering with the function of the oestrogen response element (ERE) or responses to oestrogen in general. We can use antisense sequences complementary to mRNA for the oestrogen receptor. And we can do the competition studies: overload with exogenous ERE so that the oestrogen receptor does not reach the native ERE. Our first studies (McCarthy et al 1991) were antisense experiments with glutamic acid decarboxylase. We also introduced an antisense sequence for the progesterone receptor mRNA directly into hypothalamic neurons and reduced progesterone-stimulated reproductive behaviour (Ogawa et al 1992). We are taking so much time to document exactly how that works, that we haven't got back to the study that you suggest.

Mellon: You inject the antisense sequence into the hypothalamus and you think that is quantitative enough to be sure you have reached most neurons that contain progesterone receptors?

Pfaff: We are trying simply to administer enough. It is not quantitative in that we can expect to reduce the amount of translating message by a given percentage. All we can do is hope to get enough antisense sequence into the cells in order to get a behavioural effect. We are trying to figure out how it works.

Zoeller: Would it be possible to inject the consensus sequence for the cAMP regulatory element to compete out factors that regulate enkephalin expression?

Pfaff: George Uhl did that and showed unilateral diminution in afferent-stimulated enkephalin expression.

Sawchenko: Don, can you give us more details of the gene transfer protocol? What volume you are injecting? Are they iontophoretic injections?

Pfaff: No, they are pressure injections in 1 μl. We either simply inject the stock in which the amplicon, the defective virion, was grown, which still contains the helper virus or we concentrate the vector (Kaplitt et al 1991a,b). The efficiency is low: we calculate the number of virions we are injecting and count the number of cells expressing β-galactosidase, and divide the latter by the former. The efficiency has not been higher than 1%. Altogether there are three parameters we can alter—concentration, efficiency and time. Because the efficiency has been so low, we tried increasing the concentration. This has increased the yield of expressing cells, but the pictures aren't as nice because the cells are piled on top of each other. Usually, after injection of lower concentrations of the vector, we get nice expression by scattered individual cells.

Sawchenko: Do you think it is possible to introduce these genes discretely into a substantial fraction of a cell population, like the ventrolateral part of the ventromedial nucleus?

Pfaff: The cells express splendidly by themselves, but if you asked whether we could drive some physiological function, I think the answer would be not yet. We hope to have functional effects with the concentrated preparation.

Sawchenko: What about the mundane constraints imposed by the irregular topographies of hypothalamic cell groups? It would seem that in order to introduce the material into enough cells to see an effect, you are necessarily going to introduce the gene into a number of extraneous cells and this could confound the results. Has this been a problem?

Pfaff: This kind of molecular approach, when looking at neuronal functions, will have the disadvantages of every neuropharmacological approach. We hope that the gene we are introducing will be physiologically neutral in the cells that don't have the connections to affect the function we are interested in.

Sawchenko: Is subversion of axonal transport mechanisms a conceivable trick with which to achieve more precise targeting?

Pfaff: The wild-type herpes virus is transported. We thought that since the amplicon has, as far as we know, the same protein coat as a wild-type virus, it should be transported retrogradely. We certainly have looked (Kaplitt et al 1991b), because we feared this would be a short-coming of the molecular manipulation, i.e. that the deficient virus would be transported all over.

References

Kaplitt MG, Rabkin SD, Pfaff DW 1991a Molecular alterations in nerve cells: direct manipulation and physiological mediation. Curr Top Neuroendocrinol, in press

Kaplitt MG, Pfaus JG, Kleopoulos SP, Hanlon BP, Rabkin SD, Pfaff DW 1991b Expression of a foreign gene in adult mammalian brain following in vivo transfer via a herpes simplex virus type 1 defective viral vector. Mol & Cell Neurosci 2:320–330

McCarthy MM, Schwartz-Giblin S, Pfaff DW 1991 Intracerebral administration of glutamic acid decarboxylase (GAD) antisense oligodeoxynucleotide reduces lordosis behavior in the rat. Soc Neurosci Abstr 17:497

Ogawa S, Olazabal UE, Pfaff DW 1992 Behavioral change after local administration of antisense sequence for progesterone receptor mRNA in female rat hypothalamus. In: Baserga R, Denhardt DT (eds) Antisense strategies. Ann NY Acad Sci, in press

Weisinger G, DeCristofara JD, LaGamma EF 1990 Multiple preproenkephalin transcriptional start sites are induced by stress and cholinergic pathways. J Biol Chem 265:17389–17392

Receptors and neural effects of oxytocin in the rodent hypothalamus and preoptic region

J. J. Dreifuss, E. Tribollet, M. Dubois-Dauphin and M. Raggenbass

Department of Physiology, University Medical Centre, 9 Avenue de Champel, CH-1211 Geneva 4, Switzerland

Abstract. Vasopressin and oxytocin are produced in and secreted from not only hypothalamo-hypophysial neurons which shed their products into the circulation to act as hormones or releasing factors, but also from neurons whose axons form tracts which remain within the central nervous system. Using tritiated or radioiodinated ligands, binding sites for vasopressin and for oxytocin have been detected by *in vitro* autoradiography. In the rat hypothalamus binding sites for vasopressin are present in the suprachiasmatic, stigmoid and arcuate nuclei, and oxytocin receptors in the area of the ventromedial nucleus. Electrophysiological evidence obtained using single cell recordings in slices suggests that oxytocin-binding sites present in the ventromedial hypothalamus and in the bed nucleus of the stria terminalis mostly represent functional, neuronal receptors. The expression of these receptors (but not of the vasopressin receptors) depends on gonadal steroid hormones, as does that of uterine and mammary gland oxytocin receptors. Modifications of the hormonal status associated with, for example, puberty and lactation cause 'up-regulation' of central and peripheral oxytocin receptors. The central administration of oxytocin facilitates (and the administration of oxytocin agonists inhibits) maternal behaviour and the milk ejection reflex, therefore the hormonal and neural actions of oxytocin appear to be complementary in ensuring the birth and development of the offspring.

1992 Functional anatomy of the neuroendocrine hypothalamus. Wiley, Chichester (Ciba Foundation Symposium 168) p 187–208

New-born mammals are incapable of independent life and their survival requires nursing and maternal care. In particular, an adequate supply of milk is required to provide the young with the nutrients needed for survival and growth. During nursing, maternal oxytocin—released in response to sucking by the new-born—stimulates the contraction of the myoepithelial cells of the mammary gland and milk is thus ejected. In lactating rats, in which it is possible to measure the rise in intramammary pressure at milk ejection, an abrupt rise in pressure is seen every few minutes during suckling. The pressure wave occurs simultaneously in recordings taken from different teats and is similar in amplitude from one milk ejection to the next. It is due to the release of approximately one milliunit of oxytocin once every five minutes or so.

Secretion of oxytocin from the neurosecretory axons in the neural lobe is regulated by the frequency and number of incoming action potentials. Each pulse of oxytocin released during nursing is preceded by an explosive but short-lived increase in firing rate in all, or nearly all, oxytocin-secreting hypothalamo-neurophypophysial neurons (Dreifuss et al 1981, Poulain & Wakerley 1982).

It is not yet known how the essentially *constant* suckling stimulus is transformed into a sequence of *intermittent* bursts of action potentials generated by the oxytocin-producing neurons which project to the neural lobe. Nor is it known where in the central nervous system of the nursing mother the sensory information elicited by the sucklings is integrated and then either 'gated' during the interval between two milk ejections or, if sufficient time has elapsed since the previous ejection, conveyed to the oxytocin-secreting neurons.

There is strong evidence that oxytocin itself, by a central action, determines the interval between each milk ejection: intracerebroventricular or intracerebral injection of oxytocin evokes a striking facilitation of the reflex (Freund-Mercier & Richard 1984, Moos et al 1991). Within minutes, the frequency of the bursts of action potentials increases, as does the number of action potentials in each burst. These changes are mirrored by the pulses of oxytocin released from the neural lobe, which occur more frequently and are of greater amplitude. Conversely, intracerebroventricular administration of an oxytocin antagonist temporarily slows the milk ejections, suggesting that *endogenous* oxytocin is normally involved in their timing (Freund-Mercier & Richard 1984, Belin & Moos 1986).

Rats are born immature and helpless. Their mother keeps them in a nest which she builds prior to parturition. When in the nest, she crouches over her young and by so doing she exposes her mammary region, thus allowing suckling. If one of the pups is displaced, the mother carries it back to the nest.

Intracerebroventricular injection of oxytocin can induce, after a short latent period, these and other aspects of maternal behaviour in virgin female rats, provided their oestrogen levels are high (Pedersen & Prange 1979). Vasopressin is comparatively ineffective (Pedersen et al 1982). Additional evidence supporting a role for endogenous oxytocin in stimulating the onset of maternal behaviour has been presented (Fahrbach et al 1985). An attempt was made to suppress spontaneous maternal behaviour following natural parturition in rats by injecting an oxytocin antagonist into the ventricular spaces. After delivery, the pups were removed from the mother rat for a period and thereafter, pups and nesting material were presented to her in a test cage. Saline-injected mothers started gathering the pups immediately and displayed full maternal behaviour within minutes. Antagonist-treated mothers showed a marked delay in pup retrieval (van Leengoed et al 1987).

Sheep and other ungulates differ from rodents in giving birth to precocious young, which can follow their mother soon after birth. A specific bond is formed between the mother and her own young; conversely, the mother sheep rejects young that are not her own. In female sheep, intracerebroventricular oxytocin has been shown to stimulate a rapid onset of maternal behaviour and to induce her to stay with the lamb (Kendrick et al 1987).

These studies indicate that oxytocin—including endogenous oxytocin—exerts some of its effects within the central nervous system. From the late 1970s, the existence of vasopressin- or oxytocin-containing pathways and synapses in several areas of the central nervous system has been established by immunocytochemistry (Sawchenko & Swanson 1982, Sofroniew 1985).

In our laboratory, we have accumulated autoradiographical and electrophysiological evidence compatible with the view that oxytocin acts in the rodent central nervous system. In this paper, we describe the distributions of sites binding oxytocin and, for comparison, vasopressin as revealed by light microscopic autoradiography in the hypothalamus and preoptic region (Tribollet 1992). We shall summarize the evidence that these binding sites might represent functional receptors capable of affecting bioelectrical signalling amongst neurons. A final section deals with the mechanism of action of oxytocin, which we currently study in an area of the brain outside the hypothalamus.

The distribution of binding sites for oxytocin and vasopressin

The distribution of specific binding sites for oxytocin in the hypothalamus, in the preoptic area and in adjacent parts of the rat brain is shown in Fig. 1 (*see photographs on the left hand*). Specific binding was defined as that which was displaced in the presence of an excess of non-radioactive oxytocin.

The ventrolateral part of the ventromedial hypothalamic nucleus is particularly rich in oxytocin-binding sites, as is the adjacent neuropil (Fig. 1D,E). Binding was also observed in the medial tuberal nucleus and in the mamillary complex (Fig. 1F,G). More anteriorly, weak labelling was present throughout the medial preoptic area (Fig. 1C). No specific binding sites were detected in the hypothalamic supraoptic and paraventricular nuclei, the main sites of oxytocin synthesis (Fig. 2). There was also some 'mismatch' between the distribution of binding sites for oxytocin and the fields of termination of oxytocin-containing axons revealed by immunocytochemical methods (Sofroniew 1985, Raggenbass et al 1987a).

Areas in the vicinity of the hypothalamus which showed faint, but specific labelling included the lateral septum, the dorsomedial part of the caudoputamen, the paraventricular nucleus of the thalamus, as well as the medial and basolateral amygdaloid nuclei. The central amygdaloid nucleus showed intense labelling, as did the bed nucleus of the stria terminalis.

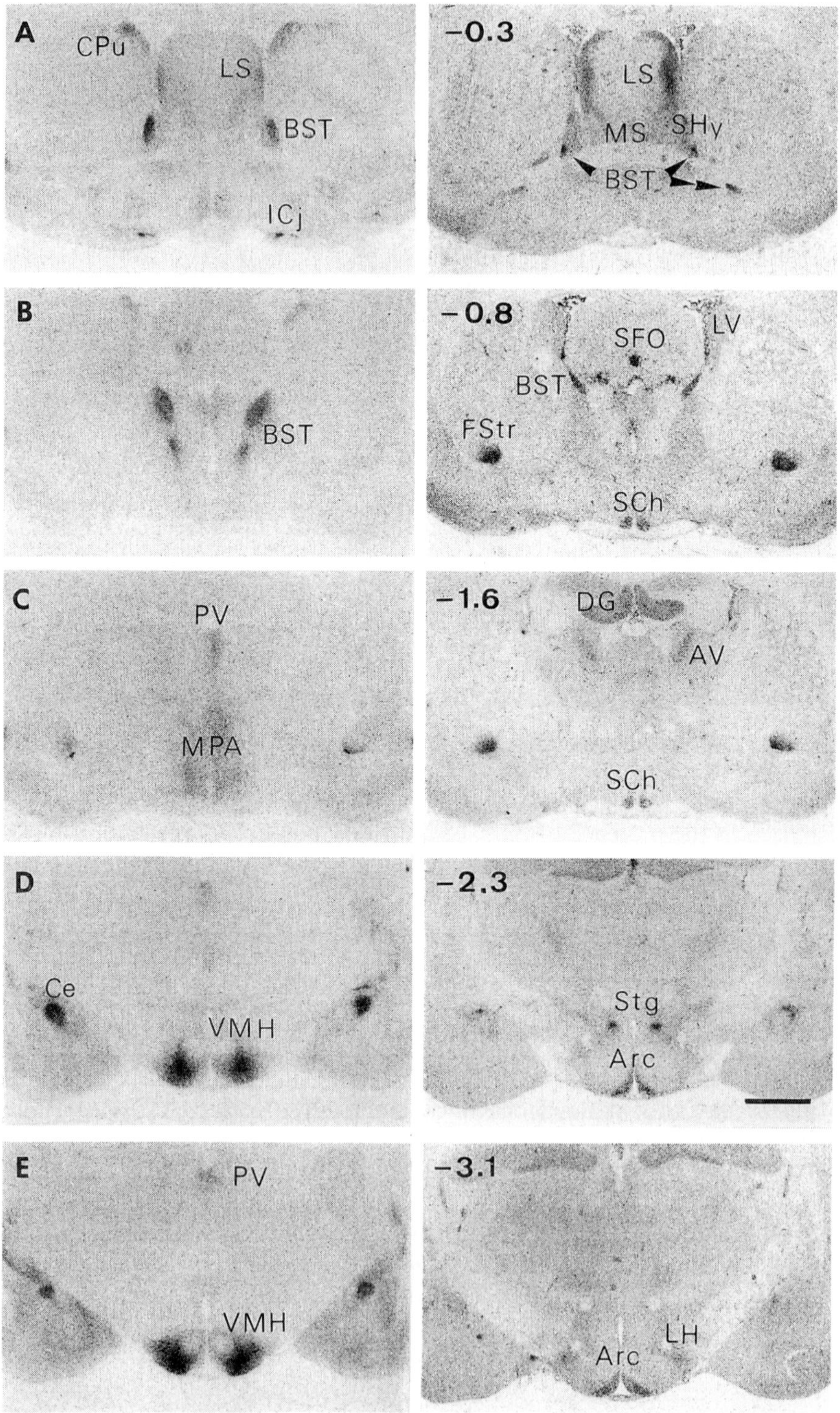
A
CPu
LS
BST
ICj
−0.3
LS
MS
SHy
BST
B
BST
−0.8
SFO
LV
BST
FStr
SCh
C
PV
MPA
−1.6
DG
AV
SCh
D
Ce
VMH
−2.3
Stg
Arc
E
PV
VMH
−3.1
LH
Arc

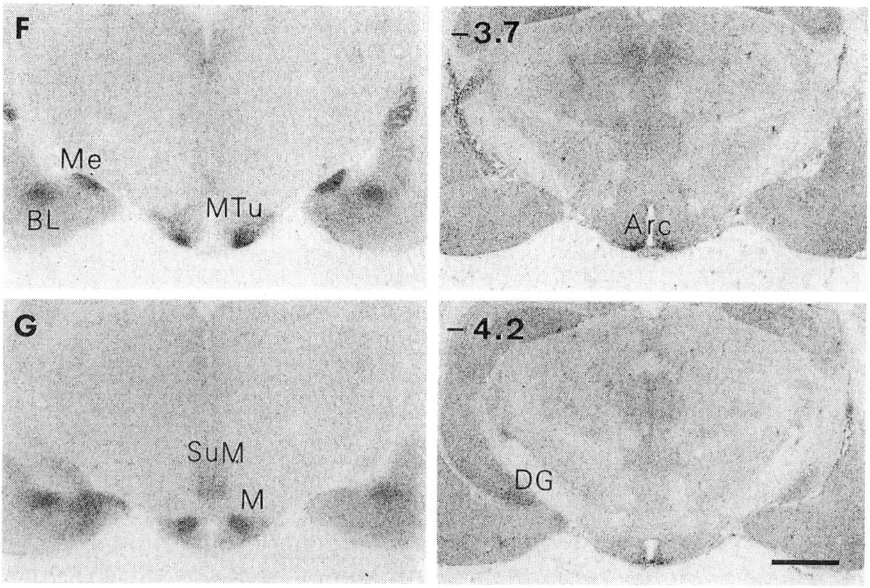

FIG. 1. Distribution of binding sites for neurohypophysial peptides in the rat hypothalamus and preoptic region. Coronal sections, 15 μm thick, were obtained from the brains of adult male rats; nine consecutive rostrocaudal levels (A–G) are shown. At each level, the number on the top of the photograph on the right indicates the approximate distance in mm from the bregma. At all levels illustrated, two sections from the same brain are shown which are separated by no more than a few tens of micrometers: the section on the left shows the binding sites for oxytocin, the section on the right those for vasopressin. To label the former, sections were incubated with 0.05 nM of an ^{125}I-labelled oxytocin antagonist (specific activity *ca.* 2000 Ci/mmol) (Elands et al 1988); to label the latter, sections were incubated with 2 nM [^{3}H]vasopressin (specific activity *ca.* 50 Ci/mmol). Bar = 2 mm.

Arc, arcuate hypothalamic nucleus; AV, anteroventral thalamic nucleus; BL, basolateral amygdaloid nucleus; BST, bed nucleus of the stria terminalis; Ce, central amygdaloid nucleus; CPu, caudate putamen; DG, dendate gyrus; FStr, fundus striati; ICj, islands of Calleja; LH, lateral hypothalamic area; LS, lateral septal nucleus; LV, lateral ventricle; M, medial mammillary nucleus; Me, medial amygdaloid nucleus; MPA, medial preoptic area; MS, medial septal nucleus; MTu, medial tuberal nucleus; PV, paraventricular thalamic nucleus; SCh, suprachiasmatic nucleus; SFO, subfornical organ; SHy, septohypothalamic nucleus; Stg, stigmoid hypothalamic nucleus; SuM, supramammillary nucleus; VMH, ventromedial hypothalamic nucleus.

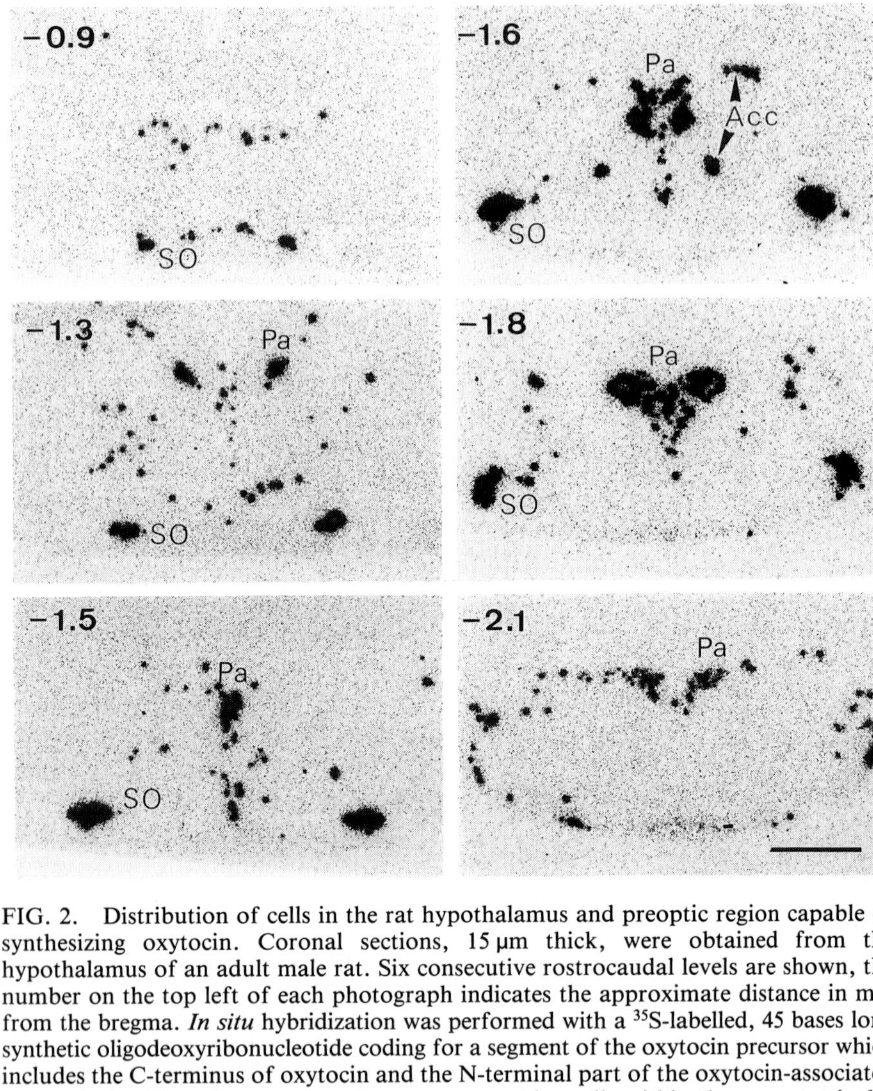

FIG. 2. Distribution of cells in the rat hypothalamus and preoptic region capable of synthesizing oxytocin. Coronal sections, 15 µm thick, were obtained from the hypothalamus of an adult male rat. Six consecutive rostrocaudal levels are shown, the number on the top left of each photograph indicates the approximate distance in mm from the bregma. *In situ* hybridization was performed with a ^{35}S-labelled, 45 bases long synthetic oligodeoxyribonucleotide coding for a segment of the oxytocin precursor which includes the C-terminus of oxytocin and the N-terminal part of the oxytocin-associated neurophysin. Note the presence of numerous reactive cells within the paraventricular (Pa) and supraoptic (SO) nuclei, or forming the so-called accessory nuclei (Acc) or occurring in isolation. Bar = 1 mm.

No major sex-specific differences in the distribution and amount of oxytocin receptors were found when comparing the brains of intact adult male and female rats. On the other hand, experiments on castrated animals suggested that in either sex, in some regions of the brain, the density of oxytocin receptors was dependent upon the circulating level of gonadal steroid hormones (Tribollet et al 1990).

Gonadectomy induced the virtual disappearance of binding sites for oxytocin in the hypothalamic ventromedial nucleus and a marked reduction in the bed nucleus of the stria terminalis. This effect could be reversed by treating the castrated animals with gonadal steroid hormones. The density of binding sites for oxytocin in the ventromedial hypothalamus and in the bed nucleus of the stria terminalis is affected by physiological modifications of the hormonal status of the animal, such as those associated with puberty (Tribollet et al 1989) and with parturition (Insel 1990).

Comparison of the left-hand and right-hand autoradiograms depicted in Fig. 1 shows that there is little overlap in the anatomical distributions of receptors for vasopressin and for oxytocin. Three hypothalamic nuclei are shown to contain vasopressin receptors: the suprachiasmatic nucleus (Fig. 1B,C), the stigmoid nucleus and the area adjacent to it (Fig. 1 D,E), and the arcuate nucleus (Fig. 1D–F). Neither the paraventricular nor the supraoptic nuclei were labelled.

Extrahypothalamic areas endowed with binding sites for vasopressin include several nuclei in the septal region, restricted areas in the bed nucleus of the stria terminalis, the subfornical organ, the fundus striati, the dentate gyrus and the anteroventral thalamic nucleus (Fig. 1). [^{3}H]Vasopressin also labelled the choroid plexus in the lateral ventricle and in the dorsal third ventricle (Fig. 1B). Otherwise no vascular labelling could be demonstrated.

Are binding sites functional receptors?

The presence of vasopressin- and oxytocin-binding sites in the hypothalamus and preoptic region does not prove that these peptides play a role in interneuronal communication in these parts of the brain. Binding sites may be located on glial and endothelial cells or, if present on neurons, may not be involved in neuronal signalling. It was thus useful to assess whether the presence of vasopressin- or oxytocin-binding sites in a given brain area correlates with the existence, in the same area, of neuronal sensitivity to these neurohypophysial peptides.

Evidence regarding the actions of oxytocin and vasopressin on the firing of single neurons in various regions of the central nervous system has been obtained using the *in vitro* slice preparation (Mühlethaler et al 1985). This experimental method has the advantage that a thin piece of nervous tissue can be kept alive and perfused with medium to which the peptide to be tested is added at known concentrations.

Approximately half of the spontaneously active neurons located in the bed nucleus of the stria terminalis in slices obtained from lactating rats responded to oxytocin with an increase in firing (Ingram et al 1990). This effect was concentration related and probably receptor mediated, since it could be reversibly suppressed by a synthetic structural analogue acting as an antagonist. Moss et al (1991) proposed that oxytocin-sensitive neurons located in the bed nucleus of the stria terminalis might mediate the facilitatory effect exerted by oxytocin

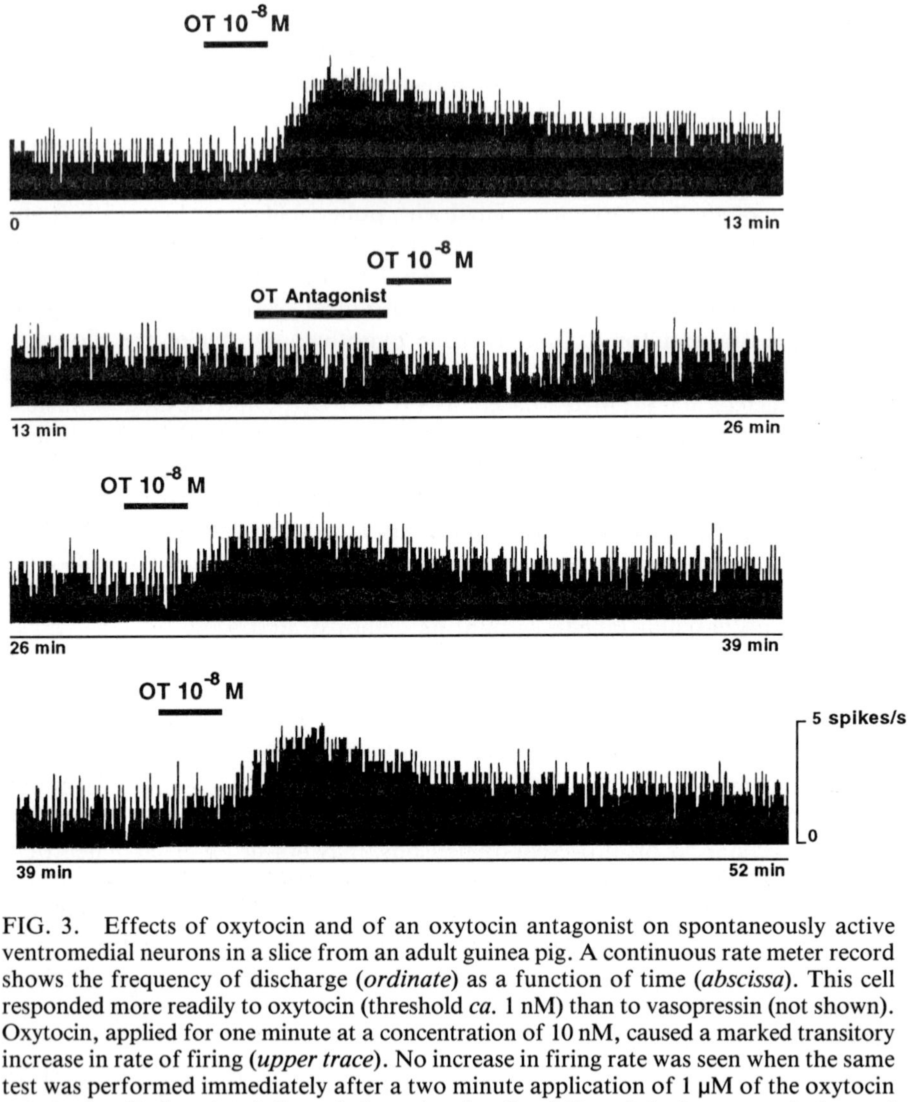

FIG. 3. Effects of oxytocin and of an oxytocin antagonist on spontaneously active ventromedial neurons in a slice from an adult guinea pig. A continuous rate meter record shows the frequency of discharge (*ordinate*) as a function of time (*abscissa*). This cell responded more readily to oxytocin (threshold *ca.* 1 nM) than to vasopressin (not shown). Oxytocin, applied for one minute at a concentration of 10 nM, caused a marked transitory increase in rate of firing (*upper trace*). No increase in firing rate was seen when the same test was performed immediately after a two minute application of 1 μM of the oxytocin antagonist, [1-deamino 2-O-methyltyrosine,4-valine,8-D-arginine] vasopressin (*second trace*). Oxytocin progressively regained its excitatory effect following washout of the antagonist (*third and fourth traces*).

on milk ejection. They showed that microinjections of low amounts of oxytocin into the bed nucleus of suckled lactating rats facilitated the milk-ejection reflex by increasing the frequency and the amplitude of the bursts of action potentials occurring periodically in presumptive oxytocin neurons.

The hypothalamic ventromedial nucleus contains binding sites for oxytocin in the rat (Fig. 1 D,E) and in the guinea pig. Inenaga et al (1991) reported that oxytocin applied in nanomolar concentrations excited about half of the spontaneously active ventromedial neurons tested in slices from guinea pigs; no inhibition was observed. The response was concentration dependent and was probably exerted postsynaptically, since it persisted under conditions of synaptic blockade. Moreover, the effect was deemed to be specific, because it could be mimicked by a selective oxytocin receptor agonist and suppressed by a synthetic structural analogue known to act as an antagonist (Fig. 3).

Similar findings have been obtained in hypothalamic slices from rats. The use of selective agonists and antagonists suggested again that the action of oxytocin was receptor mediated. Because the density of binding sites for oxytocin in the area of the ventromedial nucleus of the rat hypothalamus is markedly reduced by castration and can be restored to normal levels by the injection of ovarian steroids, a further experiment compared the responsiveness to oxytocin of ventromedial neurons in slices prepared from ovariectomized rats treated or not with oestrogen. Kow et al (1991) found that a greater percentage of neurons were responsive in slices obtained from the rats given the steroid. These findings support the idea that oestrogens, which are known to potentiate the facilitatory action of oxytocin on female reproductive behaviour, may do so by increasing the number of functional oxytocin receptors in neurons of the ventromedial hypothalamus.

These results suggest that binding sites for oxytocin revealed by autoradiography, or some of them, represent functional receptors and that centrally released oxytocin may therefore participate in interneuronal communication. Similarly, electrophysiological evidence for a neuronal action of vasopressin complements the autoradiographical demonstration of binding sites for vasopressin in the lateral septum (Raggenbass et al 1987b) and in the hypothalamic suprachiasmatic nucleus (Liou & Albers 1989, Dubois-Dauphin et al 1990). This positive correlation established between neuronal sensitivity and the presence of binding sites (Raggenbass et al 1989) contrasts with the apparent 'mismatch' seen when one compares the distribution of immunoreactivity with the map of binding sites revealed by autoradiography.

Mechanism of action of oxytocin

In the rat, a tract of oxytocin-containing axons arises in the hypothalamic paraventricular nucleus and innervates the dorsal vagal complex (Sawchenko & Swanson 1982), where synapses containing vesicles immunoreactive for neurohypophysial peptides have been detected (Siaud et al 1989). A role of oxytocin in the dorsal vagal complex of the brainstem is suggested by the observation that local microinjection of oxytocin affects gastric acid secretion. This effect could be mimicked by electrical stimuli applied to the hypothalamic

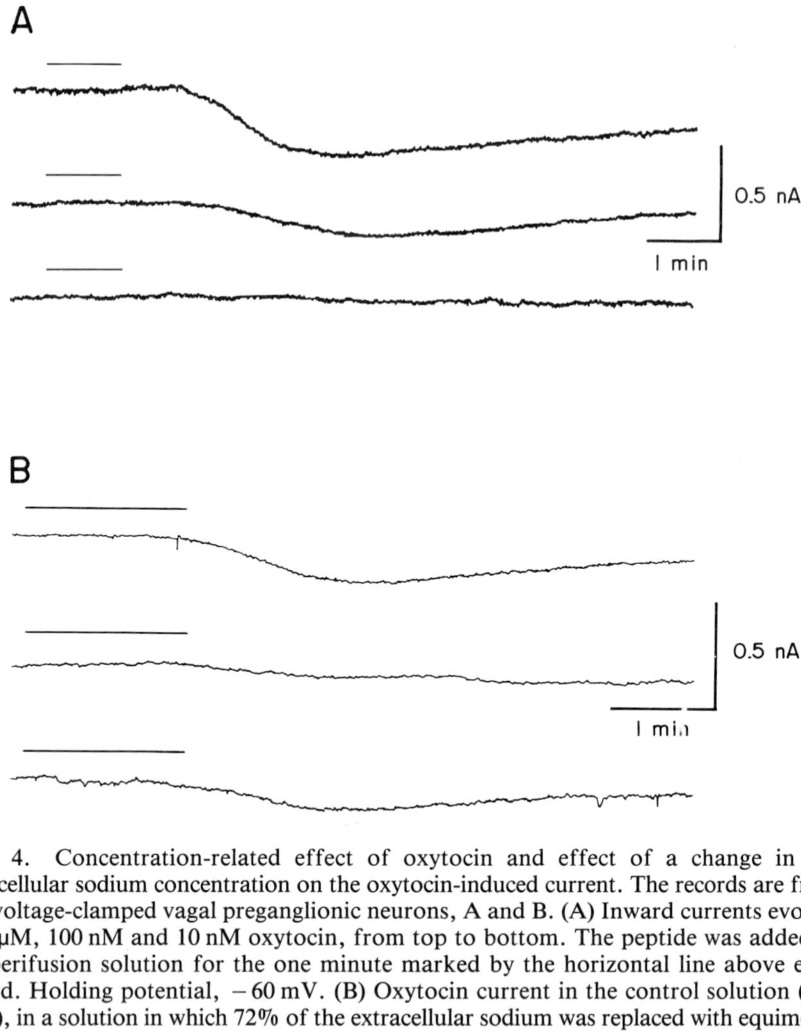

FIG. 4. Concentration-related effect of oxytocin and effect of a change in the extracellular sodium concentration on the oxytocin-induced current. The records are from two voltage-clamped vagal preganglionic neurons, A and B. (A) Inward currents evoked by 1 μM, 100 nM and 10 nM oxytocin, from top to bottom. The peptide was added to the perifusion solution for the one minute marked by the horizontal line above each record. Holding potential, −60 mV. (B) Oxytocin current in the control solution (*top trace*), in a solution in which 72% of the extracellular sodium was replaced with equimolar amounts of the impermeant cation N-methyl-D-glucamine (*middle trace*), and back again in the control solution (*bottom trace*). The peptide concentration was always 1 μM, the duration of application 1.5 minutes. Holding potential, −62 mV.

paraventricular nucleus. Moreover, microinjection of an oxytocin antagonist into the dorsal vagal complex blocked the effect of the hypothalamic stimulation on gastric secretion (Rogers & Herman 1986).

Stable intracellular recordings can be obtained from the large vagal preganglionic neurons while brainstem slices from rats are perfused with oxytocin. The peptide reversibly depolarized a majority of neurons and caused a transient increase in their firing frequency. The excitatory action of oxytocin was due to an interaction with oxytocin-type receptors, as indicated by the observation that a selective oxytocin agonist mimicked the effect of oxytocin and that a potent antagonist could suppress it (Dreifuss et al 1988).

Oxytocin acted postsynaptically. The oxytocin-induced membrane depolarization and increase in the discharge frequency persisted when the normal perfusion solution was replaced with one containing reduced calcium and increased magnesium concentrations, i.e. under conditions in which transmembrane calcium currents and synaptic transmission were markedly reduced.

When vagal neurons were clamped at or near their resting membrane potential, oxytocin generated a transient inward current. Its amplitude was concentration related (Fig. 4A). In addition, in voltage-clamped neurons, the oxytocin-induced currents recorded in the normal solution and in the presence of tetrodotoxin had similar peak amplitudes and time courses. These results suggest that the current generated by oxytocin is probably not carried by calcium ions to any significant extent or by sodium diffusing across a tetrodotoxin-sensitive sodium channel (Dubois-Dauphin et al 1992).

An indication that sodium ions might nevertheless be involved was obtained when responses to oxytocin in the normal perfusion solution were compared with those recorded in media containing reduced concentrations of sodium. A typical result is illustrated in Fig. 4B. In the normal solution, in this cell oxytocin evoked an inward current of nearly 200 pA. The response to the same amount of peptide was markedly reduced in a solution in which some of the extracellular sodium was replaced by N-methyl-D-glucamine, and this reduction was partially reversible. We are thus led to conjecture that oxytocin can enhance the excitability of vagal motoneurons by generating a tetrodotoxin-resistant sodium current. Whether oxytocin acts by a similar mechanism in other areas of the central nervous system remains unknown.

Oxytocin exerts its main known hormonal effects in the female mammal at parturition and during lactation, by interacting with specific receptors present on the surface of cells in the uterus and in the mammary gland. It has been shown that the oxytocin receptors present in the myometrium and decidua of the uterus and on myoepithelial cells of the mammary gland are regulated by gonadal steroid hormones and that their number increases markedly shortly before labour (Fuchs et al 1983). Receptor numbers remain elevated in the mammary gland throughout lactation.

Central effects exerted by oxytocin have recently been much investigated (Richard et al 1991). In this paper, we have given special attention to those neural effects of oxytocin which may also be mediated via steroid hormone-dependent oxytocin receptors, and which contribute, in conjunction with the hormonal effects, to the survival and growth of the offspring by triggering maternal behaviour and by facilitating nursing.

Acknowledgements

Work by the authors is supported in part by the Swiss National Science Foundation (grant 31.28624-90). M. Dubois-Dauphin gratefully acknowledges receipt of a Career Development Award of the Professor Dr Max Cloëtta Foundation. We thank Dr M. Manning (Toledo, Ohio, USA) for the gift of oxytocin structural analogues, Drs S. Jard and C. Barberis (Montpellier, France) for performing the purification of [^{3}H]vasopressin and the radioiodination of the oxytocin analogue, and Dr Zhenhui Li for participating in the experiment illustrated in Fig. 3.

References

Belin V, Moos F 1986 Paired recordings from supraoptic and paraventricular oxytocin cells in suckled rats: recruitment and synchronization. J Physiol (Lond) 377:369–390

Dreifuss JJ, Tribollet E, Mühlethaler M 1981 Temporal patterns of neural activity and their relation to the secretion of posterior pituitary hormones. Biol Reprod 24:51–72

Dreifuss JJ, Raggenbass M, Charpak S, Dubois-Dauphin M, Tribollet E 1988 A role of central oxytocin in autonomic functions: its action in the motor nucleus of the vagus nerve. Brain Res Bull 20:765–770

Dubois-Dauphin M, Pévet P, Tribollet E, Dreifuss JJ 1990 Vasopressin in the brain of the golden hamster: the distribution of vasopressin binding sites and of immunoreactivity to the vasopressin-related glycopeptide. J Comp Neurol 300:535–548

Dubois-Dauphin M, Raggenbass M, Widmer H, Tribollet E, Dreifuss JJ 1992 Morphological and electrophysiological evidence for postsynaptic localization of functional oxytocin receptors in the rat dorsal motor nucleus of the vagus nerve. Brain Res 575:124–131

Elands J, Barberis C, Jard S et al 1988 ^{125}I-labelled d$(CH_2)_5$[Tyr(Me)^2,Thr^4,Tyr(NH$_2$)^9]-OVT: a selective oxytocin receptor ligand. Eur J Pharmacol 147:197–207

Fahrbach SE, Morrell JI, Pfaff DW 1985 Possible role of endogenous oxytocin in estrogen-facilitated maternal behavior in rats. Neuroendocrinology 40:526–532

Freund-Mercier MJ, Richard P 1984 Electrophysiological evidence for facilitatory control of oxytocin neurones by oxytocin during suckling in the rat. J Physiol (Lond) 352:447–466

Fuchs AR, Periyasami S, Alexandrova M, Soloff MS 1983 Correlation between oxytocin receptor concentration in pregnant rat myometrium: effects of ovarian steroids. Endocrinology 113:742–749

Inenaga K, Kannan H, Yamashita H, Tribollet E, Raggenbass M, Dreifuss JJ 1991 Oxytocin excites neurons located in the ventromedial nucleus of the guinea-pig hypothalamus. J Neuroendocrinol 3:569–573

Ingram CD, Cutler KL, Wakerley JB 1990 Oxytocin excites neurones in the bed nucleus of the stria terminalis of the lactating rat in vitro. Brain Res 527:167–170

Insel TR 1990 Regional changes in brain oxytocin receptors post-partum: time course and relationship to maternal behaviour. J Neuroendocrinol 2:539–545

Kendrick KM, Keverne EB, Baldwin BA 1987 Intracerebro-ventricular oxytocin stimulates maternal behaviour in the sheep. Neuroendocrinology 46:56–61

Kow LM, Johnson AE, Ogawa S, Pfaff DW 1991 Electrophysiological actions of oxytocin on hypothalamic neurons in vitro: neuropharmacological characterization and effects of ovarian steroids. Neuroendocrinology 54:526–535

Liou SY, Albers HE 1989 Single unit response of suprachiasmatic neurons to arginine vasopressin (AVP) is mediated by a V_1-like receptor in the hamster. Brain Res 477:336–343

Moos F, Ingram CD, Wakerley JB, Guerné Y, Freund-Mercier MJ, Richard P 1991 Oxytocin in the bed nucleus of the stria terminalis and lateral septum facilitates bursting of hypothalamic oxytocin neurons in suckled rats. J Neuroendocrinol 3:163–171

Mühlethaler M, Raggenbass M, Dreifuss JJ 1985 Oxytocin and vasopressin. In: Rogawski A, Barker JL (eds) Neurotransmitter actions in the vertebrate nervous system. Plenum Publishing Corporation, New York, p 439–458

Pedersen CA, Prange AJ Jr 1979 Induction of maternal behavior after intracerebroventricular administration of oxytocin. Proc Natl Acad Sci USA 76:6661–6665

Pedersen CA, Ascher JA, Monroe YL, Prange AJ Jr 1982 Oxytocin induces maternal behavior in virgin female rats. Science (Wash DC) 216:648–650

Poulain DA, Wakerley JB 1982 Electrophysiology of hypothalamic magnocellular neurons secreting oxytocin and vasopressin. Neuroscience 7:773–808

Raggenbass M, Dubois-Dauphin M, Charpak S, Dreifuss JJ 1987a Neurons in the dorsal motor nucleus of the vagus nerve are excited by oxytocin in the rat but not in the guinea pig. Proc Natl Acad Sci USA 84:3926–3930

Raggenbass M, Tribollet E, Dreifuss JJ 1987b Electrophysiological and autoradiographical evidence of V_1 vasopressin receptors in the lateral septum of the rat brain. Proc Natl Acad Sci USA 84:7778–7782

Raggenbass M, Tribollet E, Dubois-Dauphin M, Dreifuss JJ 1989 Correlation between oxytocin neuronal sensitivity and oxytocin receptor binding: an electrophysiological and autoradiographical study comparing rat and guinea pig hippocampus. Proc Natl Acad Sci USA 86:750–754

Richard P, Moos F, Freund-Mercier MJ 1991 Central effects of oxytocin. Physiol Rev 71:331–370

Rogers RC, Herman GE 1986 Hypothalamic paraventricular nucleus stimulation-induced gastric acid secretion and bradycardia suppressed by oxytocin antagonist. Peptides (NY) 7:695–700

Sawchenko PE, Swanson LW 1982 Immunohistochemical identification of neurons in the paraventricular nucleus of the hypothalamus that project to the medulla or to the spinal cord in the rat. J Comp Neurol 205:260–272

Siaud P, Denoroy L, Assenmacher I, Alonso G 1989 Comparative immunocytochemical study of the catecholaminergic and peptidergic afferent innervation to the dorsal vagal complex in rat and guinea pig. J Comp Neurol 290:323–335

Sofroniew MV 1985 Vasopressin, oxytocin and their related neurophysins. In: Björklund A, Hökfelt T (eds) Handbook of chemical neuroanatomy, vol 4: GABA and neuropeptides in the CNS. Part 1. Elsevier Science Publishers, Amsterdam, p 93–165

Tribollet E 1992 Vasopressin and oxytocin receptors in the rat brain. In: Björklund A, Hökfelt T, Kuhar MJ (eds) Handbook of chemical neuroanatomy, vol 10: Neuropeptide receptors in the CNS. Part III. Elsevier Science Publishers, Amsterdam p 289–320

Tribollet E, Charpak S, Schmidt A, Dubois-Dauphin M, Dreifuss JJ 1989 Appearance and transient expression of oxytocin receptors in fetal, infant, and peripubertal rat brain studied by autoradiography and electrophysiology. J Neurosci 9:1764–1773

Tribollet E, Audigier S, Dubois-Dauphin M, Dreifuss JJ 1990 Gonadal steroids regulate oxytocin receptors but not vasopressin receptors in the brain of male and female rats. An autoradiographical study. Brain Res 511:129–140

van Leengoed E, Kerker E, Swanson HH 1987 Inhibition of post-partum maternal behaviour in the rat by injecting an oxytocin antagonist into the cerebral ventricles. J Endocrinol 112:275–282

DISCUSSION

Pfaff: What is known about second messenger systems or third messenger systems that might be involved in responses to oxytocin?

Dreifuss: The peripheral oxytocin receptors are linked to membrane phosphoinositide metabolism, which leads to intracellular calcium mobilization (see Flint et al 1986). I am not aware of similar studies done in the central nervous system.

Pfaff: One of the striking things about oxytocin is the projection to brainstem and spinal cord. Has anybody speculated about the physiological meaning of the terminations, either in the dorsal motor nucleus of the vagus or in the spinal cord, where some of the terminations are in the intermediolateral column?

Dreifuss: There is evidence that oxytocin affects visceral functions by modulating the parasympathetic and sympathetic efferents of the central nervous system. In my paper, I mentioned that microinjection of oxytocin into the dorsal vagal complex caused gastric acid secretion. Rogers & Herman (1986) also observed a slowing of the heart rate and Siaud et al (1991) reported a reduction in pancreatic insulin secretion following similar microinjection of oxytocin. There is some evidence that sympathetic preganglionic neurons located in the intermediolateral column of the spinal cord are also responsive to oxytocin (Gilbey et al 1982). On the other hand, in the spinal cord oxytocin could modulate the afferent sensory information that originates from the mammary gland or from the vagina and gets to the dorsal horn.

Poulain: We have done electron microscopy of the spinal cord and found a number of oxytocin-containing synapses, right in the terminal field of the afferent input from the mammary gland (Rousselot et al 1990). This is highly suggestive of an interaction between central oxytocin and the incoming afferent input during suckling, but we have no physiological evidence for this.

Palkovits: Dr Dreifuss, could you say a little more about the intracellular recordings from the vagal preganglionic neurons?

Dreifuss: The large preganglionic vagal neurons are antidromically invaded by action potentials set up by stimuli applied ventrolaterally to the dorsal motor nucleus, along the course followed by the axons arising from this nucleus. We have injected Lucifer Yellow into their perikarya and verified that oxytocin-responsive cells were located within the boundaries of the nucleus. In the nucleus of the solitary tract, which is located just dorsally to the dorsal motor nucleus of the vagus, approximately half the cells responded to neurohypophysial peptides, but more readily to vasopressin than oxytocin (Raggenbass et al 1989). These electrophysiological results correlate well with the autoradiographical data, because we detected oxytocin binding in the dorsal motor nucleus of the vagus and vasopressin binding in the nucleus of the solitary tract. There is, however, 'mismatch' when relating these results to the distribution of oxytocin immunoreactive axon terminals. Although oxytocin terminals are present in both nuclei, they are more numerous in the nucleus of the solitary tract than in the dorsal motor nucleus of the vagus.

Palkovits: Is there a vasopressin response in the dorsal vagal nucleus?

Dreifuss: Vasopressin can excite vagal motoneurons, but less powerfully than oxytocin at the same concentration (Charpak et al 1984). Because vasopressin can bind to oxytocin receptors and thereby mimic the action of oxytocin, we believe the effects of vasopressin in the dorsal motor nucleus might result from its action on oxytocin receptors.

Sawchenko: If you look at the topography of those oxytocin terminals in the dorsal motor nucleus, they are preferentially distributed to one aspect of it, the lateral column, which happens to be the source of vagal preganglionic inputs to the caecum. On the basis of the immunohistochemistry, I would predict that at various mediolateral levels of the dorsal motor nucleus, you should see differential responses to oxytocin.

This mismatch between neuronal sensitivity and the presence of binding sites is particularly acute in the forebrain. For example, in the hippocampal formation and in the amygdala, the number of oxytocin fibres is vanishingly small. It is not like the situation with vasopressin where, under certain hormonal conditions, the gene can be expressed at several extrahypothalamic sites. The only places we know oxytocin to be expressed in cell bodies are in the paraventricular and supraoptic nuclei and in the accessory magnocellular cells. So how is it getting to your receptors? Would you consider the relatively large and differentially regulated oxytocin content of the cerebrospinal fluid to be a possibility?

Dreifuss: No. I would rather consider diffusion in nervous tissue over a longer distance than what is normally believed; for example, from the nucleus of the solitary tract to the dorsal motor nucleus of the vagus.

Makara: You might be able to explain the mismatch by invoking the cerebrospinal fluid or diffusion in the mouse, but it will not work for the elephant! We have to consider the size of the brain and the distances the peptides have to travel. I wonder whether the mismatch might show simply that those peptides don't have a function at that site.

Hatton: If you project the dendritic trees in those facial nucleus cells, do they extend further up, close to the terminal area?

Dreifuss: As I said, the autoradiographical data and the immunocytochemical data don't coincide, so it cannot just be a question of shape or extension of dendrites.

Hatton: You have done similar studies with these peptides in the hippocampus, where there are many receptors for oxytocin. Have you seen the same kind of tetrodotoxin-resistant curve in the hippocampus?

Dreifuss: In the hippocampus, only a small fraction of all neurons that we tested responded to oxytocin: we believe they are a class of inhibitory neurons (Mühlethaler et al 1984). It turned out to be difficult to obtain stable intracellular recordings from these cells and we therefore have not been able to study the mechanism by which oxytocin affects them. On the other hand, we have obtained long-term recordings from motoneurons in the facial nucleus of young rats. Virtually all these neurons responded to vasopressin by generating a sustained voltage-dependent inward current (Raggenbass et al 1991), which strongly resembles the sodium current induced by oxytocin in the dorsal motor nucleus of the vagus nerve.

Palkovits: What about oxytocin receptors in the olfactory region?

Dreifuss: We haven't done any electrophysiological experiments in that part of the brain so far.

Nagy: Autoradiographic studies performed some years ago by several groups showed binding of oxytocin and vasopressin in the paraventricular nuclei and the supraoptic nuclei of the rat hypothalamus. I was not surprised by these data, partially because vasopressin may bind to neurophysin and give false labelling.

Dreifuss: Those studies used relatively high concentrations of radioligand (see Freund-Mercier et al 1987). Recently, it has been shown that this represents most probably binding of the ligand to neurophysin rather than to receptors (Freund-Mercier et al 1991).

Nagy: Have you looked at the pituitary gland for binding sites?

Dreifuss: We have looked at the pituitary gland and found sites that specifically bind [^{3}H]vasopressin in rat, human and sheep anterior lobes. These sites are located mainly, and possibly exclusively, on corticotrophs (Du Pasquier et al 1991a,b).

Robinson: It is very difficult to state what is binding to neurophysin and what is not, partly because neurophysin binding sites are also heterogeneous. They vary with different pH values and different concentrations of the ligands. We tried to show that there was a distinct vasopressin-binding site on the pituitary gland (Dashwood & Robinson 1988). One of the odd things is the absence of the binding in the pituitary of the Brattleboro rat where the neurophysin remains: the oxytocin neurophysin is still present but the binding sites are not. Thus, I don't think the problem of binding to neurophysin versus receptors is totally resolved.

Dreifuss: It's not totally resolved for another reason. Japanese authors have provided electrophysiological evidence using hypothalamic slices that vasopressin and oxytocin can affect the firing of supraoptic and paraventricular neurons (Inenaga & Yamashita 1986, Yamashita et al 1987). The possibility remains that autoradiography fails to reveal all oxytocin receptor subtypes. We may be missing rapidly inactivating receptors or receptors with very low affinity.

Robinson: When you get oxytocin release or vasopressin release, of course, there is also neurophysin release. Because none of us can find any extracellular function for neurophysin, we like to ignore it, but especially in a confined synaptic cleft the concentrations of this binding protein will be very high. Perhaps the peptide–neurophysin complex is active in a different way from the peptide alone administered by iontophoretic or pressure injections. Many people, including yourself, have looked for activities of neurophysin, and perhaps it's premature to discount activities of peptide–neurophysin or peptide–neurophysin–glycopeptide complexes, especially within the confines of the synaptic cleft.

Lightman: Although the maternal effects of oxytocin are similar in many different species, the sites of oxytocin receptors are extremely different. In the sheep no one has ever been able to find any oxytocin receptors at all, but oxytocin still has powerful maternal effects. Would you like to comment on that?

Dreifuss: Kendrick et al (1988) in Cambridge not only reported that intracerebroventricular oxytocin caused a rapid onset of maternal behaviour, they also showed that during parturition oxytocin was released in some parts of the brain, for example the olfactory bulb. Frédéric Lévy and Eliane Tribollet in my laboratory have not been able to detect sites that bind oxytocin in the sheep brain. However, vasopressin-binding sites are present in the glomerular layer and in the external plexiform layer of the sheep olfactory bulb. The possibility must therefore be considered that vasopressin or oxytocin released in the olfactory bulb acts on vasopressin receptors rather than on oxytocin receptors when triggering some aspects of maternal behaviour in the sheep. Both peptides have some biological effect on either receptor.

Hatton: In the olfactory bulb there is a real mismatch because there are no oxytocin fibres there either, so where does the oxytocin come from? When you put retrograde tracers into the olfactory bulb, at least of a rat, no oxytocin cells retrogradely transport the tracers, yet there's a lot of oxytocin in the bulb when you analyse the dialysate.

Poulain: The question of the mismatch between function and the anatomical findings is crucial. Although there are no oxytocin receptors in the supraoptic nucleus, there is release of oxytocin within the supraoptic nucleus during suckling (Moos et al 1989). There are also synapses containing oxytocin (Theodosis 1985) and there is facilitation of the electrical activity of oxytocin neurons when you inject oxytocin into the supraoptic nucleus itself (Moos & Richard 1989).

Dreifuss: We can, for the time being, reveal two classes of oxytocin receptor, one oestrogen-dependent and the other oestrogen-independent. They both have a high affinity and they closely resemble the receptors known peripherally. Whether these are the only receptors in the central nervous system, we cannot say.

Poulain: From the infusion experiments by Moos et al (1991), in which they injected oxytocin into various areas of the brain, it is clear that the time course of action is very long; it takes minutes from the beginning of a local infusion to an effect on the pattern of milk ejection. So there appear to be different classes of receptor for oxytocin.

Theodosis: Many authors have described mismatch between the presence of a neuropeptide and its receptor in the brain. One review concluded that there is such mismatch in most cases (Herkenham 1987). Either one finds peptide

neurosecretory axons and terminals. They also have processes that terminate in the perivascular space, where they occupy about 50% of the area, the rest being occupied by neurosecretory terminals. During stimulation, there is about a 15% increase in the extent of neurovascular contact resulting from an increase in number of terminals and/or in the size of individual terminals impinging on the perivascular space (Wittkowski & Brinkman 1974, Tweedle & Hatton 1980, 1987, Luckman & Bicknell 1991); a concomitant decrease in the number of pituicyte processes results in a corresponding reduction in glia–vascular contact. These changes are rapid: in neurohypophyses maintained *in vitro*, altered neuronal–glial conformations are detectable even after one hour of stimulation (Luckman & Bicknell 1991). Together with these modifications at the pericapillary region, there is a significant reduction in the number of neurosecretory axons enclosed by pituicytes (Tweedle & Hatton 1980). As in the hypothalamic nuclei, the changes are reversible and pituicyte processes reoccupy the perivascular space upon cessation of stimulation and again surround neurosecretory axons at a rate which depends on the duration of stimulation (Tweedle & Hatton 1987). It is not yet determined whether these changes are specific to one or the other of the neurosecretory systems (see Luckman & Bicknell 1991).

Similar neuronal-glial changes occur in another neurohaemal structure, the external layer of the median eminence where neurosecretory terminals from parvocellular neurons, together with astrocytic and tanycyte processes, abut onto a perivascular space next to fenestrated capillaries. The ultrastructure of this area has not been analysed under different physiological states but under certain experimental conditions, such as bilateral adrenalectomy, there is a 20% increase in the extent of neurovascular contact, accompanied by a corresponding decrease in glia-vascular contact. Here also, the changes are reversible (Wittkowski & Scheuer 1974).

Elsewhere in the hypothalamus

Reversible changes in the conformation of astrocytes have been detected in the arcuate nucleus, which participates in the control of the release of pituitary gonadotrophins. Ultrastructural analyses show altering proportions of neuronal soma surface covered by astrocytes and changes in the number of presynaptic terminals ensheathed by astrocytic processes throughout the oestrous cycle (Olmos et al 1989).

Neuronal surface juxtapositions have been noted between somatostatin-immunoreactive elements in the periventricular region extending from the suprachiasmatic nucleus to the median eminence (Alonso et al 1985). They are reminiscent of those seen between stimulated oxytocinergic neurons and suggest that neuronal-glial remodelling also occurs in this area of the hypothalamus. However, the observations should be interpreted with caution because juxtapositions between somatostatin profiles were not quantified or assessed in terms of different experimental or physiological conditions.

Reorganization of synaptic inputs

Neurosecretory neurons in the magnocellular and arcuate nuclei are richly innervated and these innervations undergo important alterations during the same physiological conditions which result in neuronal-glial conformational changes. Synaptic changes are readily detected in the magnocellular nuclei because of changes in the incidence of a particular synaptic configuration, that of axonal terminals synapsing onto two or more postsynaptic elements in the same plane of section ('shared' synapses, Fig. 2). Such terminals are extremely rare in the nuclei of unstimulated animals but their incidence significantly increases during stimulation (Theodosis et al 1981, Hatton & Tweedle 1982, Theodosis & Poulain 1984b). Once stimulation ceases, they again diminish in frequency (Fig. 3B), at a rate that depends on the duration of stimulation. For example, after a normal lactation, at least one month is necessary for their incidence to return to 'virgin' values (Theodosis & Poulain 1984b); if the young are removed immediately after birth, these 'shared' terminals are no longer visible within 10 days (Montagnese et al 1987). From our ultrastructural analyses on identified profiles (Figs 2 & 4B), we found that they contact esssentially oxytocinergic somata and dendrites (Theodosis & Poulain 1984a, 1989, Chapman et al 1986, Theodosis et al 1986a,c, Montagnese et al 1987).

In the arcuate nucleus of the rat, changes in the density of axosomatic synapses occur throughout the oestrous cycle. The number of axosomatic synapses significantly decreases during the 24 hours between pro-oestrus and oestrus and increases again between oestrus and metoestrus (Olmos et al 1989).

Possible cellular mechanisms

When one considers all the ultrastructural observations, it becomes readily apparent that the structural reorganization of each hypothalamic centre involves both neurons and glial cells. The active participation of neurons is obvious, for several reasons: (i) magnocellular cell bodies hypertrophy under stimulation (see for example, Chapman et al 1986) and it is likely that this is accompanied by a change in their geometry and in that of their dendrites, as has now been visualized *in situ* in certain adult neuronal systems of living animals (Purves & Voyvodic 1987). (ii) Neurosecretory axons change their morphology, as seen from the changes in length and number of individual nerve terminals in the neurohypophysis and median eminence under different conditions. (iii) Magnocellular neurons produce and destroy attachment plates in relation to the extent of their surface in juxtaposition (Theodosis et al 1981, 1986a, Chapman et al 1986, Theodosis & Poulain 1984b). (iv) Rearrangement of synaptic

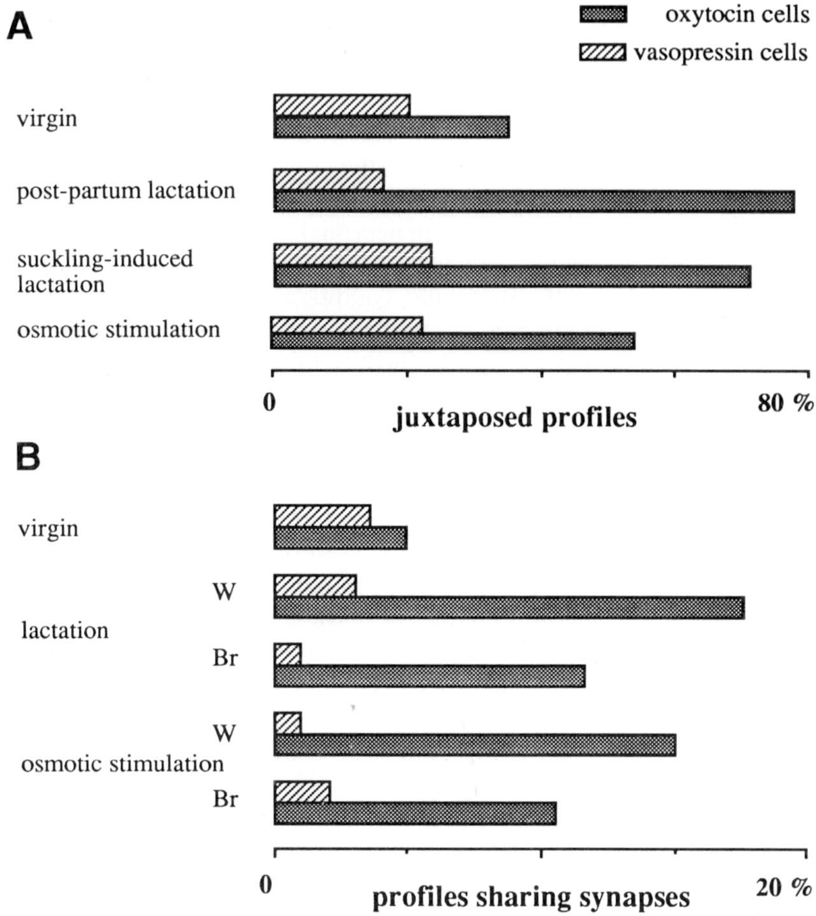

FIG. 4. Specificity of the neuronal–glial (A) and synaptic (B) changes to the oxytocinergic system in the supraoptic nucleus. The data are derived from quantitative analyses in which neurosecretory profiles were identified at the electron microscopic level by morphological means (in Brattleboro homozygote (Br) rats) or immunocytochemical methods in Wistar (W) or Br rats (for further details see Theodosis et al 1986a, Chapman et al 1986, Montagnese et al 1987). Similar changes are seen during lactation (postpartum or induced in pseudogestating females by continuously suckling litters) and prolonged osmotic stimulation (or chronic dehydration). Dehydration was induced experimentally in Wistar rats by giving them hypertonic water to drink for 10 days; Brattleboro homozygotes suffer genetically from diabetes insipidus and are in a constant state of dehydration, even with water provided *ad libitum*. The increase in the number of profiles juxtaposed and sharing the same synapse during both stimuli concerns only the oxytocinergic profiles. The proportion of vasopressinergic profiles in direct juxtaposition or sharing the same synapse remains low and no different from that in unstimulated animals (virgin).

inputs requires the formation and destruction of postsynaptic densities. (v) In the arcuate nucleus, freeze-fracture analyses reveal a reorganization of the perikaryal plasmalemma which parallels the fluctuations in oestrogen levels during the oestrous cycle (Garcia-Segura et al 1988).

How astrocytes participate in the structural reorganization of each region is less obvious. Juxtapositions between magnocellular somata could simply result from astrocytic processes being passively squeezed away when the somata hypertrophy under stimulation. However, this cannot be the case since both oxytocin- and vasopressin-containing somata hypertrophy when chronically stimulated, yet only the former show a reduction in glial coverage (Chapman et al 1986). Also, the important increase in dendritic juxtapositions cannot be explained by mere volume changes. A more attractive hypothesis is active retraction of glial processes. Such a mechanism may explain some of the phenomena, such as the decreased number of glial processes in the perivascular spaces of the stimulated neurohypophysis and median eminence and the reduction of astrocytic coverage of arcuate neurons at metoestrus, but, it cannot account for all the observations. For example, glial coverage of both oxytocinergic and vasopressinergic somata actually increases in absolute values under stimulation (Chapman et al 1986, Theodosis et al 1986a); what is peculiar to oxytocinergic somata is that the proportion of their plasmalemma covered by glial elements diminishes. In these nuclei then, neuronal juxtapositions result not so much from active glial retraction, but rather from insufficient glial coverage. A more likely hypothesis for glial participation in the changing anatomy of each structure is that glial processes move, both extending and retracting. More importantly, such glial movements are coordinated with the altering morphology of the neurons with which they are associated.

The cellular mechanisms underlying the synaptic changes also need to be resolved. In the hypothalamic nuclei, the remodelling of synaptic inputs invariably occurs in concert with the neuronal-glial conformational changes. Indeed, in both the stimulated supraoptic and paraventricular nuclei, it is quite common to see a cluster of directly juxtaposed oxytocinergic profiles sharing one or more synaptic terminal(s). A close association of neuron-glia conformational changes has also been noted in the peripheral nervous system where synaptic rearrangements are coincident with alterations in the positions and forms of associated neurons and glia, visualized with video-enhanced microscopy for periods of over 100 days in adult living mice (Pomeroy & Purves 1988). Studies of the developing central nervous system indicate that the formation and stabilization of synaptic connections require not only interactions between pre- and postsynaptic partners but also interaction between these and the astrocytes with which they are associated. In view of the close relationship between synaptic and glial changes in the different adult neuronal systems described in this review, such interaction may also be important in adult synaptic rearrangements. Assuming that this is the case, many questions remain.

For example, does the changing incidence of synapses in the hypothalamus represent *de novo* formation of axonal terminals with consequent degeneration, or do boutons already in place produce and destroy synaptic specializations during different physiological conditions? Degenerating terminals have not been detected in the magnocellular or arcuate nuclei. It is then possible that the number of boutons remains unchanged but sites of interaction with postsynaptic elements become inoperative because of intervening astrocytic processes. The validity of this hypothesis requires rigorous analysis of synaptic densities but it is noteworthy that in the arcuate nucleus the reduction in synapses from pro-oestrus to oestrus is coincident with an increase in the incidence of presynaptic terminals ensheathed by astrocytic processes (Olmos et al 1989).

If the absence of glial coverage favours synapse formation, then we still need to determine how. Synaptogenesis involves the active participation of pre- and postsynaptic partners and, as suggested from other examples of synaptic plasticity (see for example, Cotman & Nieto-Sampedro 1984), the rate of synapse formation may depend on the intensity of stimulation in the presynaptic element. Shared synapses in the activated oxytocinergic system may therefore be extensions of pre-existent terminals which become hyperactive and induce an adjacent postsynaptic element, at a site of the neuronal surface not covered by astrocytes, to create a postsynaptic density. A similar mechanism may operate in the arcuate nucleus, where active presynaptic terminals, in close proximity to sites of neuronal somata not covered by astrocytes, induce postsynaptic density formation.

Factors implicated in the structural changes

Role of adhesion molecules

Many different molecules are currently attributed a role in neuronal–glial interactions and synaptogenesis. Of these, the neural cell adhesion molecules (N-CAMs) appear to be critical because they participate in cell–cell adhesion and cell–substrate interactions, thereby regulating tissue architecture. N-CAMs are expressed by neurons and glial cells and exist in a variety of isoforms, differing in their proteic and carbohydrate moieties. In developing tissues, N-CAM contains more than 30% polysialic acid (PSA), an oligosaccharide of α-2-8-linked sialic acid residues. This isoform is gradually replaced by several 'adult' isoforms with much less PSA. In the rodent brain, the conversion takes place during the late perinatal or early postnatal periods but its timing differs in different regions, as they acquire their definite structure. The timing of the conversion and the fact that the ability of N-CAM to promote cell adhesion is attenuated by PSA have suggested that N-CAM and especially its PSA may serve as overall regulators of contact-dependent cell–cell interactions. Indeed, PSA-N-CAM expression may be a characteristic of motile cells or of cells undergoing reshaping (for review and further discussion see Rutishauser et al 1988).

We recently showed that neurons and glial cells of the hypothalamo-neurohypophysial system continue to express high levels of PSA-N-CAM immunoreactivity in adult animals (Fig. 5) (Theodosis et al 1991). We have also found that PSA-N-CAM immunoreactivity persists in other central systems capable of morphological plasticity, including the arcuate nucleus and median eminence. This immunoreactivity is seen in male and female rats and the expression of the molecule, at least within the hypothalamo-neurohypophysial system, does not appear to vary significantly in relation to the physiological state of the animal. It would appear then that polysialylation is a permanent feature of certain neurons and astrocytes, retained throughout the animal's life.

N-CAM and, in particular, its PSA may therefore be involved in tissue reorganization not only in developing tissues but also in the adult central nervous system, in areas that can manifest morphological plasticity. A causal relationship between PSA expression and activity-dependent plasticity has been demonstrated in the peripheral nervous system, in the establishment of intramuscular nerve branching (Landmesser et al 1990). If it is the continued expression of PSA-N-CAM that allows certain adult neurons and astrocytes to change their morphology and connections under different conditions, then we need to determine the signals that make this possible. As we will see below, several different factors appear to be implicated in the induction of the structural reorganization in each of the systems. It may be that although neurons and glial cells in some systems maintain certain 'embryological' features that allow them to be plastic in adulthood, the signals inducing them to do so are specific to each of the systems.

Role of oxytocin

Because the morphological changes in the magnocellular nuclei involve exclusively oxytocinergic neurons and because they occur consistently whenever the oxytocinergic system is activated, an obvious signal that may be involved in their genesis is oxytocin itself. When oxytocin is infused into the third ventricle of normally hydrated, non-gestating, non-lactating female rats for one week, the oxytocinergic system undergoes the same structural changes as those seen when it is physiologically activated (Theodosis et al 1986b, Montagnese et al 1990). This action of oxytocin is very specific; it could be mimicked only by the potent oxytocin agonist, 4-threonine-oxytocin (Montagnese et al 1990). Vasopressin had little effect, as did cholecystokinin, a peptide which facilitates the activity of oxytocin neurons (Verbalis et al 1986). We still do not know whether oxytocin acts directly on the cells to change their morphology but it is of interest that processes of postnatal hypothalamic astrocytes maintained in culture are highly motile (Madarasz et al 1991) and this motility is modified

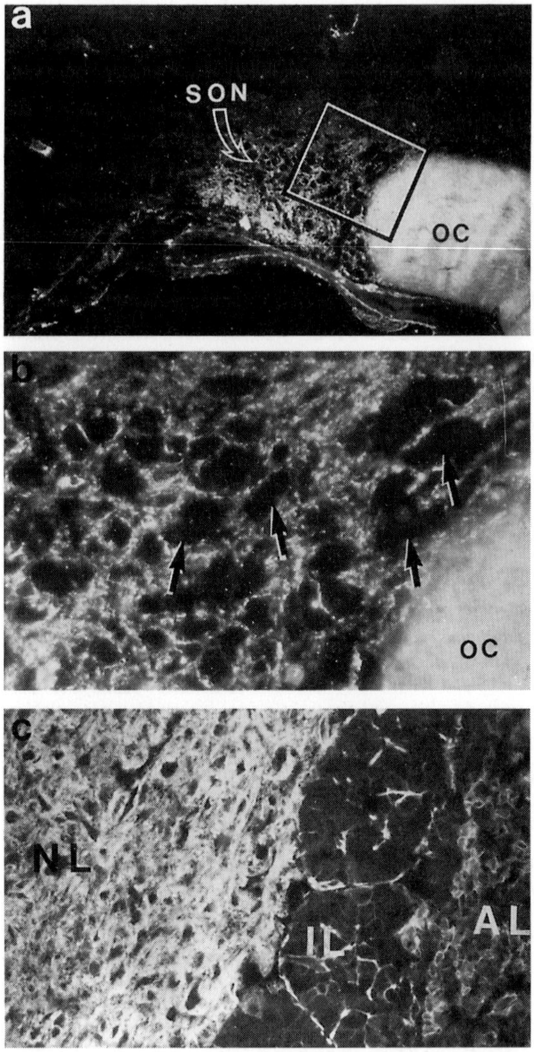

FIG. 5. Light microscopic localization of immunoreactivity for the polysialylated isoform of the neural cell adhesion molecule (N-CAM) in the hypothalamo-neurohypophysial system of adult rats. After reaction of sections with an antibody specific for this N-CAM isoform, a striking immunoreactivity is seen in (a,b) the supraoptic nucleus (SON) and (c) neural lobe (NL) of the hypophysis. In the SON, the immunoreactivity is particularly strong in the ventral portions of the nucleus and in the neuropil around immunonegative neurosecretory somata. Note the absence of immunoreactivity in the adjacent hypothalamus. In lactating animals, there is less reaction in the dorsal regions of the nucleus where there are numerous clusters of closely apposed neurosecretory somata (arrows in b). In the hypophysis, there is strong reaction in the NL, masking cellular

when oxytocin is included in the culture milieu (E. Madarasz & D. T. Theodosis unpublished work, 1989).

Influence of steroid hormones

There are now many examples of the participation of steroid hormones in the restructuring of adult nervous tissue. Steroid-accumulating neurons exist in a number of central systems and they have been implicated in dendritic growth and synaptic plasticity (Bottjer & Arnold 1984, Leedy et al 1987) and even neurogenesis in the adult (Nottebohm 1985). As described in this review, the influence of oestrogen in inducing neuronal–glial and synaptic plasticity is particularly clear in the arcuate nucleus. The effect of oestrogen is visible not only *in vivo* but also *in vitro* where application of the steroid to cultured hypothalamic astrocytes dramatically changes their morphology (Garcia-Segura et al 1989).

In the magnocellular nuclei, oxytocin does not act alone but requires a particular steroid milieu to induce structural plasticity. In our experiments, central infusion of oxytocin had an effect only in rats undergoing a prolonged dioestrus and in castrated females that had been given progesterone injections for four days followed by 48 hours of oestrogen (Montagnese et al 1990), an endocrine state similar to that of parturition and lactation. Steroids alone are not sufficient to change the ultrastructure of the magnocellular nuclei since this remains unmodified during most of gestation (Fig. 3) and pseudogestation (Montagnese et al 1987), when plasma sex steroid levels are elevated for several days. High levels of progesterone or oestrogen, alone or in combination, without the concomitant central infusion of oxytocin also do not affect the anatomy of the system (Montagnese et al 1990). The steroids implicated may not be ovarian or placental. For example, prolonged dehydration, which induces changes in the oxytocinergic system similar to those seen during lactation (Fig. 4), is accompanied by changes not in circulating sexual steroids but in corticosteroids (Johnson & Levine 1973), which may exert the same function. On the other hand, testosterone can induce the neuronal–glial changes in the neurohypophysis (Tweedle et al 1988). It is possible then that the nature and peripheral source of the steroids vary but their ultimate central action is the same.

Role of neurotransmitters

Many factors are implicated in the restructuring of each system; these may include not only neuropeptides and steroid hormones but also more classical

details; some reaction is also seen in the intermediate (IL) and anterior (AL) lobes. Immunoperoxidase staining, dark field optics. OC, optic chiasma, appearing bright because of dark field illumination of its myelinated fibres. Adapted from Theodosis et al (1991).

neurotransmitters. Recent observations suggest that noradrenaline plays such a role, because incubation of isolated neural lobes in media containing noradrenaline or its agonists produced the same changes as those seen *in vivo* when the system is stimulated; the action is presumably via β-adrenergic receptors (Luckman & Bicknell 1991). The effects appear direct, since application of the neurotransmitter did not induce an increase in peptide release. Also, application of the transmitter to pituicytes *in vitro* induced rapid, reversible changes in their morphology (Bicknell et al 1989).

Functional significance of neuron–astrocyte and synaptic plasticity

That the relationship of particular neurons and astrocytes varies in several hypothalamic centres under different physiological conditions strongly suggests that a dynamic regulation of the structure of neurons and their associated glial cells is closely related to normal neuronal function, not only during development but also in the adult. One possibility is that the neuronal-glial changes allow remodelling of synapses to take place; another is that they result in changes in the ionic microenvironment. In either case, there would be important consequences for neuronal activity. Owing to the paucity of experimental evidence, we will here focus our discussion on the hypothalamo-neurohypophysial system.

Shared synapses allow several postsynaptic elements to receive the same information simultaneously and an obvious consequence is that they influence coordination of neuronal firing. Such coordination appears critical in the function of the oxytocinergic system under certain conditions. During suckling, for example, oxytocin neurons display, every 3–10 minutes, brief high frequency discharges of action potentials that last for 3–4 seconds. They occur simultaneously in all the population and are responsible for the release of a pulse of oxytocin which acts on the mammary gland and provokes milk ejection. The neurons display a similar electrical activity at parturition (for review see Poulain & Wakerley 1982). As in other neuronal systems, GABA (γ-aminobutyric acid) is a major inhibitory transmitter in the magnocellular nuclei (see for example Theodosis et al 1986c) and we have shown that the majority of shared synapses, at least within the supraoptic nucleus, are GABAergic (Theodosis et al 1986c). Thus, an increased inhibitory input to oxytocin neurons may coordinate their accelerated firing by permitting activity to occur only at particular intervals, as is the case during the milk ejection reflex or the expulsion of each fetus at parturition. On the other hand, some shared synapses are dopaminergic (Buijs et al 1984), which would imply some shared excitatory input that could further facilitate coordinated firing.

One possible consequence of extensive neuronal plasmalemma juxtapositions is that they lead to electrical field interactions, thereby affecting the excitability of contiguous neurons (see for example Dudek et al 1983). They may also permit the accumulation of ions, particularly potassium, which is normally cleared away by intervening astrocytes. Recordings with potassium-sensitive microelectrodes have shown that during suckling-induced milk ejection, high frequency discharges are indeed accompanied by a transient increase in the concentration of extracellular potassium, the amplitude of which could eventually contribute to synchronization of the onset of a burst in adjacent oxytocin neurons that are directly juxtaposed (Coles & Poulain 1991). However, the rise in potassium is very localized and becomes negligible at a short distance, so that it is clear that it cannot make any significant contribution to synchronizing the onset of bursts between oxytocin cells that are not contiguous. Moreover, comparison of potassium concentration variations in the supraoptic nucleus during antidromic stimulation and reflex milk ejection indicates that potassium is cleared from around oxytocin cells more effectively than from around vasopressin cells, which avoids significant increases in other parts of the nucleus or generation of slow potentials.

The electrical activity generated in neurosecretory somata propagates down their axons to the terminals in the neurohypophysis where it induces release of the respective peptides into the extracellular space. Such release is calcium dependent and is facilitated at high frequency of firing (for review see Poulain & Theodosis 1988). Even a slight accumulation of potassium in the extracellular space can facilitate release. During electrical stimulation of the pituitary stalk and during milk ejection, there is a significant increase in extracellular potassium (Leng & Shibuki 1986) and this may be the result of the withdrawal of pituicyte processes from around the neurohypophysial axons and terminals.

Acknowledgements

We thank R. Bonhomme and S. Senon for their constant support and expert technical assistance. The authors' work referred to in this review was partially supported by grants from the Conseil Régional d'Aquitaine (EPR 9003045) and the Fondation pour la Recherche Médicale.

References

Alonso G, Tapia-Arancibia L, Assenmacher I 1985 Electron microscopic immunocytochemical study of somatostatin neurons in the periventricular nucleus of the rat hypothalamus with special reference to their relationships with homologous neuronal processes. Neuroscience 16:297–306

Bicknell RJ, Luckman SM, Inenaga K, Mason WT, Hatton GI 1989 Beta-adrenergic and opioid receptors on pituicytes cultured from adult rat neurohypophysis: regulation of cell morphology. Brain Res Bull 22:1–10

Bottjer SW, Arnold AP 1984 Hormones and structural plasticity in the adult brain. Trends Neurosci 7:168–171

Buijs RM, Geffard M, Pool CW, Hoorneman EMD 1984 The dopaminergic innervation of the supraoptic and paraventricular nucleus. A light and electron microscopical study. Brain Res 323:65–72

Chapman DB, Theodosis DT, Montagnese C, Poulain DA, Morris JF 1986 Osmotic stimulation causes structural plasticity of neurone–glia relationships of the oxytocin but not vasopressin secreting neurones in the hypothalamic supraoptic nucleus. Neuroscience 17:679–686

Coles JA, Poulain DA 1991 Extracellular K^+ in the supraoptic nucleus of the rat during reflex bursting activity by oxytocin neurones. J Physiol (Lond) 439:383–409

Cotman CW, Nieto-Sampedro M 1984 Cell biology of synaptic plasticity. Science (Wash DC) 225:1287–1294

Dudek FE, Andrew RD, MacVicar BA, Snow RW, Taylor CP 1983 Recent evidence for and possible significance of gap junctions and electrotonic synapses in the mammalian brain. In: Jasper HH, van Gelder NM (eds) Basic mechanisms of neuronal hyperexcitability. Alan R Liss, New York p 31–73

Garcia-Segura LM, Hernandez P, Olmos G, Tranque PA, Naftolin F 1988 Neuronal membrane remodeling during the oestrus cycle: a freeze fracture study in the arcuate nucleus of the rat hypothalamus. J Neurocytol 17:377–383

Garcia-Segura LM, Torres-Aleman I, Naftolin F 1989 Astrocytic shape and glial fibrillary acidic protein immunoreactivity are modified by estradiol in primary rat hypothalamic cultures. Dev Brain Res 47:298–302

Gregory WA, Tweedle CD, Hatton GI 1980 Ultrastructure of neurons in the paraventricular nucleus of normal, dehydrated and rehydrated rats. Brain Res Bull 5:301–306

Hatton GI, Tweedle CD 1982 Magnocellular peptidergic neurons in hypothalamus: increases in membrane apposition and number of specialized synapses from pregnancy to lactation. Brain Res Bull 8:197–204

Johnson TJ, Levine S 1973 Influence of water deprivation on adrenocortical rhythms. Neuroendocrinology 11:268–273

Landmesser L, Dahm L, Tang J, Rutishauser U 1990 Polysialic acid as a regulator of intramuscular nerve branching during embryonic development. Neuron 4:655–667

Leedy MG, Beattie MS, Bresnahan JC 1987 Testosterone-induced plasticity of synaptic inputs to adult mammalian motoneurons. Brain Res 424:386–390

Leng G, Shibuki K 1986 Extracellular potassium changes in the rat neurohypophysis during activation of the magnocellular neurosecretory system. J Physiol (Lond) 392:97–111

Luckman SM, Bicknell RJ 1991 Morphological plasticity that occurs in the neurohypophysis following activation of the magnocellular neurosecretory system can be mimicked in vitro by beta-adrenergic stimulation. Neuroscience 39:701–709

Madarasz E, Theodosis DT, Poulain DA 1991 *In vitro* formation of type 2 astrocytes derived from postnatal rat hypothalamus or cerebral cortex. Neuroscience 43:211–221

Montagnese C, Poulain DA, Vincent JD, Theodosis DT 1987 Structural plasticity in the rat supraoptic nucleus during gestation, post-partum lactation and suckling-induced pseudogestation and lactation. J Endocrinol 115:97–105

Montagnese C, Poulain DA, Theodosis DT 1990 Influence of ovarian steroids on the ultrastructural plasticity of the adult supraoptic nucleus induced by central administration of oxytocin. J Neuroendocrinol 2:225–231

Nottebohm F 1985 Neuronal replacement in adulthood. Ann NY Acad Sci 457:143–161

Olmos G, Naftolin F, Perez J, Tranque PA, Garcia-Segura LM 1989 Synaptic remodeling in the rat arcuate nucleus during the estrous cycle. Neuroscience 32:663–667

Peters A 1974 Synapses. In: Peters A, Palay SL, Webster H (eds) The fine structure of the nervous system. Harper & Row, New York p 132–163

Pomeroy SL, Purves D 1988 Neuron/glia relationships observed over intervals of several months in living mice. J Cell Biol 107:1167–1175

Poulain DA, Theodosis DT 1988 Coupling of electrical activity and hormone release in mammalian neurosecretory neurons. Curr Top Neuroendocrinol 9:73–104
Poulain DA, Wakerley JB 1982 Electrophysiology of hypothalamic magnocellular neurones secreting oxytocin and vasopressin. Neuroscience 7:773–808
Purves D, Voyvodic JT 1987 Imaging mammalian nerve cells and their connections over time in living animals. Trends Neurosci 10:398–404
Rutishauser U, Acheson A, Hall AK, Mann DM, Sunshine J 1988 The neural cell adhesion molecule (NCAM) as a regular of cell–cell interactions. Science (Wash DC) 240:53–57
Theodosis DT, Poulain DA 1984a Evidence that oxytocin-secreting neurones are involved in the ultrastructural reorganisation of the rat supraoptic nucleus apparent at lactation. Cell Tissue Res 235:217–219
Theodosis DT, Poulain DA 1984b Evidence for structural plasticity in the supraoptic nucleus of the rat hypothalamus in relation to gestation and lactation. Neuroscience 11:183–193
Theodosis DT, Poulain DA 1989 Neuronal-glial and synaptic plasticity in the adult rat paraventricular nucleus (BRE 23448). Brain Res 484:361–366
Theodosis DT, Poulain DA, Vincent JD 1981 Possible morphological bases for synchronisation of neuronal firing in the rat supraoptic nucleus during lactation. Neuroscience 6:919–929
Theodosis DT, Chapman DB, Montagnese C, Poulain DA, Morris JF 1986a Structural plasticity in the hypothalamic supraoptic nucleus at lactation affects oxytocin- but not vasopressin-secreting neurones. Neuroscience 17:661–678
Theodosis DT, Montagnese C, Rodriguez F, Vincent JD, Poulain DA 1986b Oxytocin induces morphological plasticity in the adult hypothalamo-neurohypophysial system. Nature (Lond) 322:738–740
Theodosis DT, Paut L, Tappaz ML 1986c Immunocytochemical analysis of the GABAergic innervation of oxytocin- and vasopressin-secreting neurones in the rat supraoptic nucleus. Neuroscience 19:207–222
Theodosis DT, Rougon G, Poulain DA 1991 Retention of embryonic features by an adult neuronal system capable of plasticity: polysialylated N-CAM in the hypothalamo-neurohypophysial system. Proc Natl Acad Sci USA 88:5494–5498
Tweedle CD, Hatton GI 1976 Ultrastructural comparisons of neurons of supraoptic and circularis nuclei in normal and dehydrated rats. Brain Res Bull 1:103–121
Tweedle CD, Hatton GI 1977 Ultrastructural changes in rat hypothalamic neurosecretory cells and their associated glia during minimal dehydration and rehydration. Cell Tissue Res 181:59–72
Tweedle CD, Hatton GI 1980 Evidence for dynamic interactions between pituicytes and neurosecretory axons in the rat. Neuroscience 5:661–667
Tweedle CD, Hatton GI 1987 Morphological adaptability at neurosecretory axonal endings on the neurovascular contact zone of the rat neurohypophysis. Neuroscience 20:241–246
Tweedle CD, Modney BK, Hatton GI 1988 Ultrastructural changes in the rat neurohypophysis following castration and testosterone replacement. Brain Res Bull 20:33–38
Verbalis JG, McCann MJ, McHale CM, Stricker EM 1986 Oxytocin secretion in response to cholecystokinin and food: differentiation of nausea from satiety. Science (Wash DC) 232:1417–1419
Wittkowski W, Brinkman H 1974 Changes in extent of neuro-vascular contacts and number of neuro-glial synaptoid contacts in the pituitary posterior lobe of dehydrated rats. Anat Embryol 146:157–165
Wittkowski W, Scheuer A 1974 Functional changes of the neuronal and glial elements at the surface of the external layer of the median eminence. Z Anat Entwicklungsgesch 143:255–262

DISCUSSION

Palkovits: What type of glial cells are there in the supraoptic nucleus? Are they the same as those in the magnocellular paraventricular glial element?

Theodosis: Most are astrocytes and their morphology is the same in the supraoptic and paraventricular nuclei. Whether this means that they are the same is another matter. I have just shown you that even in the same nucleus astrocytes next to vasopressin cells don't behave the same way as astrocytes next to oxytocin cells.

Palkovits: What is the lifetime of the glial cells?

Theodosis: I don't know. We tend to think they are like neurons; that once in place, they remain there. However, astrocytes in the hypothalamus can divide, for example after dehydration (Paterson & Leblond 1977).

Palkovits: Yes, they found the mitotic rate of the astrocytes to be up to 5%. One other question, have you checked the glial–neuronal connections in those animals where you have a specific stimulus for glial cells? I mean after transecting the axons the degenerating terminals are cleaned by glial cells. What has happened after transection of the axons which resulted in retrograde degeneration of the neuronal cell body? That should be a very strong stimulus for the glial cells. It would be interesting to study, because it's a very rapid change—the degeneration occurs in hours.

Theodosis: We have not used such a model, but we should.

Hatton: We have done this in slices. We looked at horizontal slices of hypothalamus by electron microscopy one, two and five hours after making the slice. We get more of these multiple synapses with time.

Halász: Is the oxytocin plasticity evident both in males and in females?

Theodosis: Yes. Under osmotic stimulation, the male supraoptic nucleus undergoes the same plastic changes and the neurons involved are oxytocinergic (see text).

Nagy: I am surprised that you got the same results with dehydration and lactation.

Theodosis: We all tend to think that dehydration stimulates only vasopressin release. However, oxytocin is released as well and the oxytocin system is very active, especially over long periods of dehydration.

Nagy: What would you expect to see if you dehydrated a lactating female?

Theodosis: We have examined the nuclei of dehydrated, lactating animals and we find the same changes as during lactation; that is, the stimuli are not additive.

Liposits: What's the percentage of oxytocinergic cells that receive oxytocinergic fibres? How many labelled boutons can you see on one oxytocin neuron?

Theodosis: We don't have figures for the number of oxytocinergic terminals. One expects to find at least one or two per section of the nucleus examined with the electron microscope—they are rare.

Gainer: What is moving? Is it the terminals, is it the glia, is it the cells?

Theodosis: I think that both neurons and astrocytes change their conformation. Probably, the neurons change first and their associated astrocytes follow.

Mellon: Could the specificity be on the oxytocin neuron? The oxytocin neuron might be either repelling the glia by expressing something that repels them, or the glia might be passively making contact and the oxytocin neuron expressing something that causes those neurons to zip up and make more and more contacts and squeeze out the glia.

Theodosis: No. I think the results favour the hypothesis of active neuron–glia changes: oxytocin neurons change in morphology, as do their associated astrocytes. Vasopressin cells are not affected by the changes and therefore their astrocytes are not either.

Palkovits: If you remove the neurons from the arcuate nucleus by treatment with monosodium glutamate or from the ventromedial nucleus by injections of gold thioglucose, do all glial cells disappear?

Theodosis: I don't know, we haven't studied this.

Epelbaum: If you treat with oestradiol valerate, you will eliminate only the β-endorphin neurons, not the others.

Theodosis: But I don't want to get rid of the neurons, I'm saying that they *are* important. I think that the glial cells are responding to the conformational changes of the neurons.

Plant: Could the apparent changes in synaptic density in the arcuate nucleus be accounted for just by swelling and shrinking of the cells in this nucleus?

Theodosis: No. In the arcuate nucleus, neuronal somata do not hypertrophy.

Barraclough: Do you know what kind of synapses there are in the arcuate nucleus?

Theodosis: No, Olmos et al (1989) did not examine the neurochemistry of the synapses showing plasticity.

Gainer: Why is it good for lactation to have many more GABA synapses?

Poulain: I don't know. My feeling is that during lactation, one of the main activities of the oxytocin system is related to the milk ejection reflex. Therefore there should be some sort of inhibitory system that prevents oxytocin cells being involved in all other regulations in which they normally take part. There is circumstantial evidence for this. Several groups have shown that the responsiveness of the oxytocin system, for instance to dehydration, is diminished during lactation compared to that in normal animals.

The other thing is that during the milk ejection reflex, oxytocin cells have a very peculiar pattern of electrical activity, which is highly synchronized with a burst occurring every few minutes. It may be that the rich GABAergic innervation of these neurons is one of the means to synchronize the population by permitting the excitatory wave within these cells to occur at a very specific moment when the cells are fairly well hyperpolarized.

Gainer: Have you recorded or counted the number of spikes per second during the non-ejection phase, the non-burst phase, and compared it to normal circumstances?

Poulain: It is not changed.

Gainer: So the GABA isn't reducing the basal activity?

Poulain: We have done a lot of statistical analysis and comparisons with normal animals or with animals before suckling starts. What we call the background activity between two milk ejection bursts doesn't seem to be affected at all by suckling. We are doing pharmacological experiments at the moment with GABA and it is very complicated. When we infuse a GABA agonist, the reflex slows down. If we infuse a GABA antagonist, the reflex also slows down. In both cases there is an inhibitory effect on the electrical activity of the cells. My feeling is that you need to keep the cell membrane depolarized within a range where eventually bursting activity will occur.

Liposits: Are these oxytocinergic cells oestrogen receptive in the supraoptic nucleus?

Theodosis: There is no evidence—again the lack of receptors.

Liposits: What about the GABAergic cells that innervate the supraoptic neurons? Do they synthesize oestrogen receptors?

Theodosis: The origin of the GABAergic innervation is unknown.

Pfaff: Do the GABA cells synapse on other GABA cells? Is there any evidence for glycine? For instance, could there be Renshaw-like cells around the supraoptic nucleus?

Theodosis: Around the supraoptic nucleus there is no morphological evidence for this. Because we don't know where the cell bodies of these GABAergic terminals are, I can't really answer your question.

Young: Do you know if there are morphological changes in the parvicellular portion of the paraventricular nucleus?

Theodosis: There are no changes there.

Sawchenko: Could you summarize just what juxtaposition means? To what extent is this indicative of some functional coupling between neurons?

Theodosis: In the 1950s everyone was convinced that the brain communicated that way, via electrical interactions, field effects, across neuronal membranes. The only morphological correlates of such interactions, at least within the mammalian brain, are gap junctions. The juxtapositions in the magnocellular nuclei represent neuronal membranes, not separated by glial elements, between which there is a normal extracellular space. What may be of physiological consequence is that the amount of neuronal membrane juxtaposed is very extensive. I have to ask the electrophysiologists if this means something functionally. For instance, in one plane of section, you can sometimes see up to 17 μm of neuronal surface juxtaposed.

Poulain: One of the first ideas that comes to the mind of an electrophysiologist is that the juxtapositions facilitate synchronization of the electrical activity.

The first papers describing neuronal juxtapositions stressed this idea. There are several ways this may lead to synchronization. One is by field effects, but then you need a very well organized topography of the neurons, as in the hippocampus. Another way is linked to potassium: when one cell is active it extrudes potassium which can influence the excitatability of another cell. With Jonathan Coles, we have done a series of experiments using potassium-sensitive electrodes. We found that at the time of a milk ejection burst, when recording an oxytocin cell closely enough that there are beautiful extracellular spikes, you also see a rise in extracellular potassium (Coles & Poulain 1991). But the concentration of potassium is not very high. The basal level is about 2.5 mM: the increase can be 0.5 mM. We calculated that in the best cases, within a patch of neuronal apposition, the potassium concentration may reach about twice the basal level. This might influence the neighbouring cell. However, if you move the electrode slightly away from the point where you are recording, the potassium concentration falls rapidly. If you record around a vasopressin cell at the time of a milk ejection, you see no change whatsoever. Our conclusion was that the increase in extracellular potassium would be able to facilitate the synchronization of only a few neighbouring cells, but it would have no influence at a distance from the cell because it doesn't diffuse across the nucleus very well.

Hatton: How big were the tips of your electrodes?

Poulain: All these are standard microelectrodes with a tip in the range 0.8–3 μm. This has been addressed in detail by Ransom et al (1987). They concluded that potassium changes in the extracellular space are measured faithfully with such electrodes. I don't think the size of the tip would account entirely for the lack of a rise in potassium throughout the nucleus (see Coles & Poulain 1991).

Hatton: Considering that the synaptic cleft is 10 nm and your electrode is 3 μm, the size of the hole that the electrode makes can have a profound influence on the potassium concentration one might record. I would guess that the true potassium concentration might be a thousand times higher.

Poulain: No, I think this supports our conclusion. If we are that far away from the cleft and we don't detect a rise in potassium concentration away from the cell, it means that the rise doesn't spread. I don't discount that in a highly localized space there may be effects on neighbouring cells. In the nucleus, there are clusters of oxytocin cells that are quite far apart. The localized rise in extracellular potassium concentration may be a facilitatory mechanism, but it doesn't really account for the gross physiology of the system.

Hatton: Probably not in the burst, but maybe in the overall excitability of the cells. The ephaptic interaction may happen around the dendrites the way Taylor & Dudek (1982, 1984a,b) showed for the hippocampus. They get electrical fields around the dendrites; I don't think you are recording the dendrites. It could be that small groups of neurons are effectively synchronized.

Theodosis: Oxytocin neurons do occur in clusters. This is particularly striking in the stimulated paraventricular nucleus, where one often finds groups of 3–7 directly juxtaposed somata, sharing the same synapses. There is also extensive dendritic bundling.

Hatton: Not only are the cell body membranes very close, there are also long lengths of dendritic membrane that are closely apposed. In the lactating rat, you can find bundles of as many as 13 dendrites without glia between them. This really sets the stage for electrical interactions. Electronic coupling also occurs and increases during lactation. It may even increase during the milk ejection reflex; we don't know how dynamic it really is, because one can't study the milk ejection reflex in a slice.

Poulain: We should be aware that when we talk about synchronization, it depends entirely on the level of accuracy we are thinking of. For instance, in electrotonic coupling, the degree of synchronization is very high: when there is a spike in one cell, there is an immediate change in the other cell. For the milk ejection reflex, extracellular recordings allow us to estimate that the whole population of oxytocin cells throughout the four nuclei, that is about 9000 to 10 000 oxytocin cells, fires within 0.5 seconds, but there is no spike-to-spike synchronization (Belin & Moos 1986).

So it depends on what we are talking about in terms of synchronization. If we are talking about minute phenomena, such as interactions through gap junctions, they cannot account for the gross physiology of the milk ejection reflex. I am convinced, as you are, that they may constitute additional mechanisms, not necessarily secondary ones, to the one that triggers the activity throughout the four nuclei. But they do not constitute the primary mechanism for such synchronization.

Gainer: So you think the primary mechanism for the synchronization is synaptic? Which synaptic transmitter do you think is involved in this primary mechanism? Is it GABA? There is some evidence that $GABA_B$ receptors can be depolarizing under some circumstances.

Theodosis: Another candidate is glutamate, since there are glutamate synapses in these nuclei, as we can see with electron microscopic immunocytochemistry. The question is, do they also undergo plasticity? Could glutamate be the factor that sets off this system?

Hatton: There are glutamate receptors on both cell types.

Poulain: All attempts so far to find an external trigger of the electrical activity of the oxytocin cells have failed. There are several regions, lesions of which stop the milk ejection reflex at once. There is no evidence for a structure that facilitates the reflex. The only way one can manipulate the pattern of milk ejection is by injecting oxytocin intracerebrally in certain regions.

There is also evidence for interactions between the individual supraoptic and paraventricular nuclei. If you inject oxytocin or acetylcholine into one supraoptic nucleus, there is an immediate excitatory response in the other supraoptic nucleus. There is also anatomical evidence suggesting that there might be interactions between the paraventricular nucleus and the supraoptic nucleus (see Moos & Richard 1989).

We do not know whether there is an external clock, which would set off all the oxytocin cells, or whether the clock is within the cells. I would speculate that during suckling all the cells receive a certain amount of neurotransmitters, which, little by little, change the membrane properties of the cells. Then when they are 'ripe', one cell starts, another set starts somewhere, and because they are interconnected (through additional facilitatory mechanisms) the whole system bursts. After that, the mechanism starts again. But, as yet, we have no evidence for an intracellular mechanism responsible for the activation of oxytocin cells.

Palkovits: When you inject oxytocin unilaterally and the contralateral nucleus responds, is it a systemic effect or does it occur only through neuronal connections?

Poulain: Moos & Richard (1989) showed this using electrical recordings: the connection is neuronal but not direct. However, after splitting the brain sagittally through the third ventricle, the synchronization of the four nuclei is retained.

Pfaff: Is there evidence from other systems that polysialic moieties on N-CAM make the cells 'slippery' rather than 'sticky'?

Theodosis: A beautiful example is the neuromuscular junction. When fibres grow into the muscle, there is high expression of PSA-N-CAM. Endosialidase, which cleaves off polysialic residues, changes fasciculation and innervation patterns.

Pfaff: That doesn't distinguish between guiding or adhesion functions on the one hand and 'slipperiness' on the other. In the gonadotropin-releasing hormone (GnRH) neuronal migration system, there is N-CAM on cells that grow out first; we call them pioneer cells. We could not say that the pioneer cells are making things 'slippery' so that the GnRH cells can migrate, as opposed to N-CAM providing adhesion or a guiding function.

Theodosis: The hypothesis is that in the nervous system PSA-N-CAM is expressed at high levels at certain times when connections are being made and the tissue architecture is being established. PSA-N-CAM expression then diminishes and adult isoforms of N-CAM are expressed. Because of the timing of the conversion, it is thought that polysialylation is important for motility.

Pfaff: That does not prove that the adhesion molecules work through slipperiness rather than through guidance.

Theodosis: In *in vitro* models, it is clear that the ability of N-CAM to promote cell adhesion depends on its degree of sialylation. Adhesion is attenuated by PSA and loss of PSA increases adhesiveness (Hoffman & Edelman 1983, Rutishauser et al 1988).

References

Belin V, Moos F 1986 Paired recordings from supraoptic and paraventricular oxytocin cells in suckled rats: recruitment and synchronization. J Physiol (Lond) 377:369–390

Coles JA, Poulain DA 1991 Extracellular K^+ in the supraoptic nucleus of the rat during reflex bursting activity by oxytocin neurones. J Physiol (Lond) 439:383–409
Hoffman S, Edelman GM 1983 Kinetics of homophilic binding by embryonic and adult forms of the neural cell adhesion molecule. Proc Natl Acad Sci USA 80:5762–5766
Moos F, Richard P 1989 Paraventricular and supraoptic bursting oxytocin cells in rat are locally regulated by oxytocin and functionally related. J Physiol (Lond) 408:1–18
Olmos G, Naftolin F, Perez J, Tranque PA, Garcia-Segura LM 1989 Synaptic remodeling in the rat arcuate nucleus during the estrous cycle. Neuroscience 32:663–667
Paterson JA, Leblond CP 1977 Increased proliferation of neuroglia and endothelial cells in the supraoptic nucleus and hypophysial neural lobe of young rats drinking hypertonic sodium chloride solution. J Comp Neurol 175:373–390
Ransom BR, Carlini WG, Yamata CL 1987 Tip size of ion exchanger based K^+-selective microelectrodes. II. Effects on measurements of evoked $[K^+]_o$ transients. Can J Physiol Pharmacol 65:894–897
Rutishauser U, Acheson A, Hall AK, Mann DM, Sunshine J 1988 The neural cell adhesion molecule as a regulator of cell–cell interactions. Science (Wash DC) 240:53–57
Taylor CP, Dudek FE 1982 Synchronous neural after-discharges in rat hippocampal slices without active chemical synapses. Science (Wash DC) 218:810–812
Taylor CP, Dudek FE 1984a Excitation of hippocampal pyramidal cells by an electrical field effect. J Neurophysiol 52:126–142
Taylor CP, Dudek FE 1984b Synchronization without active chemical synapse during hippocampal after-discharges. J Neurophysiol 52:143–155

Neural control of the synthesis and release of luteinizing hormone-releasing hormone

Charles A. Barraclough

Center for Studies in Reproduction, Department of Physiology, School of Medicine, University of Maryland, 655 West Baltimore Street, Baltimore, MD 21201, USA

Abstract. Preovulatory surges of luteinizing hormone (LH) depend upon neurotransmitter activation of neurons that secrete LH-releasing hormone (LHRH, gonadotropin-releasing hormone GnRH) and noradrenaline plays a pivotal role in this critical event. The interactio ıs amongst noradrenaline and other neurotransmitters such as GABA (γ-aminobutyric acid), opiates, serotonin and excitatory amino acids (N-methyl-D-aspartate, NMDA) on LHRH neuronal activity are complex. GABA and opiates suppress the presynaptic release of noradrenaline but only GABA also directly affects the responsiveness of LHRH neurons to noradrenaline. Morphine induces the release of serotonin which either directly or indirectly via other neurotransmitters (e.g. dopamine) sensitizes LHRH neurons to the stimulatory effects of noradrenaline. NMDA rapidly induces LH release but whether this drug directly affects the activity of LHRH neurons is not known. The neuronal release of LHRH is modulated by the action of oestrogen on these various neurotransmitter systems. Antioestrogens, when placed into the medial preoptic area of otherwise completely oestrogenized rats, block LH surges; LHRH mRNA levels in such animals resemble those in 9-day castrated rats. Normally, LHRH message levels increase about the time of increased noradrenaline secretion just before the LH surge. NMDA rapidly releases LH and LHRH mRNA levels are significantly raised within 15 minutes and remain so over the next 45 minutes. Thus, it seems that stimuli which evoke LHRH release also increase LHRH mRNA transcription to replenish the hormone released during the LH surge.

1992 Functional anatomy of the neuroendocrine hypothalamus. Wiley, Chichester (Ciba Foundation Symposium 168) p 233–251

One of the most important endocrine events of the mammalian reproductive cycle is the midcycle surge of luteinizing hormone (LH). It is this hormone which induces ovulation, thereby making oocytes available for fertilization in the Fallopian tubes.

The role of noradrenaline in the release of LH-releasing hormone

Preovulatory LH surges in rats depend upon a complex interplay amongst various neurotransmitter systems and neurons that secrete LH-releasing hormone (LHRH). Of the neurotransmitters, noradrenaline has a critical and pivotal role in the stimulation of the activity of LHRH neurons (Barraclough & Wise 1982). In earlier studies we observed that turnover of noradrenaline (an index of secretion) markedly increased in selective hypothalamic regions containing LHRH perikarya (medial preoptic area) and axon terminals (median eminence) concomitant with pro-oestrus or steroid-induced LH surges in ovariectomized rats (Rance et al 1981a, Wise et al 1981). Moreover, blockade of LH surges with phenobarbitone (Rance & Barraclough 1981) or sequential inhibition[1] of LH release by progesterone (Rance et al 1981b) abolished the increase in noradrenaline turnover normally associated with LH surges. From these studies we concluded that an increase in hypothalamic noradrenaline release is a critical event in triggering the surge release of LHRH and LH.

The primary source of hypothalamic noradrenaline is medullary A1 and A2 noradrenergic neurons (Moore & Bloom 1979). We reasoned that if noradrenaline is one of the major neurotransmitters which activates LHRH neurons, then, by electrically stimulating A1 neurons, we should be able to evoke a rise in plasma LH. However, regardless of the stimulation parameters used in these studies, we were unable to alter basal plasma LH concentrations. In contrast, if we first partially depolarized LHRH neurons with an electrochemical stimulus and then electrically stimulated A1 neurons, we observed an amplification of LH release (Fig. 1). This augmented LH response could be blocked by a specific α_1-adrenoreceptor antagonist or by other drugs that suppress the synthesis of noradrenaline (Gitler & Barraclough 1988), indicating that noradrenaline was released in response to electrical stimulation of A1 neurons. More recently, Herbison et al (1990) have used a microdialysis procedure to measure endogenous release of hypothalamic noradrenaline. They observed that when A1 neurons are activated using electrical stimulation parameters similar to those described by us, an increase in noradrenaline release could be measured in the medial preoptic area. These observations led us to pose the following question: if noradrenaline activates LHRH neurons and electrical stimulation of A1 neurons releases noradrenaline into the medial preoptic area, a region which contains LHRH perikarya, why doesn't A1 stimulation elicit the release of LH? To resolve this question we reinvestigated the effects of noradrenaline on the function of LHRH neurons. When noradrenaline was infused into the third ventricle of oestrogen-treated ovariectomized rats, only a modest rise in plasma LH was obtained (Hartman et al 1989). However, if LHRH neurons were first

[1]When progesterone is injected into oestrogen-primed rates, it initially amplifies LH release and induces mating behaviour. 24 hours later, a similar injection blocks oestrogen-induced LH release and mating behaviour. These secondary effects are referred to as sequential inhibition.

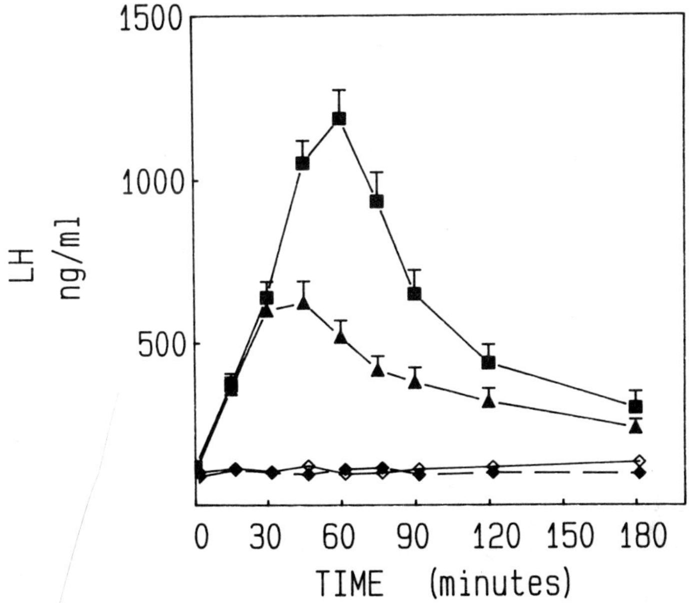

FIG. 1. The effect of electrical stimulation of A1 neurons on plasma concentrations of luteinizing hormone (LH) in rats anaesthetized with chloral hydrate (CH). ◆, controls; ◇, electrical stimulation of A1 neurons; ▲, bilateral electrochemical stimulation of the medial preoptic area; □, bilateral electrochemical stimulation of the medial preoptic area at time 0 then electrical stimulation of A1 neurons at 30 minutes for 15 minutes. From Gitler & Barraclough (1988).

depolarized electrochemically and then noradrenaline was infused into the third ventricle, a marked enhancement in LH release occurred. These enhanced responses to noradrenaline were not due to the effects of anaesthesia or to alterations in pituitary responsiveness to LHRH in this animal model. Rather, they were the result of the electrochemical stimulus rendering LHRH neurons more responsive to noradrenaline. From these studies, we concluded that LHRH neurons in their normal resting state are relatively unresponsive to noradrenaline. These observations raised the possibility that LHRH neurons fail to respond to noradrenaline because their activity is being modulated by local inhibitory interneurons. Depolarization of LHRH neurons by electrochemical stimulation bypasses these inhibitory controls and thereafter these peptidergic cells readily respond to exogenous or endogenous noradrenaline.

Role of GABA interneurons in the response of LHRH neurons to noradrenaline

A major candidate for neurons that alter the responsiveness of LHRH neurons to noradrenaline is the GABA (γ-aminobutyric acid) interneurons. LHRH neurons are innervated by GABA neurons (Jennes et al 1983, Leranth et al 1985) and GABA receptor agonists suppress LH surges in steroid-treated ovariectomized rats (Adler & Crowley 1986). Also, at the time of an LH surge, push–pull perfusion data suggest that GABA release rates decline and noradrenaline secretion increases (Demling et al 1985, Jarry et al 1988). Because both GABA and noradrenaline neurons contain oestrogen receptors (Flugge et al 1986, Heritage et al 1977), changes in plasma oestradiol levels could affect LHRH release by modulating activity in either or both of these neurotransmitter systems. Evidence has been presented that GABA tonically inhibits the presynaptic release of noradrenaline (see Hartman et al 1990 for references). Thus, this ubiquitous inhibitory neurotransmitter could affect the activity of LHRH neurons both via its direct synaptic connections and indirectly by modulating the presynaptic release of noradrenaline. We attempted to reduce some of these variables by evaluating whether responsiveness of LHRH neurons to noradrenaline is altered when specific GABA receptor antagonists are used to block endogenous GABA effects on LHRH cells. We suggested that perhaps the reason LHRH neurons normally are not responsive to noradrenaline is that GABA maintains resting membrane potentials of these cells in a hyperpolarized state. Thus, excitatory signal inputs (e.g. noradrenaline) are relatively ineffective in activating LHRH neurons during the time GABA influences are dominant.

In these studies, ovariectomized oestradiol-treated rats were anaesthetized with chloral hydrate prior to use. All rats in this and subsequent experiments were studied between 08.00 and 11.00 hours to avoid the time when spontaneous LH surges normally occur in our animals (13.00–15.00 hours).

When a $GABA_A$ receptor antagonist, bicuculline, was infused into the third ventricle or directly into the medial preoptic area (0 time) and noradrenaline was infused into the ventricle (+15 minutes), LH release was increased significantly compared to after noradrenaline alone (Fig. 2). Similar amplified responses were obtained if a $GABA_B$ receptor antagonist, phacofen, was first infused intraventricularly or into the medial preoptic area and noradrenaline was given (intraventricularly) 15 minutes later. When the preoptic area was electrochemically stimulated at 0 time and was followed by intraventricular injection of bicuculline or bicuculline + noradrenaline, there was no further increase in the peak secretion of LH (Hartman et al 1990). The results clearly indicate that GABA modulates LHRH neuronal responsiveness to noradrenaline and suggest that electrochemical stimulation improves responsiveness of LHRH neurons by bypassing inhibitory GABA influences. Demling et al (1985) and Jarry et al (1988) have provided evidence which suggests that at the time of the LH surge, GABA secretion rates decline and noradrenaline turnover rates increase. The present results broaden these earlier conclusions to include the LHRH neuron itself. Thus, withdrawal of GABA influences would not only allow a greater presynaptic release of noradrenaline but would also result in a substantial increase in the responsiveness of LHRH neurons to this secreted neurotransmitter.

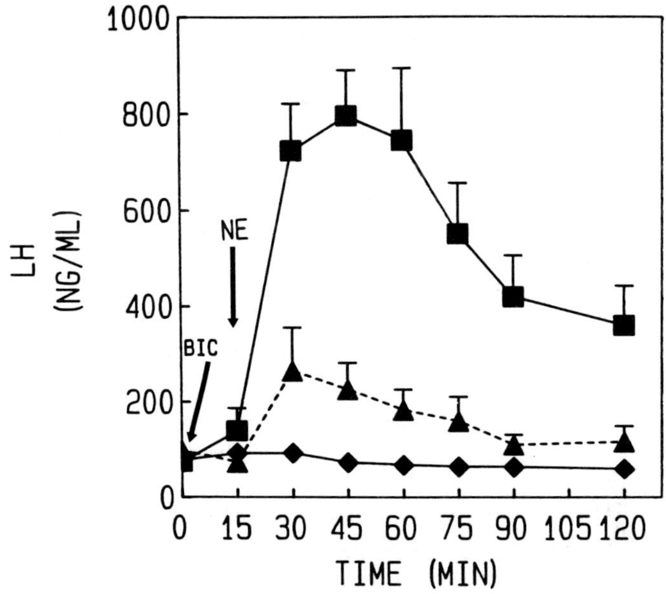

FIG. 2. The effect of intraventricular infusion of noradrenaline and bicuculline on plasma concentrations of luteinizing hormone (LH) in oestrogen-treated ovariectomized rats. ◆, bicuculline; ▲, noradrenaline; ■, bicuculline + noradrenaline. From Hartman et al (1990).

Role of opiate and serotonin neurons in the response of LHRH neurons to noradrenaline

Other neurotransmitter systems whose axons synapse directly on LHRH perikarya include those that secrete serotonin (5-HT, 5-hydroxytryptamine) (Kiss & Halász 1985) and opiates (Leranth et al 1988). Each of these neurotransmitters has been shown by many investigators to alter the secretion of LH (for reviews see Barraclough & Wise 1982, Petersen et al 1989b). Whether their effects are mediated via direct synaptic connections or indirectly by modulating the presynaptic release of other neurotransmitters (or both) is still being examined. In an attempt to resolve this question, we have examined whether opiate synapses affect the response of LHRH neurons to noradrenaline. Endogenous noradrenaline secretion was suppressed by the chloral hydrate anaesthesia used in these studies. Consequently, we could evaluate the LH responses elicited by intraventricular noradrenaline without interference from any endogenous noradrenaline which otherwise might be released in response to drugs that affect presynaptic opiate receptors. Treatment of these animals with naloxone (a μ opiate receptor antagonist) had no effect on basal LH release. The intraventricular infusion of noradrenaline resulted in a modest rise in plasma LH similar to that observed previously (Hartman et al 1989). When rats were treated with naloxone 15 minutes before the intraventricular infusion of noradrenaline, the LH release evoked by noradrenaline was unchanged. Similarly, rats in which the medial preoptic area was stimulated electrochemically to partially depolarize LHRH neurons showed no effect of naloxone on the amplified responses obtained to noradrenaline (He & Barraclough 1992). Thus, in contrast to GABA synaptic connections, which dramatically affect the activity of LHRH neurons, those made by opiate terminals do not alter the response of these neurons to noradrenaline.

In the course of these studies, we made a series of surprising observations. When morphine (an opiate receptor agonist) was given subcutaneously and noradrenaline was infused into the third ventricle 30 minutes later, LH release was markedly enhanced (531 ± 72 ng/ml peak value) compared to the responses obtained after only intraventricular noradrenaline (259 ± 31 ng/ml). Further, if rats were treated with morphine 15 minutes before electrochemical stimulation of the medial preoptic area, then noradrenaline was infused into the ventricle 30 minutes after that, LH secretion reached peak values of 2392 ± 222 ng/ml. In earlier studies, we observed that morphine blocks LH surges and this effect could be attributed to inhibition of the presynaptic release of noradrenaline (Akabori & Barraclough 1986). The question therefore arose, why does morphine treatment enhance the response of LHRH neurons to noradrenaline? One neurotransmitter that is dramatically affected by morphine is 5-HT (Johnson & Crowley 1984). We considered the possibility that the stimulatory effects of morphine were being exerted indirectly via activation of central 5-HT neurons. To resolve this question, in the next series of studies we restricted the site of action of morphine to 5-HT neurons by infusing the drug directly into the dorsal raphe nucleus (DRN). This nucleus contains neurons whose axons contribute to the rich serotonergic innervation of the hypothalamus (Conrad et al 1974, Azimita & Segal 1978, Petersen et al 1989b). Infusions of either saline or morphine alone did not affect basal LH concentrations. When morphine was infused into the DRN and noradrenaline was given 30 minutes later, plasma LH release was significantly enhanced compared to after only noradrenaline. In a final group of rats, morphine was infused into the DRN, the medial preoptic area was stimulated electrochemically and noradrenaline was infused intraventricularly. LH release was enhanced markedly and reached peak values that were significantly higher than those obtained in any of the other treatment groups (Fig. 3). While we have not resolved the mechanisms or sites of action by which morphine improves the responsiveness of LHRH neurons to noradrenaline, several possibilities are being considered: (1) morphine releases 5-HT which acts directly on LHRH neurons to sensitize these cells to the subsequent action of noradrenaline; (2) 5-HT acts indirectly on other inhibitory neurotransmitter systems (e.g. dopamine, GABA) and by suppressing their

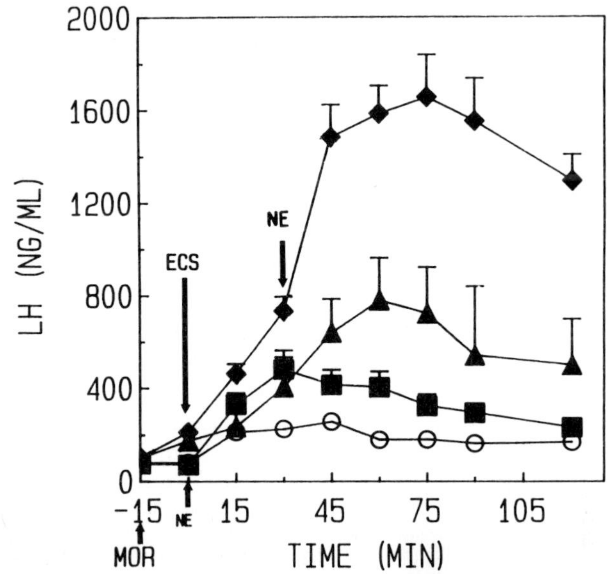

FIG. 3. The effect of morphine infusion into the dorsal raphe nucleus on the release of luteinizing hormone in response to noradrenaline and electrochemical stimulation (ECS) of the medial preoptic area. ○, intraventricular noradrenaline; ■, ECS; ▲, morphine + ECS; ◆, morphine infusion before noradrenaline + ECS. From He & Barraclough (1992).

activity allows greater release of LHRH in response to noradrenaline (He & Barraclough 1992); (3) morphine stimulates the release of neurotransmitters which act synergistically with noradrenaline to enhance LHRH secretion.

Regulation of LHRH mRNA transcription

LHRH perikarya do not contain oestrogen receptors (Shivers et al 1983), therefore this steroid presumably exerts its feedback effects by modulating the activity of neurotransmitter neurons that communicate with LHRH neurons (e.g. noradrenaline, GABA, 5-HT). We have used *in situ* histochemistry and quantitative image analysis methods to evaluate cellular LHRH mRNA levels and thereby obtain an index of neuronal activity. Evidence exists suggesting that levels of mRNAs encoding neuropeptides reflect levels of neuropeptide synthesis (Young & Zoeller 1987).

Recently, we identified a population of LHRH neurons that are involved in the oestrogen-induced LH surge. In oestrogen-treated ovariectomized rats, we

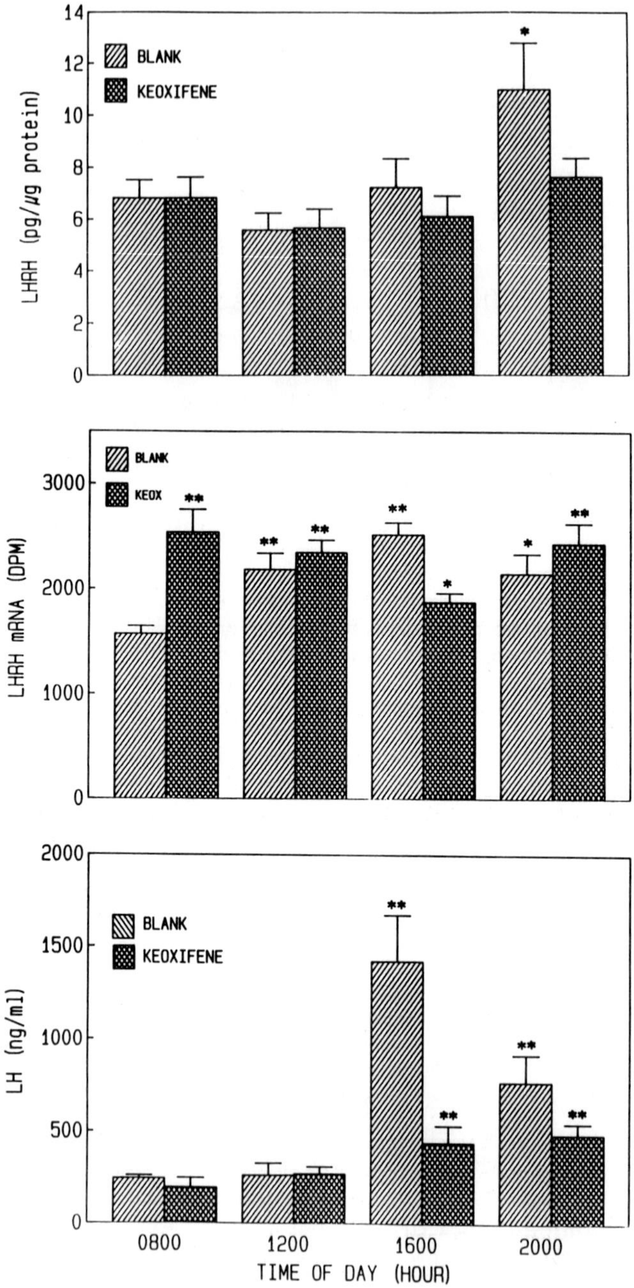
14
12
10
8
6
4
2
0
LHRH (pg/μg protein)
BLANK
KEOXIFENE
*
3000
2000
1000
0
LHRH mRNA (DPM)
BLANK
KEOX
**
*
2000
1500
1000
500
0
LH (ng/ml)
BLANK
KEOXIFENE
0800
1200
1600
2000
TIME OF DAY (HOUR)

FIG. 4.

observed that LH surges were blocked when microimplants of the non-steroidal antioestrogen, Keoxifene (Eli Lilly & Co), were placed into the medial preoptic area (Petersen & Barraclough 1989). We have used this model to examine changes in cellular levels of LHRH mRNA in perikarya located in the region of the Keoxifene implant (Petersen et al 1989a). In addition, because LHRH synthesis and release are dynamic events, we compared temporal changes in the amount of LHRH mRNA in the medial preoptic area, LHRH concentrations in the median eminence and plasma LH levels in rats having spontaneous LH surges and in rats in which LH release was suppressed by Keoxifene. The results of this study are shown in Fig. 4. Animals receiving blank cannulae had LH surges at 16.00 hours whereas LH levels in rats with Keoxifene implants were markedly suppressed. In the same animals, LHRH concentrations did not differ between 08.00 and 16.00 hours. The only difference was a significant rise in LHRH at 20.00 hours in control rats. Statistical comparisons of LHRH mRNA values at 08.00, 12.00, 16.00 and 20.00 hours in control rats revealed that they were significantly elevated at the later three times in rats with blank micro-cannulae. The levels in Keoxifene-treated rats were significantly higher at 08.00 hours than in control rats and were comparable to those in 9-day ovariectomized rats, which agrees with reports by others. Zoeller et al (1988) found that between 08.00 and 09.00 hours, LHRH mRNA levels in the area of the organum vasculosum of the lamina terminalis were significantly higher in 9-day ovariectomized females without hormone replacement than in those treated for two days with oestrogen. Similarly, Toranzo et al (1989) observed that LHRH mRNA levels had increased 14 days after ovariectomy or castration of male rats and that oestradiol or dihydrotestosterone treatment for 14 days prevented this rise. In contrast to these observations are those of others who report a decrease in LHRH mRNA levels following ovariectomy and an increase in response to oestrogen replacement (Pfaff 1986, Roberts

FIG. 4. Temporal changes in median eminence LHRH, medial preoptic area LHRH mRNA and plasma LH two days after implanting blank or Keoxifene-containing microcannulae into the medial preoptic area of oestrogen-treated ovariectomized rats. Animals receiving blank cannulae had afternoon LH surges. In control rats, median eminence LHRH concentrations were significantly higher at 20.00 h compared to other times, whereas in animals receiving Keoxifene no changes in LHRH values occurred. In control rats (blank), LHRH mRNA levels increased significantly between 08.00 and 12.00 h and remained elevated until 20.00 h. In contrast, LHRH mRNA values in Keoxifene-treated rats were significantly higher at 08.00 h and significantly lower at 16.00 h than controls at the same times. At all times examined, LHRH mRNA levels in Keoxifene-treated rats were significantly higher than those in controls at 08.00 h. In rats receiving Keoxifene, plasma LH concentrations were increased significantly at 16.00 and 20.00 h compared with 08.00 and 12.00 h but never reached surge values. $^{*}P<0.05$; $^{**}P<0.01$ compared with 08.00 h control group (Petersen et al 1989a).

et al 1989). These discrepancies may, in part, be due to the time after castration and to the number of days of steroid replacement before examination of the brains. The gonadectomized rat may be a poor model in which to study LHRH gene expression because steroid withdrawal results in gross perturbations of the neurotransmitter regulation of the activity of LHRH neurons. It would seem more appropriate to evaluate how trans-synaptic stimuli that evoke preovulatory-like LHRH surges also affect LHRH gene transcription. Recently, Petersen et al (1991) observed that after the injection of NMDA, plasma concentrations of LH in male rats rose rapidly to peak values within 15 minutes. In the same animals, LHRH mRNA levels were significantly raised at 15 minutes and remained high at 60 minutes. These observations support reports by others that stimuli which activate the release of neuropeptides also produce rapid changes in the mRNAs encoding these peptides (Uhl & Nishimori 1990, Mézey & Palkovits 1991). Consequently, it is not unreasonable to suggest that the oestrogen-induced rise in LHRH mRNA we observed immediately prior to the LH surge in our antioestrogen studies (Petersen et al 1989a) could be the result of the stimulatory effects of neurotransmitters which activate the surge release of LHRH.

Conclusions

A summary of some of the neurotransmitter events that regulate neuronal release of LHRH to evoke preovulatory LH surges is presented in Fig. 5. (1) Noradrenaline secretion increases in the synaptic cleft to activate the release of LHRH. This event requires an increase in activity within noradrenergic neurons coupled with a decline in the presynaptic inhibition of noradrenaline release by local GABA and opiate neurons. (2) With the decrease in direct inhibitory GABA influences on LHRH neurons, their responsiveness to noradrenaline markedly increases. (3) Direct opiate synapses on LHRH neurons do not affect the latter's responsiveness to noradrenaline. However, opiates suppress the presynaptic release of noradrenaline (Akabori & Barraclough 1986). (4) An increase in 5-HT secretion before the LH surge may sensitize LHRH neurons to noradrenaline. This may occur via direct 5-HT synapses or indirectly by a suppression of inhibitory neurotransmitters (e.g. dopamine, GABA). (5) Opiate agonists increase 5-HT release, which suppresses tuberoinfundibular secretion of dopamine (Alper et al 1980). Whether this also affects LHRH release is not known. Opiates also may affect activity of dopamine neurons within zona incerta or local preoptic dopamine neurons that might synapse with and modulate the activity of LHRH neurons. Tyrosine hydroxylase-containing axons have been identified synapsed on LHRH neurons and, presumably, some of these are dopamine-containing neurons (Chen et al 1989). (6) There is evidence that the same stimuli that activate the release of LHRH also increase transcription of LHRH mRNA.

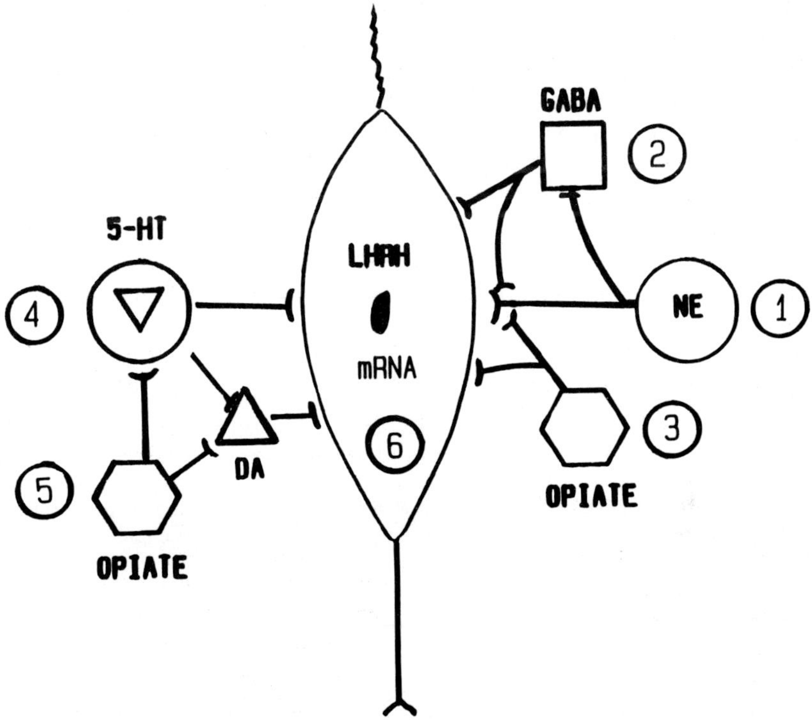

FIG. 5. Summary of neurotransmitter events affecting the activity of neurons that secrete luteinizing hormone-releasing hormone (LHRH). See text for details. DA, dopamine; GABA, γ-aminobutyric acid; NE, noradrenaline; 5-HT, serotonin.

Thus, the occurrence of preovulatory LH surges requires a complex interplay amongst stimulatory and inhibitory neurotransmitter systems. Perhaps most of these events are set into motion by the rise in plasma oestradiol which occurs as ovarian follicles mature to reach preovulatory size (Graafian follicles). It is via this mechanism that the ovary signals the brain and pituitary gland that it is competent to respond to a surge release of LH.

Acknowledgement

This research was supported by NIH Grant HD-02138 to Dr. Charles A. Barraclough.

References

Adler BA, Crowley WR 1986 Evidence of γ-aminobutyric acid modulation of ovarian hormonal effects on luteinizing hormone secretion and hypothalamic catecholamine activity in the female rat. Endocrinology 118:91–97

Akabori A, Barraclough CA 1986 Effects of morphine on LH secretion and catecholamine turnovers in the hypothalamus of estrogen-treated rats. Brain Res 362:221–226

Alper RH, Demarest KT, Moore KE 1980 Morphine differentially alters synthesis and turnover of dopamine in central neuronal systems. J Neural Transm 48:157–165

Azimita EC, Segal M 1978 An autoradiographic analysis of differential ascending projections of the dorsal and median raphe nuclei in the rat. J Comp Neurol 179:641–668

Barraclough CA, Wise PM 1982 The role of catecholamines in the regulation of pituitary luteinizing hormone and follicle stimulating hormone secretion. Endocr Rev 3:91–119

Chen W-P, Witkin JW, Silverman A-J 1989 Gonadotropin releasing hormone (GnRH) neurons are directly innervated by catecholamine terminals. Synapse (NY) 3:288–290

Conrad LCAK, Leonard CM, Pfaff DW 1974 Connections of the median and dorsal raphe nuclei in the rat: an autoradiographic and degeneration study. J Comp Neurol 156:179–205

Demling J, Fuchs E, Baumert M, Wuttke W 1985 Preoptic catecholamine, GABA and glutamate release in ovariectomized and ovariectomized estrogen-treated rats utilizing a push-pull cannula technique. Neuroendocrinology 41:212–218

Flugge G, Oertel WH, Wuttke W 1986 Evidence for estrogen-receptive GABAergic neurons in the preoptic/anterior hypothalamic area of the rat brain. Neuroendocrinology 43:1–5

Gitler MA, Barraclough CA 1988 Stimulation of the medullary A1 noradrenergic system augments luteinizing hormone release induced by medial preoptic nucleus stimulation. Neuroendocrinology 48:351–359

Hartman RD, Petersen S, Barraclough CA 1989 Limited responsiveness of LHRH neurons to norepinephrine may account for failure of locus coeruleus or medullary A1 electrical stimulation to increase plasma LH in estrogen-treated ovariectomized rats. Brain Res 476:35–44

Hartman RD, He J-J, Barraclough CA 1990 γ-aminobutyric acid-A and -B receptor antagonists increase luteinizing hormone-releasing hormone neuronal responsiveness to intracerebroventricular norepinephrine in ovariectomized estrogen-treated rats. Endocrinology 127:1336–1345

He J-J, Barraclough CA 1992 Morphine but not naloxone enhances LHRH neuronal responsiveness to norepinephrine. J Neuroendocrinol 4:91–99

Herbison AE, Heavens RP, Dyer RG 1990 Oestrogen modulation of excitatory A1 noradrenergic input to rat medial preoptic gamma aminobutyric acid neurons demonstrated by microdialysis. Neuroendocrinology 52:161–168

Heritage AS, Grant LD, Stumpf WE 1977 ^{3}H Estradiol in catecholamine neurons in rat brain stem: combined localization by radioautoradiography and formaldehyde-induced fluorescence. J Comp Neurol 176:607–630

Jarry H, Perschl A, Wuttke W 1988 Further evidence that preoptic hypothalamic GABAergic neurons are part of the GnRH pulse and surge generator. Acta Endocrinol 118:573–579

Jennes L, Stumpf WE, Tappez ML 1983 Anatomical relationships of dopaminergic and GABAergic systems with GnRH system in the septohypothalamic area. Exp Brain Res 50:91–99

Johnson DM, Crowley WR 1984 Effects of opiate antagonists on serotonin turnover and on luteinizing hormone and prolactin secretion in estrogen- or morphine-treated rats. Neuroendocrinology 38:322–327

Kiss J, Halász B 1985 Demonstration of serotonergic axons terminating on luteinizing hormone-releasing hormone neurons in the preoptic area of the rat using a combination of immunocytochemistry and high resolution autoradiography. Neuroscience 14:69–78

Leranth C, MacLusky NJ, Aakamoto H, Shanabrough M, Naftolin F 1985 Glutamic acid decarboxylase-containing axons synapse on LHRH neurons in the rat medial preoptic area. Neuroendocrinology 40:536–539

Leranth C, MacLusky NJ, Shanabrough M, Naftolin F 1988 Immunocytochemical evidence for synaptic connections between pro-opiomelanocortin-immunoreactive axons and LHRH neurons in the preoptic area of the rat. Brain Res 449:167–176

Mézey E, Palkovits M 1991 Time dependent changes in CRF and its mRNA in the neurons of the inferior olive following surgical transection of the olivocerebellar tract in the rat. Mol Brain Res 10:55–60

Moore RY, Bloom FE 1979 Central catecholamine neuron systems: anatomy and physiology of the norepinephrine and epinephrine systems. Annu Rev Neurosci 2:113–168

Pfaff DW 1986 Gene expression in hypothalamic neurons: luteinizing hormone-releasing hormone. J Neurosci Res 16:109–115

Petersen SL, Barraclough CA 1989 Suppression of spontaneous LH surges in estrogen-treated ovariectomized rats by microimplants of antiestrogen into the preoptic brain. Brain Res 484:279–289

Petersen SL, Cheuk C, Hartman RD, Barraclough CA 1989a Medial preoptic microimplants of the antiestrogen, Keoxifene, affect luteinizing hormone-releasing hormone mRNA levels, median eminence luteinizing hormone-releasing hormone concentrations and luteinizing hormone release in ovariectomized, estrogen-treated rats. J Neuroendocrinol 1:279–283

Petersen SL, Hartman RD, Barraclough CA 1989b An analysis of serotonin in hyypothalamic regions based on a 5-hydroxytryptophan accumulation or push-pull perfusion. Effect of mesencephalic raphe or locus coeruleus stimulation and correlated changes in plasma luteinizing hormone. Brain Res 495:9–19

Petersen SL, McCrone S, Keller M, Gardner E 1991 Rapid increase in LHRH mRNA levels following NMDA. Endocrinology 129:1679–1681

Rance N, Barraclough CA 1981 Effects of phenobarbital on hypothalamic LHRH and catecholamine turnover rates in proestrous rats. Proc Soc Exp Biol Med 166:425–431

Rance N, Wise PM, Selmanoff MK, Barraclough CA 1981a Catecholamine turnover rates in discrete hypothalamic areas and associated changes in median eminence luteinizing hormone-releasing hormone and serum gonadotropins on proestrus and diestrous day 1. Endocrinology 108:1795–1802

Rance N, Wise PM, Barraclough CA 1981b Negative feedback effects of progesterone correlated with changes in hypothalamic norepinephrine and dopamine turnover rates, median eminence luteinizing hormone-releasing hormone, and peripheral plasma gonadotropins. Endocrinology 108:2194–2199

Roberts JL, Dutlow CM, Jakubowski M, Blum M, Millar RP 1989 Estradiol stimulates preoptic area-anterior hypothalamic proGnRH-GAP gene expression in ovariectomized rats. Mol Brain Res 6:127–134

Shivers BD, Harlan RE, Morrell JI, Pfaff DW 1983 Absence of oestradiol concentration in cell nuclei of LHRH-immunoreactive neurones. Nature (Lond) 304:345–347

Toranzo D, Dupont E, Simard J et al 1989 Regulation of pro-gonadotropin-releasing hormone gene expression by sex steroids in the brain of male and female rats. Mol Endocrinol 3:1748–1756

Uhl GR, Nishimori T 1990 Neuropeptide gene expression and neural activity: assessing a working hypothesis in nucleus caudalis and dorsal horn neurons expressing preproenkephalin and preprodynorphin. Cell Mol Neurobiol 10:73–98

Wise PM, Rance N, Barraclough CA 1981 Effects of estradiol and progesterone on catecholamine turnover rates in discrete hypothalamic regions in ovariectomized rats. Endocrinology 108:2186–2193

Young WS III, Zoeller RT 1987 Neuroendocrine gene expression in the hypothalamus: *in situ* hybridization histochemical studies. Cell Mol Neurobiol 7:353–365

Zoeller RT, Seeburg PH, Young WS III (1988) In situ hybridization histochemistry for messenger ribonucleic acid (mRNA) encoding gonadotropin-releasing hormone (GnRH): effect of estrogen on cellular levels of GnRH mRNA in female rat brain. Endocrinology 122:2570–2577

DISCUSSION

Palkovits: What do you think about noradrenaline fibres in the preoptic area? Are there specific preoptic area noradrenaline fibres or are they just collaterals of ascending noradrenergic fibres from the lower brainstem? Using a double tract tracing technique, it has been shown that noradrenaline-containing exons in the preoptic area are collaterals of the fibres that terminate in the paraventricular nucleus or other forebrain areas. Are there any noradrenaline terminals specific to the preoptic area?

Barraclough: I think this would be almost impossible to determine, mainly because as the central noradrenergic tract enters the hypothalamus via the medial forebrain bundle, it distributes thousands of neurofibrils throughout the hypothalamus. There is evidence that the principle source of hypothalamic noradrenaline is the medullary A1 neurons. But how these axons are distributed once they reach the hypothalamus is not known.

Palkovits: That would be important—to know whether it is a general effect or specific to the LHRH group.

Barraclough: We looked at noradrenaline turnover only in the medial preoptic area, suprachiasmatic nuclei, arcuate nuclei and median eminence, not in other brain regions. In all these areas, there was an increase in noradrenaline turnover concomitant with the preovulatory LH surge on pro-oestrus. I don't know if noradrenaline turnover also increases in other hypothalamic areas or in other parts of the brain. To my knowledge, during the interval when LH surges are occurring, there is no concomitant increase in plasma concentrations of thyroid-stimulating hormone, adrenocorticotropic hormone (ACTH) or growth hormone. Yet we know that noradrenaline will trigger the release of these other pituitary hormones. Moreover, we demonstrated that blockade of noradrenaline turnover by drugs prevents LH surges and in animals in which LH surges do not occur (dioestrous rats, androgen-sterilized rats), noradrenaline turnover does not increase during the afternoon.

Palkovits: We have some indirect evidence from microdialysis studies in stressed rats. There was a beautiful stress-induced release of noradrenaline in the paraventricular nucleus, and exactly the same release in the preoptic area and in the amygdala.

When you talk about opiates, which do you mean?

Barraclough: I presume we are evaluating the role of POMC neurons on LHRH neuronal function, because Leranth et al (1988) have described ACTH immunopositive synapses on LHRH dendritic shafts. Moreover, naloxone is a μ opiate receptor antagonist and morphine is an agonist of this receptor. We have not looked at whether dynorphin affects LHRH release.

Moore: We now have the pharmacological tools to separate the opiate receptors. β-funaltrexamine (Research Biochemicals Incorporated, USA) is a μ antagonist and nor-binaltorphimine (RBI) is a κ-antagonist. They are very selective.

Palkovits: How about corticotropin-releasing factor (CRF)?

Barraclough: I don't know. Stressful stimuli will block preovulatory LH surges. According to Rivier & Rivest (1991) and others, the CRF neurons in the paraventricular region are not the neurons which seem to be affecting LHRH neuronal function. There are local CRF neurons in the medial preoptic area and these may have an important role in modulating LHRH neuronal activity.

Palkovits: In our hands the more frequent synaptic contacts on LHRH neurons are made by CRF-containing neurons. And there are plenty of CRF neurons in the preoptic area. We lesioned all potential inputs— paraventricular lesion, lower brainstem lesions, etc—without any morphological changes in preoptic CRF neurons. There was no sign of degeneration there. We concluded that most of the CRF in the preoptic area should be local.

One more question, how about neuropeptide Y? There is plenty of NPY in the preoptic area.

Barraclough: NPY given intraventricularly will cause a modest rise in LH of a similar magnitude to that seen with noradrenaline. If you combine NPY with the noradrenaline, there is an amplified response but still not a very good one. NPY also directly affects the pituitary response to LHRH. Consequently, when NPY is infused into the third ventricle, it can pass easily via the portal circulation to the pituitary gland. If you have an increase in pituitary responsiveness to LHRH in the presence of NPY, it makes it difficult to determine the principle site of action of NPY.

Theodosis: Besides opiates and CRF, are there any other peptidergic candidates directly innervating LHRH neurons?

Barraclough: I am not aware of additional anatomically identified synapses on LHRH neurons other than those I described. However, substance P terminals are present in the medial preoptic area and may contact LHRH cells. There are also neurotensin neurons in the preoptic area; they possess oestrogen receptors and seem to have a role in regulating preovulatory LH surges (Alexander et al 1989a,b).

Theodosis: Do you have any idea of the relative importance of each of these innervations?

Barraclough: There's no way of telling the hierarchical order of control. People quote percentages of synapses on LHRH neurons, but the relative importance of each is not known. The sequence of neurotransmitter-mediated events that has to occur before LHRH surges is shown by our studies. Noradrenaline is essential for the preovulatory release of LHRH. The release of this neurotransmitter in adequate amounts to activate LHRH neurons requires that GABA and opiate inhibitory influences on the presynaptic release of noradrenaline decline. There must also be a withdrawal of postsynaptic inhibitory influences on LHRH neurons, so that these neurons become more responsive to noradrenaline.

Jarry et al (1988) have done push–pull studies in which they measured moment-to-moment changes in GABA release in the preoptic area before and during steroid-induced LH surges. They reported that GABA declines simultaneously with an increase in noradrenaline; these observations support our results with GABA receptor antagonists.

What I find disconcerting is that the same A1 electrical stimuli that increase hypothalamic noradrenaline also increase GABA secretion in the preoptic area (Herbison et al 1990). Where this GABA is coming from and whether the GABA neurons that respond to A1 activation synapse on LHRH neurons, we don't know. The preoptic area contains many GABA neurons and, presumably, they don't all synapse with LHRH cells. Also, not all of these GABA neurons possess oestrogen receptors. Whether it is only the oestrogen receptor-bearing neurons that affect LHRH neuronal activity while other GABA cells do not remains to be resolved.

Clarke: The results that Jane Robinson has in the sheep (Robinson et al 1991) support the Wuttke data in so much as the fall in GABA that is seen during the preovulatory period coincides with the LH surge.

Barraclough: I love sheep.

Clarke: I wouldn't say that too loudly!

Moore: You are stating that the lack of effect of A1 stimulation was due to high inhibitory tone of either opiate or GABA neurons on the noradrenaline neurons. Is that correct?

Barraclough: Perhaps. Herbison et al (1990) found that similar electrical stimulation of A1 neurons releases noradrenaline within the preoptic area but only in very small amounts. They see a rise from 1 to about 3 pg/10 µl of dialysis fluid. We can't even measure such small amounts by HPLC in our laboratory. If there are LHRH neurons which have limited responsiveness and you infuse these very small amounts of noradrenaline, you won't see a response. A1 stimulation seems to release noradrenaline but not in sufficient quantities to activate resting LHRH neurons. However, if we partially depolarize GnRH neurons and repeat the experiment, the small amounts of noradrenaline are capable of amplifying the response.

Moore: You are injecting more noradrenaline than is present in the whole brain. Why didn't you inject the antagonists, bicuculline or naloxone, concurrently with stimulation of the A1 neuron rather than injecting noradrenaline?

Barraclough: The amount of noradrenaline we infuse into the ventricle and the amount which reaches LHRH neurons and its distribution within the hypothalamus are not known. However, I agree with you and we are doing those experiments at the moment.

What concerns me with the study you suggest is that we may have to treat with both opiate and GABA receptor antagonists to release adequate amounts of noradrenaline to activate LHRH neurons. I am not sure what non-specific effects such combined drug treatment would have in rats anaesthetized with chloral hydrate. It is not easy to duplicate experimentally the events that occur spontaneously in animals at the time of the LH surge.

Sawchenko: The major afferent influences on this A1 cell group appear to arise from the cardiovascular system. There is also evidence for some nociceptive influences on it, conveyed by ascending spinal systems. If the A1 cell group is providing the trigger for the surge, then what do you conceive as driving the trigger?

Barraclough: I wish I knew. It is true that these A1 neurons don't just regulate LHRH neurons, they also play an important role in regulating the cardiovascular system. I don't even know if activity in these noradrenaline cells increases at the time of increased hypothalamic noradrenaline turnover at pro-oestrus. They could fire at a steady rate and the release of noradrenaline could be modulated in the preoptic area terminal field by changes in the opiate and GABA systems. We still have to decipher some of these events.

Clarke: All of the studies you described were in ovariectomized oestrogen-treated animals and I appreciate the reasons for that. Have you got similar data from animals treated with chloral hydrate, for example, in the preovulatory period and then given these treatments?

Barraclough: We have not used pro-oestrous rats because of the expense of performing these studies. However, I really don't believe that we would see many differences. All our experiments are done in the morning, prior to the time the preovulatory LH surges occur.

Zoeller: If NMDA produces a very quick increase in LHRH message levels, suggesting a temporal link between secretory activity and LHRH transcription, why would Keoxifene, which blocks the surge, produce an increase in LHRH message?

Barraclough: The NMDA studies I described were done in male rats; we don't know if the same changes in LHRH mRNA would occur in ovariectomized oestrogen-treated females. The anti-oestrogen studies were done in 7-day ovariectomized rats. Keoxifene was placed in the medial preoptic area shortly after inserting oestrogen capsules subcutaneously. The purpose of these studies was to deprive local preoptic neurons of oestrogen in an otherwise completely oestrogenized rat. The LHRH mRNA levels two days after implantation of the Keoxifene were similar to those reported by you (Zoeller et al 1988) for 7-day ovariectomized rats that had not received oestrogen. However, rats treated with antioestrogen do not resemble those ovariectomized animals in other respects—plasma levels of LH are low in the former and high in the latter.

Zoeller: Do you think you could put Keoxifene anywhere apart from the preoptic area and have a similar effect?

Barraclough: I don't know. When we put Keoxifene into the lateral preoptic area it had no effect on steroid-induced LH surges. These effects were obtained only when the implants were placed in the medial preoptic regions. It would be interesting to see what effect Keoxifene would have in regions such as the ventral lateral medulla (A1 cells) or periarcuate nucleus (POMC neurons).

Young: Naomi Rance and I have shown that infundibular neurons in the human are oestrogen responsive—they appear to enlarge and have increased levels of substance P and neurokinin B transcripts postmenopausally (Rance & Young 1991). They also contain oestrogen receptor transcripts (Rance et al 1990). Do you think that these neurons, which are the equivalent of the arcuate nucleus in the rat, may be involved in this circuitry?

Barraclough: One could postulate an influence of these peptides on LHRH terminal fields in the median eminence rather than on the cell body. However, I believe that normally, depolarization of LHRH neurons occurs at the cell body, not at the terminal. This does not preclude the possibility that modulation of LHRH release can also occur within the median eminence.

Young: But these neurons may also project to the preoptic area. There are, as you mentioned, contacts between substance P and LHRH-containing neurons.

Plant: What part of that circuitry accounts for the circadian neurogenic trigger for the LH surge? You can give oestrogen to the rat but the surge will occur only at a particular time of day. It is keyed into the light cycle. Is that circuitry included in your model?

Barraclough: How light influences the time of surge, I don't know.

Theodosis: I find the system very complicated. I still don't understand how the oestrogen-responsive cells take part in this control.

Barraclough: The activity of these neurotransmitter systems could be regulated by oestradiol. What oestradiol does after it enters brain neurons and how it alters neuronal activity, we don't know.

Pfaff: For the GABA neurons, which Flugge et al (1986) showed to have oestrogen receptors, you can either hypothesize that oestrogen keeps up the GABAergic tone to prevent leakage, or the opposite, that in pro-oestrus disinhibition allows LHRH release.

Barraclough: But how?

Pfaff: I don't know.

Clarke: The NMDA study that you showed was such that there was an increase in LHRH mRNA 15 minutes after treatment, which surprised me. I would like to know whether that is a genomic effect or whether it is something to do with message stability or some other sort of transcriptional efficiency. Is 15 minutes considered to be a fast effect on message levels?

Barraclough: 15 minutes is fast.

Zoeller: I think it would be hard to envision changes in stability having that quick an effect.

Mellon: The only way I can think of would be if the LHRH message had an extremely short half-life. In our hands it appears to have a somewhat long half-life of at least a few hours.

References

Alexander MJ, Dobner PR, Miller MA, Bullock BP, Dorsa DM, Leeman SE 1989a Estrogen induces neurotensin/neuromedin N messenger ribonucleic acid in a preoptic nucleus essential for the preovulatory surge of luteinizing hormone in the rat. Endocrinology 125:2111–2117

Alexander MJ, Mahoney PD, Ferris CF, Carraway RE, Leeman SE 1989b Evidence that neurotensin participates in the central regulation of the preovulatory surge of luteinizing hormone in the rat. Endocrinology 124:783–788

Flugge G, Oertel WH, Wuttke W 1986 Evidence for estrogen-receptive GABAergic neurons in the preoptic/anterior hypothalamic area of the rat brain. Neuroendocrinology 43:1–5

Herbison AE, Heavens RP, Dyer RG 1990 Oestrogen modulation of excitatory A1 noradrenergic input to rat medial preoptic gamma aminobutyric acid neurones demonstrated by microdialysis. Neuroendocrinology 52:161–168

Jarry H, Perschl A, Wuttke W 1988 Further evidence that preoptic hypothalamic GABAergic neurons are part of the GnRH pulse and surge generator. Acta Endocrinol 118:573–579

Leranth C, MacLusky NJ, Shanabrough M, Naftolin F 1988 Immunohistochemical evidence for synaptic connections between pro-opiomelanocortin-immunoreactive axons and LH-RH neurons in the preoptic area of the rat. Brain Res 449:167–176

Rance NE, Young WS III 1991 Hypertrophy and increased gene expression of neurons containing neurokinin-B and substance-P messenger ribonucleic acids in the hypothalami of postmenopausal women. Endocrinology 128:2239–2247

Rance NE, McMullen NT, Smialek JE, Price DL, Young WS III 1990 Postmenopausal hypertrophy of neurons expressing the estrogen receptor gene in the human hypothalamus. J Clin Endocrinol & Metab 71:79–85

Rivier C, Rivest S 1991 Effect of stress on the activity of the hypothalamic-pituitary-gonadal axis: peripheral and central mechanisms. Biol Reprod 45:523–532

Robinson JE, Kendrick KM, Lambert CE 1991 Changes in the release of γ-aminobutyric acid and catecholamines in the preoptic/septal area prior to and during the preovulatory surge of luteinizing hormone in the ewe. J Neuroendocrinol 3:393–399

Zoeller RT, Seeburg PH, Young WS III 1988 In situ hybridization histochemistry for messenger ribonucleic acid (mRNA) encoding gonadotropin-releasing hormone (GnRH): effect of estrogen on cellular levels of GnRH mRNA in female rats. Endocrinology 100:2570–2577

The neurobiology of primate puberty

Ayesh D. Perera and Tony M. Plant

University of Pittsburgh School of Medicine, Department of Physiology, Pittsburgh, PA 15261, USA

Abstract. In higher primates the protracted prepubertal phase of development is occasioned by a mechanism that suppresses pulsatile hypothalamic gonadotropin-releasing hormone (GnRH) secretion from late infancy until the onset of puberty and thereby guarantees, in the juvenile, the quiescence of the pituitary–gonadal axis. Studies from our laboratory have employed the rhesus monkey, a representative higher primate, as an experimental paradigm. GnRH release has been measured using luteinizing hormone secretion by the *in situ* pituitary as a bioassay for the hypothalamic hormone. The nature of the prepubertal brake on pulsatile GnRH release in the monkey has been probed using physiological, neuroanatomical and neuropharmacological approaches. Such studies have led to the view that the prepubertal hiatus in pulsatile GnRH release results from a withdrawal in late infancy of a synchronized frequency-facilitated afferent neural input to the GnRH network, which in all other respects appears to exhibit properties identical to those in the postpubertal animal. The mechanism timing the onset of puberty, i.e. that responsible for the reactivation of synchronous activity in the GnRH network, is posited to be under the control of a central neural time- or growth-tracking device.

1992 Functional anatomy of the neuroendocrine hypothalamus. Wiley, Chichester (Ciba Foundation Symposium 168) p 252–267

The drive to the pituitary–gonadal axis originates within the central nervous system and is transmitted to the gonadotrophs of the adenohypophysis by an intermittent discharge of gonadotropin-releasing hormone (GnRH) into the hypophysial portal circulation. The neural mechanism responsible for generating this intermittent hypophysiotrophic signal appears to reside within the basal hypothalamus and is known as the hypothalamic GnRH pulse generator (see Plant 1986). Because the pulsatile release of GnRH results in a corresponding train of discharges of luteinizing hormone (LH) from the anterior pituitary, measurement of moment-to-moment changes in the concentration of this gonadotropin in the peripheral circulation provides a reliable index of GnRH pulse generator activity (see Plant 1986). The purpose of this paper is to examine the mechanisms that govern the ontogeny of this neuroendocrine system in higher primates and that dictate, therefore, the timing of the onset of puberty in these species.

The ontogeny of GnRH pulse generator activity

In higher vertebrates, the cells that synthesize and secrete GnRH, in contrast to other neurons in the central nervous system, originate outside the brain in the olfactory area and migrate to the hypothalamus during fetal development (Schwanzel-Fukudu & Pfaff 1989, Wray et al 1989). There are about 2000 GnRH neurons in the rhesus monkey (Silverman et al 1982) and they reach their adult pattern of distribution (the septo–preoptico–infundibular pathway) between Days 50 and 135 of gestation (Ronnekliev & Resko 1990). In the primate, unlike the rodent, the functional ontogeny of the GnRH pulse generator, as reflected by the secretion of gonadotropin, does not show a gradual maturation from early fetal life until puberty. Instead, full development of this neuroendocrine system occurs during the prenatal period. The amount of circulating LH in the primate fetus starts to rise by approximately Day 80 and reaches a peak between Days 100 and 140 (Ellinwood & Resko 1980). Jaffe (1989) has confirmed the early maturity of the GnRH pulse generator in primates by the demonstration of pulsatile LH release in the fetal rhesus monkey. During the last trimester of pregnancy, however, there is a decline in circulating gonadotropin levels in the fetus because of the inhibitory action of sex steroids, whose levels are rising during this phase of development as a result of increased production by the fetoplacental unit. After the removal of the inhibitory action of fetoplacental steroids at birth, plasma LH levels rise in the new-born to concentrations comparable to those observed in the adult (Plant 1988). The increase in gonadotropic stimulation to the infant testis elicits an adult pattern of testosterone secretion. At this stage of development, orchidectomy results in a marked hypersecretion of the gonadotropic hormones, which is characterized by frequent (hourly) LH release—a response to castration indistinguishable from that observed in the adult (Plant 1988). Such a finding underlines the view that in primates the GnRH pulse generator attains full maturity during perinatal development. (The hypothalamic GnRH pulse generator of the infantile female, in contrast to that of the neonatal male, appears not to operate at the adult circhoral frequency (Plant 1988).) The adult hypophysiotrophic drive to the gonadotrophs of the infant, however, is not sustained, and the transition from the neonatal to the juvenile stage of development is associated with a decline in gonadotropin secretion (Fig. 1). The GnRH pulse generator is held in check throughout the remainder of the juvenile phase of development and is subsequently reactivated at puberty (Watanabe & Terasawa 1989). The foregoing considerations, together with the demonstration that there are no inherent pituitary or gonadal deficits in either the prepubertal male or female monkey (see Plant 1988), has underlined the role of the hypothalamus in the control of the timing of the onset of puberty in primates.

From a comparative point of view, it is of interest to ask whether the ontogenetic pattern of the GnRH pulse generator in primates is observed in other

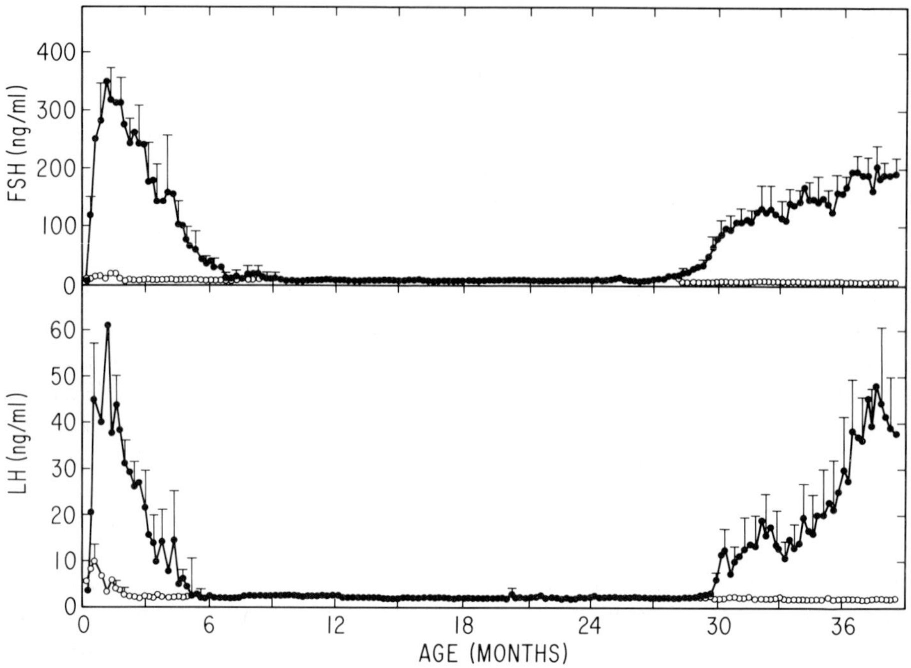

FIG. 1. The postnatal development pattern of hypothalamic gonadotropin-releasing hormone pulse generator activity in agonadal male rhesus monkeys, as reflected by the circulating gonadotropin levels, from birth until three years of age (●---●). Data for gonadally intact animals (○---○) are shown for comparison. FSH, follicle-stimulating hormone; LH, luteinizing hormone. From Plant (1985), with permission.

mammals that reach puberty at a relatively advanced age (the elephant is a species that springs to mind). Unfortunately, data on this issue are sparse. A notable exception is a recent study by Wildt and colleagues of the African lion, a species in which the male reaches puberty at three years of age (Brown et al 1991). Administration of a single GnRH injection to the prepubertal male resulted in a dramatic rise in plasma LH concentrations, without a concomitant increase in testicular tetosterone secretion. Because the ability of the pituitary to respond to GnRH indicates that the gonadotroph of the juvenile lion is exposed to a functional hypophysiotrophic stimulus, these observations suggest that the limiting factor to the onset of puberty in this species is of testicular origin.

The prepubertal hiatus in GnRH pulse generator activity

The central issue confronting those studying the control of puberty in primates is the need for a better understanding of the neurobiological basis of the prepubertal hiatus in pulsatile GnRH release. Recent results from this and other laboratories have led to the view that the limiting step underlying the prepubertal hiatus in the pulsatile release of this decapeptide is not in the GnRH neuron itself. Parameters indicative of GnRH synthesis seem to be similar during the infant, juvenile and adult stages of development. Hypothalamic GnRH content remains unchanged in the male macaque throughout postnatal life (Fraser et al 1989) and recently Vician et al (1991) have shown, using *in situ* hybridization, that there is no apparent difference in the levels of mRNA encoding this peptide in the hypothalamus of the juvenile and adult. Morphologically, we have found the distribution of GnRH neurons and their peptide content, as reflected by immunocytochemistry, to be similar in juvenile and adult rhesus monkeys (G. E. Hoffman & T. M. Plant, unpublished, Fig. 2), although Cameron et al (1985) have reported an increase in GnRH neuron cross-sectional area in the postpubertal crab-eating macaque. Preliminary examination of GnRH neurons by electron microscopy has revealed no obvious developmental difference at the ultrastructural level (A. D. Perera & T. M. Plant, unpublished, Fig. 3). Finally, the view that both GnRH biosynthetic and secretory competency are maintained throughout the juvenile period has been substantiated by the ability of the network of GnRH neurons in the prepubertal monkey to sustain a pulsatile mode of release during chronic intermittent chemical excitation with repetitive intravenous injections of N-methyl-D-aspartate (NMDA) (Gay & Plant 1988). This observation also suggests that in the juvenile there is no physical barrier, such as glial ensheathment of nerve terminals, preventing GnRH release into the hypophysial portal circulation. The latter mechanism of regulating neuroendocrine function has been established for the magnocellular system (Tweedle & Hatton 1980).

It is generally recognized that the degree of biosynthesis of a given secretory product is strongly coupled to the degree of cellular activation (e.g. Uhl & Nishimori 1990). This is certainly exemplified by the gonadotroph where in the absence of a GnRH drive in the juvenile pituitary LH content is less than 4% of that found in the infant or adult and circulating levels of LH in prepubertal animals are at least an order of magnitude lower than those observed postpubertally (Fraser et al 1989). From this argument, it must be concluded that the GnRH neuron in the hypothalamus of the juvenile monkey is being driven afferently (thereby maintaining the cell's biosynthetic potential) and that something else accounts for a mode of release that is incapable of providing a robust stimulus to the gonadotroph. Pulsatile GnRH release is not completely abolished in the prepubertal primate. In humans, ultrasensitive assays for LH have revealed the presence of low frequency, low amplitude LH pulses in prepubertal children (for example, see Wu et al 1990). Also, in the agonadal prepubertal female monkey, as in girls of comparable developmental age with Turner's syndrome, circulating gonadotropin concentrations are significant (see Plant 1988). The last finding also indicates that the prepubertal brake imposed upon pulsatile GnRH is less intense in the female.

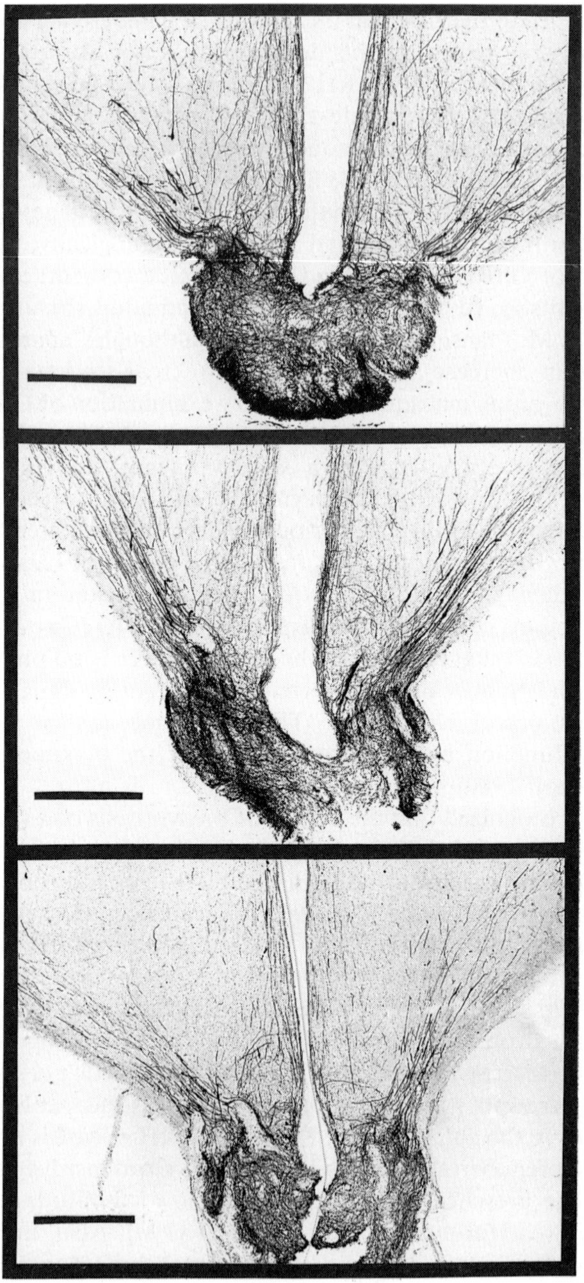

FIG. 2. Immunocytochemical identification of GnRH neurons in the mid-tuberal level of the mediobasal hypothalamus of an infant (top), juvenile (centre) and adult (bottom) male rhesus monkey. Bar = 500 μm.

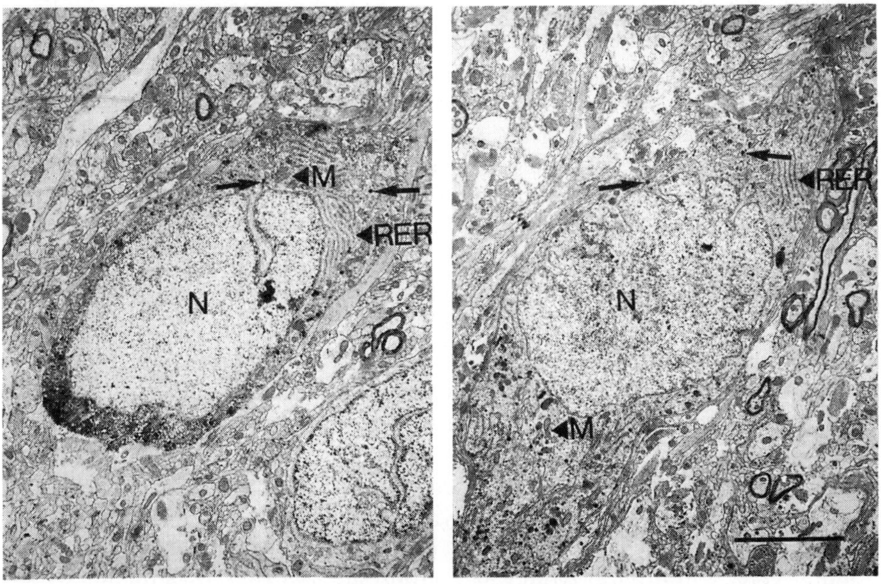

FIG. 3. A photomicrograph of a GnRH neuron from the mediobasal hypothalamus of a juvenile (left) and adult (right) castrate male rhesus monkey. Immunological reaction product (arrows) is present in the cytoplasm, but not in the nucleus (N). M, mitochondria; RER, rough endoplasmic reticulum. Bar = 2 μm.

Parenthetically, it is of interest that in other states of reversible hypogonadotropism, stimulatory afferent input to the GnRH neuronal network appears markedly compromised. In the European starling, for example, there is an almost complete disappearance of GnRH in the hypothalamus during the period of photorefractoriness that follows the breeding season (Dawson et al 1985). Such findings imply that mechanisms impairing hypophysiotropic drive to the gonadotroph in other situations may be quite different from that underlying the prepubertal hiatus in GnRH release in primates.

Two possibilities may account for the inability of the GnRH neuron in the hypothalamus of the prepubertal primate to release peptide in a mode stimulatory to the gonadotroph. Firstly, a suppression of release caused by an inhibitory synapse on the GnRH nerve terminal may be proposed. We have recently tested this hypothesis by central and peripheral administration of a variety of receptor antagonists to inhibitory neurotransmitters. Although these studies have by no means been exhaustive, receptor antagonists to γ-aminobutyric acid, serotonin, dopamine and β-endorphin have been unable to lift the prepubertal brake (Medhamurthy et al 1990, N. Mikuma, R. Medhamurthy, V. L. Gay, T. M. Plant, unpublished). We have therefore leaned towards the second possibility, namely that the hiatus is due to the removal of a pulse-generating input. The model we are proposing is directly analogous to the mechanism of synchronized frequency facilitation that accounts for the oxytocin discharge which underlies milk ejection (for review, see Bicknell 1988). According to our model, the activity of the GnRH neuron in the prepubertal state is characterized by low frequency action potentials that are able to maintain biosynthetic potential and limited release. At the onset of puberty, however, there is either an augmentation and synchronization of the existing stimulatory input or, more likely, the appearance of an additional excitatory signal coupled to the pulse generator, which leads to coordinated bursts of high frequency action potentials, resulting in the synchronized periodic release of relatively large boli of GnRH into the hypophysial portal vessels.

In the past, we have argued that the similar levels of GnRH peptide and its mRNA in the hypothalamus of juvenile and adults were indicative of comparable levels of release at these two stages of development (Gay & Plant 1988). To account for the lack of hypophysiotropic drive to the gonadotroph of the juvenile, we argued that the neurosecretory activity of the GnRH neurons of the prepubertal animal occurred asynchronously. Thus, quantitatively, the pituitary would receive a GnRH signal similar to that of the adult, but, qualitatively, stimulation would be continuous and therefore unable to elicit pulsatile release of gonadotropin. Recent evidence suggests that this hypothesis is no longer tenable. Using push–pull perfusion of the medial basal hypothalamus, Watanabe & Terasawa (1989) demonstrated that the integrated level of GnRH release in the juvenile female monkey was far less than that found in the adult: the asynchrony hypothesis called for similar integrated levels of GnRH in adults and juveniles.

The role of the NMDA receptor

The ability of certain excitatory neurotransmitter receptor agonists to elicit GnRH release in the juvenile monkey may be taken in support of the argument that the prepubertal hiatus is due to a lack of a bursting excitatory input. Repetitive administration of a specific NMDA receptor agonist (NMDA) elicits a sustained intermittent GnRH discharge in the juvenile monkey (Gay & Plant 1988), while a similar mode of administration of a non-NMDA receptor agonist (kainate) produces a progressively decrementing train of GnRH discharges (Medhamurthy et al 1992). These observations generated the idea that glutamate, acting at the NMDA receptor subtype, may be the missing component of the synchronizing facilitating pathway responsible for the drive in the adult. Furthermore, Knobil and colleagues have shown that the competitive NMDA receptor antagonist, DL-2-amino-5-phosphonopentanoic acid, can interrupt GnRH pulse generator activity in the adult monkey (Grosser et al 1989). Although there is thus strong circumstantial evidence for glutamate being the pubertal trigger, we argue that a similar line of reasoning may be put forward for other neurotransmitters.

In this regard, acute central administration of methoxamine, an α_1-adrenergic receptor agonist, can elicit a GnRH discharge in the premenarcheal rhesus monkey (Gore & Terasawa 1991), while intravenous injection of the α-adrenoreceptor antagonist, phentolamine, blocks GnRH pulsatility in the adult female (Kaufman et al 1985). Thus, a fairly strong case might also be put forward for noradrenaline being the signal responsible for the pubertal reactivation of pulsatile GnRH release.

Recently, Ojeda (1991) and Bourguignon et al (1991) have put forward models for the control of puberty in the rat that also involve excitatory amino acids. The non-competitive NMDA receptor antagonist, MK801, delays puberty in the female of this species (see Ojeda 1991). The blocking efficiency of MK801 at the NMDA receptor is increased as the receptor is activated by agonists (Honey et al 1985). This so-called 'use dependency' of MK801 led Bourguignon to suggest that glutamate tone increases at the time of puberty (Bourguignon et al 1991). Although the data for the rat are compelling, we nevertheless caution against the direct extrapolation of these findings to the primate. The non-gonadal brake imposed upon the GnRH pulse generator in the juvenile monkey—the fundamental mechanism underlying the prepubertal state in this species—is not present in the rat. This view is reflected by the finding that castration in the prepubertal rodent results in high levels of circulating gonadotropins (Ojeda et al 1990, Fraser & Plant 1989), indicating that in these species a functional frequency-facilitated and synchronized GnRH neuronal network is operational prior to puberty. We therefore suggest that the postnatal development of the GnRH pulse generator in the rat may be viewed as analogous to that occurring prenatally in primates.

Future studies

We are currently testing the frequency-facilitation hypothesis by investigating whether changes in synaptic input to the GnRH neuron occur concomitantly with the onset of puberty in the monkey. We feel that regardless of the precise mechanism controlling the onset of puberty, the reawakening of the GnRH pulse generator at this stage of development should be accompanied by some form of neural reorganization. Synaptic plasticity at the GnRH neuron has already been demonstrated in relationship to seasonal variations in hypophysiotropic drive to the gonadotroph in the ewe (Karsch et al 1987) and to changes in gonadal status in the female rhesus monkey (Witkin et al 1991).

The physiological control system that must dictate the neurobiological events discussed above is also being examined in our laboratory. That the timing of onset of puberty may be governed by a central growth-tracking device, called a 'somatometer', has been described in detail previously (see Plant 1988). An alternative model for the control of the onset of puberty based on the notion of a central time-keeping mechanism may also be envisaged. Such an idea, although intriguing, is difficult to address empirically. The recent demonstration, however, of a circannual rhythm in reproductive activity in the adult monkey (Chik et al 1989) has suggested to us the possibility that a prepubertal clock operates by 'counting' circannual rhythms: puberty in the rhesus monkey would therefore be initiated once two to three cycles had elapsed. Furthermore, Chik et al (1987) have demonstrated that seasonal cycles in reproductive activity of adults can entrain to a 32 week 'year' comprising cycles of 16 weeks of short days followed by 16 weeks of long days. If such a photoperiodic treatment represents an entrainment of their endogenous circannual rhythm to a 32 week year, the intriguing possibility exists that juveniles could be prematurely brought into puberty by exposing them to two to three abbreviated years, i.e. in less than two geophysical years. We see this as a good test to ascertain if such a clock mechanism underlies the control of puberty in the rhesus monkey. It would be necessary, of course, to establish that the circannual rhythm is present prepubertally, and that it is entrainable by photoperiodic cues at this stage of development. The expression of a circannual rhythm of prolactin secretion in the juvenile male monkey (Plant 1985) suggests that the former is true.

Summary

In primates, the hypothalamic GnRH pulse generator, which provides the drive to the pituitary–gonadal axis in adulthood, appears to be functional in the infant. An immediate transition from infancy to a state of sexual maturity, however, does not occur in these species. Instead, a brake is imposed upon pulsatile GnRH release during late infancy and this restraint is maintained throughout the prepubertal period. According to this model, the initiation of puberty in primates may be viewed as the reawakening of a mature hypothalamic GnRH pulse generator. Although the neurobiology of the prepubertal hiatus in pulsatile GnRH release remains an intriguing mystery, the withdrawal of a synchronized frequency-facilitated input to the GnRH neuron during prepubertal development may be proposed to account for the diminished hypophysiotropic drive to the gonadotroph during this phase of development. At the control systems level, the removal of the prepubertal brake on pulsatile GnRH release may be governed by a central neural time-keeping or growth-tracking device.

Acknowledgement

The work in the authors' laboratory was supported by NIH grants HD08610, HD13254 and HD16851.

References

Bicknell RJ 1988 Downstream consequences of bursting activity in oxytocin neurons. In: Leng G (ed) Pulsatility in neuroendocrine systems. CRC Press, Boca Raton, FL, p 61–74

Bourguignon J-P, Gerard A, Fawe L, Alvarez-Gonzalez ML, Franchimont P 1991 Neuroendocrine control of the onset of puberty: secretion of gonadotrophin-releasing hormone from rat hypothalamic explants. Acta Paediatr Scand Suppl 327:19–25

Brown JL, Bush M, Packer C et al 1991 Developmental changes in pituitary-gonadal function in free-ranging lions (*Panthera leo leo*) of the Serengeti Plains and Ngorongoro Crater. J Reprod Fertil 91:29–40

Cameron JL, McNeill TH, Fraser HM, Bremner WJ, Clifton DK, Steiner RA 1985 The role of endogenous gonadotropin-releasing hormone in the control of luteinizing hormone and testosterone secretion in the juvenile male monkey, *Macaca fascicularis*. Biol Reprod 33:147–156

Chik CL, Almeida OFX, Libré E, Booth JD, Renquist D, Merriam GR 1987 Photoperiod-driven changes in reproductive function in rhesus monkeys. Endocrinology (suppl) 120:abstr 881

Chik CL, Almeida OFX, Libré EA, Merriam GR 1989 Effect of continuous long days on the reproductive function of male rhesus monkeys: evidence for endogenous circannual rhythms. Endocrinology (suppl) 124:abstr 166

Dawson A, Follett BK, Goldsmith AR, Nicholls TJ 1985 Hypothalamic gonadotrophin-releasing hormone and pituitary and plasma FSH and prolactin during photostimulation and photorefractoriness in intact and thyroidectomized starlings (*Sturnus vulgaris*). J Endocrinol 105:71–77

Ellinwood WE, Resko JA 1980 Sex differences in biologically active and immunoreactive gonadotropins in the fetal circulation of rhesus monkeys. Endocrinology 107:902–907

Fraser MO, Pohl CR, Plant TM 1989 The hypogonadotropic state of the prepubertal male rhesus monkey (*Macaca mulatta*) is not associated with a decrease in hypothalamic gonadotropin-releasing hormone content. Biol Reprod 40:972–980

Fraser MO, Plant TM 1989 Further studies on the role of the gonads in determining the ontogeny of gonadotropin secretion in the guinea-pig (*Cavia porcelus*). Endocrinology 125:906–911

Gay VL, Plant TM 1988 Sustained intermittent release of gonadotropin-releasing hormone in the prepubertal male rhesus monkey induced by N-methyl-DL-aspartic acid. Neuroendocrinology 48:147–152

Gore AC, Terasawa E 1991 A role for norepinephrine in the control of puberty in the female rhesus monkey, *Macaca mulatta*. Endocrinology 129:3009–3017

Grosser PM, Williams CL, Thalabard JC, Nishihara M, Brody S, Knobil E 1989 N-methyl-D,L-aspartate (NMDA) receptor blockade of the hypothalamic gonadotropin-releasing hormone (GnRH) pulse generator activity in the rhesus monkey. Endocrinology (suppl) 124:abstr 942

Honey CR, Miljkovic Z, MacDonald JF 1985 Ketamine and phencyclidine cause a voltage dependent block of responses to L-aspartic acid. Neurosci Lett 61:135–139

Jaffe RB 1989 Fetal neuroendocrinology. In: Mancuso S (ed) Achievements in gynecology. Karger, Basle vol 17:104–110

Karsch FJ, Lehman MN, Silverman A-J 1987 Seasonal plasticity in synaptic input to luteinizing hormone-release hormone (LHRH) neurons in the sheep. Soc Neurosci Abstr 13:1527

Kaufman J-M, Kesner JS, Wilson RC, Knobil E 1985 Electrophysiological manifestation of luteinizing hormone-releasing hormone pulse generator activity in the rhesus monkey: influence of α-adrenergic and dopaminergic blocking agents. Endocrinology 116:1327–1333

Medhamurthy R, Gay VL, Plant TM 1990 The prepubertal hiatus in gonadotropin secretion in the male rhesus monkey (*Macaca mulatta*) does not appear to involve

endogenous opioid peptide restraint of hypothalamic gonadotropin-releasing hormone release. Endocrinology 126:1036–1042
Medhamurthy R, Gay VL, Plant TM 1992 Repetitive injections of L-glutamic acid, in contrast to those of N-methyl-D,L-aspartic acid, fail to elicit sustained hypothalamic GnRH release in the prepubertal male rhesus monkey (*Macaca mulatta*). Neuroendocrinology 55:660
Ojeda SR 1991 The mystery of mammalian puberty: how much more do we know? Perspect Biol Med 34:365–383
Ojeda SR, Smith-White S, Advis JP, Andrews WW, Aguado LI 1990 First preovulatory gonadotropin surge in the rodent. In Grumbach MM, Sizonenko PC, Aubert ML (eds) Control of the onset of puberty. Williams and Wilkins, Baltimore, MD, p 156–177
Plant TM 1985 A study of the role of the postnatal testes in determining the ontogeny of gonadotropin secretion in the male rhesus monkey (*Macaca mulatta*). Endocrinology 116:1341–1350
Plant TM 1986 Gonadal regulation of hypothalamic gonadotropin-releasing hormone release in primates. Endocr Rev 7:75–88
Plant TM 1988 Puberty in primates. In: Knobil E, Neill J (eds) The physiology of reproduction. Raven Press, New York, vol 2:1763–1788
Ronnekleiv OK, Resko JA 1990 Ontogeny of gonadotropin-releasing hormone-containing neurons in early fetal development of rhesus macaques. Endocrinology 126:408–511
Schwanzel-Fukuda M, Pfaff DW 1989 Origin of luteinizing hormone-releasing hormone neurons. Nature (Lond) 338:161–164
Silverman A-J, Antunes JL, Abrams GM et al 1982 The luteinizing hormone-releasing hormone pathways in rhesus (*Macaca mulatta*) and pigtailed (*Macaca nemestrina*) monkeys: new observations on thick, unembedded sections. J Comp Neurol 211:309–317
Tweedle CD, Hatton GI 1980 Evidence for dynamic interactions between pituicytes and neurosecretory axons in the rat. Neuroscience 5:661–667
Uhl GR, Nishimori T 1990 Neuropeptide gene expression and neural activity: assessing a working hypothesis in nucleus caudalis and dorsal horn neurons expressing preproenkephalin and preprodynorphin. Cell Mol Neurobiol 10:73–98
Vician L, Adams LA, Clifton DK, Steiner RA 1991 Pubertal changes in proopiomelanocortin and gonadotropin-releasing hormone gene expression in the brain of the male monkey. Mol Cell Neurosci 2:31–38
Watanabe G, Terasawa E 1989 *In vivo* release of luteinizing hormone releasing hormone increases with puberty in the female rhesus monkey. Endocrinology 125:92–99
Witkin JW, Ferin M, Popilskis SJ, Silverman A-J 1991 Effects of gonadal steroid on the ultrastructure of GnRH neurons in the rhesus monkey: synaptic input and glial apposition. Endocrinology 129:1083–1092
Wray S, Nieburgs A, Elkabes S 1989 Spatiotemporal cell expression of luteinizing hormone-releasing hormone in the prenatal mouse: evidence for an embryonic origin in the olfactory placode. Dev Brain Res 46:309–318
Wu FCW, Butler GE, Kelnar CJH, Sellar RE 1990 Patterns of pulsatile luteinizing hormone secretion before and during the onset of puberty in boys: a study using an immunoradiometric assay. J Clin Endocrinol & Metab 70:629–637

DISCUSSION

Grosvenor: Tony, could you tell us about the early nutrition of these rhesus monkeys? Were they breast-fed or bottle fed?

Plant: All the data I showed were from monkeys that were left with their mothers and breast fed. The infants were not weaned until the middle of the hiatus phase.

Grosvenor: There is a lot of LHRH (GnRH) in milk; I have often wondered what it is doing there. It is possible that the elevated LH you see neonatally is triggered by LHRH in maternal milk.

Plant: It is a possibility, but I don't think it's the case because the animals in which we examined LH time courses were temporarily taken from their mothers and had been away from the breast for as long as 6–7 hours (Plant 1982). It is difficult to see how that pulsatile release of LH could be an effect of the antecedent breast-feeding. A lot of things get into milk.

Zoeller: You concluded that there is no tonic inhibition being imposed on the GnRH neuron during the prepubertal period, what was your argument for that?

Plant: Only that GnRH content appears not to be suppressed; the GnRH gene seems to be expressed, according to Steiner's *in situ* data (Vician et al 1991). Moreover, these neurons respond immediately the first time they are excited. It seems to me less likely that there is a profound inhibitory influence on this system. If there was, one would expect the system to be down-regulated. Our experiments were done a year after the GnRH pulse generator in the hypothalamus of the prepubertal monkey had been switched off.

Zoeller: When we looked at the level of GnRH mRNA during the rat oestrous cycle, we found that it is high during dioestrus, Day 1 and Day 2, when LH levels are very low. Then, between the evening of dioestrus Day 2 and the morning of pro-oestrus, there's a tremendous drop in GnRH message level; this recovers throughout the day of pro-oestrus to go back up to higher levels by that evening. If you compare oestrogen-treated and ovariectomized animals at certain times, you see what seems to be the opposite of what you would expect—GnRH message is low when there are periods of elevated LH.

Plant: What strikes me is that fundamental difference between the gonadotroph of the prepubertal animal and the GnRH neuron at the same stage of development. The former, which is clearly not receiving any stimulus, is totally down-regulated, it has no GnRH receptors, very little LH and FSH, low levels of message for the gonadotropins and no secretion. It takes weeks to get that gonadotroph to function like a cell in the adult. To make a naive comparison, the peptide-secreting GnRH cell above the gonadotroph is totally different.

Gainer: Related to this, have you tried using bicuculline (a $GABA_A$ receptor antagonist) in these monkeys?

Plant: Yes. We can induce seizure but we can't induce any GnRH release from the hypothalamus of the prepubertal monkey with bicuculline.

Gainer: So it is not a GABA shut down.

Plant: No, nor is it endogenous opioid peptides. We have given naloxone in several modes but have not removed the prepubertal brake on GnRH release.

But one still reads in the literature that this prepubertal break is an endorphin-mediated event.

I don't mind what it is, but it would be nice to get a handle on what is responsible for the absence of the pulsatility.

Gainer: What about the synchronization taking place at the level of the median eminence?

Plant: Sure, I don't exclude that.

Gainer: How would one evaluate it? Could all these cells be somehow coupled at the level of median eminence? Then you could have decoupling. Are there gap junctions at the median eminence between these cells?

Plant: Obviously, there must be a morphological concomitant of the apulsatile mode in GnRH function during prepubertal development in primates. We are beginning to analyse this in collaboration with Paul Goldsmith using both light microscopy and electron microscopy. I am glad I am not an anatomist because we have been looking for synapses on GnRH cell bodies and we haven't found one yet.

Clarke: Tony, you say that the whole system in the primate is different to that in other species. But Doug Foster has basically shown that in the sheep in the prepubertal period there is an enhanced negative feedback response to oestrogen. In the primate, you castrate the animal and it has sustained elevated concentrations of gonadotropins. Even though in the juvenile period there's a very low level of testosterone, it may be having a very powerful negative feedback effect. Is that right?

Plant: We take the gonads and the adrenals out and the developmental pattern of gonadotropin secretion doesn't change qualitatively.

Clarke: But you still get pulsatile LH secretion in the castrate?

Plant: In infancy, yes, but not in the prepubertal monkey (6–30 months of age).

Clarke: So it's not the adrenal.

Plant: Then you might argue that it's in the diet—and I would say you prove that! Interestingly, the adrenalectomized animals, even though we treat them with cortisol and mineralocorticoids, don't grow as well, as judged by body weight measurements and some linear measurements, but they have puberty roughly at the same time as adrenally intact animals, which is curious.

Nagy: Tony, I would like to ask your opinion of the possible role of the pituitary gland itself. Which falls asleep first, the hypothalamus or the pituitary gland?

Plant: We haven't actually asked that question. I assume the hypothalamus, but we haven't measured GnRH release during that transition.

Nagy: We always regard the pituitary gland as just a passive target organ.

Plant: Certainly, I see it as a slave to the brain.

Clarke: It may not be that simple. There are very nice data from Mark Wise on bull calves. He shows quite clearly that the GnRH pulse generator starts up way before the emergence of LH pulses (Rodriguez & Wise 1989). So there is a period, at least in cattle, postnatally when there are GnRH pulses without the response at the pituitary level, for about 12 weeks.

Plant: I think that would support the idea that cattle have a different pubertal control system to the primate, where you impose a pulsatile pattern of GnRH prepubertally and everything blooms.

Clarke: Yes, it is just that not all pituitaries are slaves!

Mellon: Can one propose a model, where although the GnRH neurons are spread rather widely they do make contact with each other and the network is what is responsible for the synchrony? In that case, these contacts might be separated during the hiatus period and then re-established.

Plant: Yes, we want to look at glia in our morphological studies. In the sheep there are seasonal changes in glial ensheathment that are correlated to changes in hypophysiotropic drive (Karsch et al 1987).

Poulain: Are there any data on *in vitro* release during this prepubertal period?

Plant: We have none. I am not aware that anybody else has. It would be expensive, but anything is costly, so if it's a good question, the experiment can be done.

Poulain: You assume that the pulse generator is quiet because of a lack of drive, but that it is ready for activity. In an *in vitro* preparation, I suspect that a stimulus, such as K^+ stimulation or electrical stimulation, could be used to test this hypothesis. You could show that there is no drive for the neurons but they are ready to respond.

Plant: We would argue that NMDA-induced LH release *in vivo* makes the same point. Nevertheless, it would be nice to have a hypothalamus in a dish to look at what's happening.

Lightman: Tony, in precocious puberty in humans, particularly in males, almost any form of space-occupying lesion in the hypothalamus—it doesn't have to be in the specific area making GnRH—can bring on precocious puberty. I find it hard to imagine how such a mass could itself synchronize pulsatile GnRH release. It would be much easier and more logical to interpret these data by postulating that the tumours interrupt some inhibitory afferent input(s), which would normally be blocking the onset of puberty.

Plant: Yes, it is intuitively easier to think of a removal of an inhibition. But, if you think in terms of black boxes, you could postulate a desynchronizing black box and that a space-occupying lesion would interrupt that input.

Lightman: Pineal tumours do this reasonably commonly.

Plant: We have negative data on this issue. We have removed pineal glands from neonatal monkeys and we still see the brake (Plant & Zorub 1986). We have taken pineal glands from prepubertal animals and we still see the onset (Plant & Zorub 1982). Yet people still invoke the pineal as the controller of primate puberty.

Robinson: Tony, I would like to ask about the Cinderella of the pulse, i.e. the trough between pulses, the period without GnRH. If what is important is desynchronization, GnRH might be released all the time and its continuous presence might actually down-regulate the system. How much priming can you get with a low dose of continuous GnRH?

Plant: We have not done dose–response curves. If one accepts Dr Terasawa's data, where she has directly assessed, using push–pull perfusion, GnRH release rates in the prepubertal monkey (Watanabe & Terasawa 1989), the release rate is not equivalent to an integrated adult mean, which would be consistent with the idea of desynchrony. If each neuron in the prepubertal hypothalamus was behaving in an adult fashion in a desynchronized mode, there would be a steady integrated normal level. However, the GnRH release rate before puberty is very low, which suggests that the desynchrony is a consequence of the absence of the phasic bursting seen in the magnocellular system.

The response to the infusions of GnRH that we have given to prepubertal monkeys does not appear to be any different from those to the same infusions in an adult bearing a hypothalamic lesion that has knocked out the endogenous GnRH pulse generating system.

Robinson: What I am saying is does the pituitary like to see the absence of GnRH as well as its presence? Is it looking at some rate of change of occupation of GnRH receptor? We always focus on the high concentrations at the peak in pulsatile hormone systems. In the pulsatile release of growth hormone, for example, filling in the troughs between GH pulses has quite different effects from increasing the amplitude of the peaks. Does the pituitary just like to be without GnRH sometimes?

Plant: Not for too long.

Makara: In the animal with a lesioned hypothalamus, is there down-regulation of the responsiveness to GnRH?

Plant: If you wait long enough, yes.

Makara: Could it be steroids rather than peptides that interact with responsiveness to GnRH?

Plant: All our experiments have been done in agonadal animals. Steroids, however, will certainly interact at the pituitary level.

Makara: There are two analogies: one is the adrenal when it is deprived of ACTH. When you give ACTH to a hypophysectomized rat it takes about a week to normalize the structure of the cell as well as the function. The same doesn't happen to an ACTH cell of the rat pituitary. If we cut all the peptidergic inputs from the vasopressin and CRF cells to the median eminence and wait six weeks, fragments taken from the anterior lobe will respond normally to CRF or vasopressin. There is no 'sleeping' phenomenon in this case, but the gonadotropin cell may be completely different, even in the rat.

References

Karsch FJ, Lehman MM, Silverman A-J 1987 Seasonal plasticity in synaptic input to luteinizing hormone-releasing hormone (LHRH) neurons in the sheep. Soc Neurosci Abstr 13:425.2

Plant TM 1982 Pulsatile luteinizing hormone secretion in the neonatal male rhesus monkey (*Macaca mulatta*). J Endocrinol 93:71–74

Plant TM, Zorub DS 1982 The role of nongonadal restraint of gonadotropin secretion in the delay of the onset of puberty in the rhesus monkey (*Macaca mulatta*). J Anim Sci 55(suppl 2):43–55

Plant TM, Zorub DS 1986 Pinealectomy in agonadal infantile male rhesus monkeys (*Macaca mulatta*) does not interrupt initiation of the prepubertal hiatus in gonadotropin secretion. Endocrinology 118:227–232

Rodriguez RE, Wise ME 1989 Ontogeny of pulsatile secretion of gonadotropin-releasing hormone in the bull calf during infantile and pubertal development. Endocrinology 124:248–256

Vician L, Adams LA, Clifton DK, Steiner RA 1991 Pubertal changes in pro-opiomelanocortin and gonadotropin-releasing hormone gene expression in the brain of the male monkey. Mol Cell Neurosci 2:31–38

Watanabe G, Terasawa E 1989 *In vivo* release of luteinizing hormone releasing hormone increases with puberty in the female rhesus monkey. Endocrinology 125:92–99

Hypothalamic transplantation

H. M. Charlton

Department of Human Anatomy, Oxford University, South Parks Road, Oxford OX1 3QX, UK

Abstract. Tissue transplantation aided in formulating the neurohumoral hypothesis of anterior pituitary function. The concept of a hypophysiotropic region within the hypothalamus stemmed from experiments in which pituitary tissue was transplanted into the brain. Restoration of aberrant function of the central nervous system by transplants has been reported in two neuroendocrine models: the antidiuretic hormone-deficient Brattleboro rat and the gonadotropin-releasing hormone-deficient hypogonadal mouse. Neural transplants into the Brattleboro rat result in the survival of axons containing antidiuretic hormone but reversal of the physiological defect has not been confirmed. In the hypogonadal mouse grafts of preoptic area tissue into the third ventricle have restored pituitary hormone synthesis and secretion and gonadal activity, leading to nearly normal reproductive function. The gonadotropin-releasing hormone axons specifically innervate the median eminence of the hypothalamus, their normal target, which raises interesting questions of neurobiological graft/host interactions. The hpg model has been used to investigate factors affecting graft survival; by suitable immunosuppression it has been possible to reverse the hypogonadism with grafts of rat preoptic area tissue. Perhaps the most dramatic recent development has been the restoration of circadian rhythmicity to suprachiasmatic nucleus-lesioned hamsters by grafts of similar tissue. The rhythmicity restored is typical of the donor tissue.

1992 Functional anatomy of the neuroendocrine hypothalamus. Wiley, Chichester (Ciba Foundation Symposium 166) p 268–286

The hypothalamus acts as an integration centre controlling the autonomic, endocrine and behavioural responses essential for the survival of the individual animal and of the species. It is central to the control of water balance, thermoregulation, food intake, aggression, reproductive processes and sleep–wake cyclic activity (Swanson 1987).

The role of the hypothalamus in the modulation of neuroendocrine control mechanisms and the scientific detective work which culminated in the neurohumoral hypothesis of the control of anterior pituitary function were described in the Dale Lecture delivered by Geoffrey Harris (1972). The importance of the neurovascular link between the hypothalamus and the anterior pituitary was demonstrated by transplanting pituitary tissue to a variety of sites within the brain: only the grafts that received a blood supply directly from the median eminence of the hypothalamus could reverse the physiological effects of hypophysectomy.

Hypothalamic peptides that control the synthesis and release of growth hormones—the gonadotropic hormones, thyroid-stimulating hormone and adrenocorticotropic hormone—have all been isolated and their respective genes cloned. With the advent of immunoassays, immunocytochemistry and mRNA hybridization, our knowledge of the sites of production of these hormones and of factors controlling their synthesis and release has advanced enormously over the last decade.

Several early methods were developed to investigate the areas of the hypothalamus involved in pituitary control, for example electrical stimulation, electrolytic lesioning and deafferentation techniques using knife cuts, but several authors chose intraparenchymal transplantation of anterior pituitary tissue as a model system. Transplantation into the brain has also been used to investigate neuroendocrine mutations and recently to gain information on the role of the suprachiasmatic nucleus as a circadian pacemaker.

The hypophysiotropic area

In some of the earliest transplantation studies in which material was grafted into the central nervous system (brain), Knigge (1962) demonstrated that reproductive function could be restored in hypophysectomized rats by grafting anterior pituitary tissue into the tuberoinfundibular region of the third ventricle. Halász et al (1962, 1963, 1965) extended this work to demonstrate that in hypophysectomized rats in which pituitary tissue had been transplanted into the medial basal hypothalamus, gonadotropic hormone secretion was in the normal range, with full stimulation of the ovaries and uteri. The histological evidence suggested that direct vascularization from capillary loops of the pituitary portal vessels was not essential for the restoration of reproductive function. Apparently, the maintenance of normal thyroid and adrenal responses required contact with the median eminence. It was from these studies that the concept of a hypophysiotropic area within the medial basal hypothalamus arose.

With regard to gonadotropic hormone control, Halász et al (1965) suggested that local diffusion of releasing factors adjacent to the grafts was the probable explanation for pituitary activation. However, in studies where the median eminence was also ablated, it has been suggested that the integrity of the median eminence and of fenestrated vessels within the medial basal hypothalamus is essential for the stimulation of pituitary grafts (Setalo et al 1977).

In contrast, Shechter et al (1987) have claimed that pituitary grafts into the cerebral cortex possess active gonadotrophs. They postulate that hypophysectomy could result in excessive hypothalamic secretion of gonadotropin-releasing hormone (GnRH), sufficient to activate the cortical grafts. They did not investigate the reproductive status of the recipient animals. This area of brain tissue transplantation is still unresolved and perhaps deserves further investigation with more modern methodology.

The Brattleboro rat

The roles of the posterior pituitary gland hormones, oxytocin and vasopressin, in the control of milk ejection and water balance, respectively, are well documented (Swanson 1987). The Brattleboro rat was discovered and characterized by Valtin (1976, 1982); it is an animal model for vasopressin-deficient diabetes insipidus. In this mutant, a single G residue is missing in the glycoprotein-coding region of the gene, which results in a frame shift and the production of an aberrant vasopressin messenger RNA that cannot be translated into the mature hormone (Schmale & Richter 1984). There is little or no immunoreactive antidiuretic hormone (ADH) within the posterior pituitary gland and the mutants excrete large volumes of hypotonic urine. The potential for neural transplants to alleviate the physiological disorder was investigated by Gash et al (1980). The survival of ADH-containing cells and axons within the grafts was readily detected using immunohistochemical methods and the axons grew towards the median eminence. These authors reported an alleviation of the diabetes insipidus in their animals. Since then, several reports have described ADH axonal outgrowth from grafts but the physiological results have not been confirmed. Marciano et al (1989) demonstrated that fetal hypothalamic transplants can promote survival and functional regeneration of axotomized adult supraoptic vasopressinergic neurons in normal rats.

The hypogonadal mouse

The hypogonadal (hpg) mouse was first described by Cattanach et al (1977). It has a deficiency in hypothalamic GnRH, which controls the synthesis and release of luteinizing hormone (LH) and follicle-stimulating hormone (FSH). Consequently, pituitary LH and FSH content is drastically reduced and there is a failure of postnatal gonadal development. The genetic cause of this mutation was shown by Mason et al (1986) to be a deletion of two exons in the GnRH gene, resulting in an inability to transcribe and translate the GnRH gene.

This animal model lends itself to a study of the effects of transplanting neural tissue containing normal GnRH cell bodies into the brain. Surviving GnRH neurons and their axons can be detected by immunostaining; actions upon the pituitary gland can be detected by radioimmunoassay of hormone content, by the measurement of hormone-specific mRNA levels and by alterations in the ultrastructure; the action of pituitary hormones upon the gonads can also be followed.

Krieger et al (1982) grafted normal fetal hypothalamic preoptic area (POA) tissue into the third ventricle of hpg male mice and the physiological results of such grafts were remarkable. The hypogonadism of the mutant males was completely reversed, with increased pituitary LH and FSH synthesis and secretion to the normal range and the stimulation of full spermatogenesis within 60 days of grafting. Pituitary GnRH receptor and testicular LH receptor levels were also restored to the normal range (Young et al 1985).

In female hpg mice, grafts of POA tissue restored pituitary and ovarian function with uterine stimulation (Gibson et al 1984, Young et al 1985); however, the females did not demonstrate oestrous cycles and did not ovulate. The ovaries contained numerous ripe follicles and vaginal smears indicated long periods of oestrus. The gonadal steroids in the grafted females stimulated female sexual behaviour and many of them mated with normal males. One of the most surprising results in this study was that a significant proportion of females became pregnant (Gibson et al 1984). Such mated hpg females with grafts of POA tissue show a reflex release of LH (Gibson et al 1987). The failure of spontaneous oestrous cycles and therefore spontaneous GnRH surges every four or five days indicates that the grafts have not become fully integrated with host mechanisms controlling GnRH release. Nevertheless, Gibson et al (1991) have demonstrated that in many cases, release of LH in grafted animals is pulsatile. Exactly where such pulsatility is generated remains unknown: it could be inherent in the grafted neurons themselves, be transduced from the host pulse generator at the level of the median eminence, or be due to innervation of donor GnRH cells by host axons. The last suggestion seems the least probable because there is little interchange of axons between host and graft across the ventricular wall (Hodgkiss & Kelly 1990).

These physiological results indicate that, whilst grafts survive and release GnRH, in the vast majority of cases the neurons within the grafts are not capable of transducing much of the information that would impinge upon GnRH neurons within the preoptic area in normal mice. Extending these studies, the hpg mouse has proved of use in assessing possible sites of negative feedback of gonadal steroids upon LH and FSH synthesis and secretion. In normal male mice, subcutaneous implants of testosterone cause a massive reduction in pituitary LH and FSH content. If the feedback were primarily at the level of the pituitary gland, testosterone implants in hpg mice would prevent the synthesis of gonadotropic hormones elicited by GnRH injections. In fact, such injections increase pituitary gonadotropic hormone content in untreated and testosterone-treated hpg male mice, indicating a major site of negative feedback above the level of the gonadotroph (Charlton et al 1983). In the males with grafts of POA tissue, if feedback acted directly upon the GnRH neurons in the graft, then testosterone treatment would prevent, or drastically reduce, LH and FSH synthesis. But, subcutaneous testosterone pellets have no effect upon the graft-induced increase in gonadotropic hormone synthesis, thus feedback may reside in other neurons which normally impinge upon the GnRH cells in the preoptic area. Dissection of the graft tissue and its transplantation to the third ventricle presumably disrupts this input (Charlton et al 1987, 1988).

It might be considered that the grafts were merely acting as a minipump within the third ventricle, releasing GnRH into the brain. However, when POA tissue was grafted into the lateral ventricle, GnRH axons survived but did not grow towards the median eminence and there was no restoration of pituitary and gonadal function. In all cases where third ventricular grafts elicited reproductive activity, axon terminals were found innervating the median eminence; more than 95% of axons leaving the graft terminate close to the pituitary portal vessels (Silverman et al 1985). This guidance of GnRH nerve endings does not depend upon the presence of neurons within the arcuate nucleus because GnRH axons still innervate the median eminence in hpg mice in which these neurons have been destroyed by neonatal injections of the excitatory neurotoxin, monosodium glutamate (Silverman et al 1990).

What factors might be involved in this specificity of outgrowth to the basal hypothalamus? The axons terminate exactly where the portal vessels leading from the hypothalamus to the pituitary gland are found, and associated with these vessels is the pars tuberalis of the pituitary gland. Wray et al (1988) have demonstrated that GnRH axons from the POA grow towards anterior pituitary tissue when slice preparations of GnRH neurons are co-cultured with pituitary tissue. However, Daikoku et al (1990) have shown that when pituitary and POA tissue are grafted into the third ventricle together, GnRH axons tend to enclose pituitary grafts and not to infiltrate between cells. We transplanted POA and anterior pituitary tissue into the lateral ventricle of hpg mice and found that even after nine months there was no evidence of a robust innervation of the anterior pituitary tissue (H. M. Charlton & M. J. Wood, unpublished 1991). Of course, stimulation of pituitary function would require only innervation of fenestrated capillaries reaching the glandular tissue. Daikoku et al (1990) claimed that when POA tissue was grafted with medial basal hypothalamic tissue, median eminence-like structures could be visualized within the grafts and GnRH axons terminated on these structures. They did not extend their analysis to electron microscopy to demonstrate that the blood vessels associated with these structures were fenestrated. A.-J. Silverman & M. J. Gibson (personal communication) have found that exuberant outgrowth of GnRH axons towards the median eminence occurs even in hypophysectomized hpg mice, which further argues against a major role for the pituitary gland in the guidance mechanisms. In Daikoku's study and in that of Silverman & Gibson, the pars tuberalis is likely to be adhered to the median eminence. However, GnRH axons also innervate the organum vasculosum of the lamina terminalis and the subfornical organ and at both of these sites there is no adjacent pituitary tissue. Kadota et al (1990) have described the regrowth of damaged neurosecretory axons to fenestrated vessels of implanted peripheral tissues. They grafted three different types of tissue containing fenestrated capillaries—pineal, adrenal medulla and pituitary neural lobe—to the retrochiasmatic region of the hypothalamus and determined the degree of innervation of these tissues by regenerating neurosecretory nerve terminals. There was no obvious specificity of innervation of the neural lobe compared to the pineal gland and adrenal medulla. It may be that fenestrated vessels themselves possess neurotropic properties. Recently, using monoclonal antibody injections to depress helper T cell function, Honey et al (1991) stimulated full gonadal activity in hpg mice with rat POA xenografts. The rat GnRH axons also specifically innervated the mouse median eminence.

The suprachiasmatic nucleus

The suprachiasmatic nucleus (SCN) receives afferent inputs conveying environmental information which allows the animal to synchronize a variety of diurnal events, including neuroendocrine control mechanisms, with the dark–light cycle (Meijer & Rietveld 1989). For example, the activity rhythm of a hamster is normally precisely entrained to the light–dark regime under which it is housed. If the animal is then placed in constant darkness, the daily activity pattern becomes free running with a periodicity of about 24 hours and the animal becomes active at an earlier or later time each successive day. This periodicity is endogenous to the SCN; when the nucleus is destroyed there is completely random locomotor activity. Several reports have been published in which transplantation of SCN tissue resulted in the restoration of a circadian pattern of running activity (see Ralph & Lehman 1991). In these grafted animals, the locomotor rhythm did not become synchronized with light–dark cycles, therefore the SCN tissue had not become fully integrated with the hosts' normal afferent inputs. Nevertheless, the transplants imposed some regularity upon the recipient locomotor system.

There is a mutant strain of hamster in which the endogenous circadian clock is set to about 20 hours. Ralph et al (1990) grafted SCN tissue from the mutant strain into normal hamsters in which the SCN had been ablated and which were therefore arrhythmic. These animals were placed in running wheels after the operation and in 80% of the arrhythmic hosts a circadian running pattern was generated typical of the donor tissue. This use of mutant and normal strains has proved conclusively that the endogenous rhythm of the SCN is maintained in the transplants and that the host is driven by the donor tissue. There is evidence that transplants with no recognizable neural connections can still elicit a circadian response, which argues for some communication mechanism based on a humoral route. Even cell suspensions of the disrupted SCN remain capable of restoring behavioural rhythms (see Ralph & Lehman 1991). This will be an exciting area for neural transplantation research over the next few years.

Acknowledgements

I would like to acknowledge the collaboration of all those colleagues cited in joint publications and the financial assistance from The Wellcome Trust, The Medical Research

Council, The International Spinal Research Trust, The EPA Cephalosporin Trust and the National Institutes of Health, Bethesda, USA.

References

Cattanach BM, Iddon CA, Charlton HM, Chiappa SA, Fink G 1977 Gonadotropin-releasing hormone deficiency in a mutant mouse with hypogonadism. Nature (Lond) 269:338–340

Charlton HM, Halpin DMG, Iddon CA et al 1983 The effects of daily administration of single and multiple injections of gonadotrophin-releasing hormone on pituitary and gonadal function in the hypogonadal (hpg) mouse. Endocrinology 113:535–544

Charlton HM, Jones AJ, Ward BJ, Detta A, Clayton RN 1987 The effects of castration or testosterone upon pituitary function in hypogonadal (hpg) mice bearing normal fetal preoptic area (POA) grafts. Neuroendocrinology 45:376–380

Charlton HM, Porter Goff AE, Cox BS et al 1988 The hypogonadal mouse: a model for brain–pituitary–gonadal interactions. In: Imura H, Shizume K, Yoshida S (eds) Progress in endocrinology. Excerpta Medica, Netherlands vol 1:525–530

Daikoku-Ishido H, Okamura Y, Yanaihara N, Daikoku S 1990 Development of the hypothalamic luteinising hormone-releasing hormone-containing neuron system in the rat: in vivo and in transplantation studies. Dev Biol 140:374–387

Gash DM, Sladek JR Jr, Sladek CD 1980 Functional development of grafted vasopressin neurons. Science (Wash DC) 210:1367–1369

Gibson MJ, Krieger DT, Charlton HM, Zimmerman EA, Silverman AJ, Perlow MJ 1984 Mating and pregnancy can occur in genetically hypogonadal mice with preoptic area brain grafts. Science (Wash DC) 225:949–951

Gibson MJ, Moscovitz HC, Kokoris GJ, Silverman AJ 1987 Plasma LH rises rapidly following mating in hypogonadal female mice with preoptic area (POA) brain grafts. Brain Res 424:133–137

Gibson MJ, Miller GM, Silverman AJ 1991 Pulsatile luteinizing hormone secretion in normal female mice and in hypogonadal female mice with preoptic area implants. Endocrinology 128:965–971

Halász B, Pupp L, Uhlarik S 1962 Hypophysiotrophic area in the hypothalamus. J Endocrinol 25:147–159

Halász B, Pupp L, Uhlarik S, Tima L 1963 Growth of hypophysectomised rats bearing pituitary transplants in the hypothalamus. Acta Physiol Acad Sci Hung 23:287–292

Halász B, Pupp L, Uhlarik S, Tima L 1965 Further studies on the hormone secretion of the anterior pituitary transplanted into the hypophysiotrophic area of the rat hypothalamus. Endocrinology 77:343–353

Harris GW 1972 Humors and hormones. J Endocrinol 53:2–23

Honey CR, Charlton HM, Wood KJ 1991 Rat brain xenografts reverse hypogonadism in mice immunosuppressed with anti-CD4 monoclonal antibody. Exp Brain Res 85:149–152

Hodgkiss JP, Kelly JS 1990 An intracellular study of grafted and *in situ* preoptic area neurones in brain slices from normal and hypogonadal mice. J Physiol (Lond) 423:111–135

Kadota Y, Pettigrew K, Brightman MW 1990 Regrowth of damaged neurosecretory axons to fenestrated vessels of implanted tissues. Synapse (NY) 5:175–189

Knigge KM 1962 Gonadotrophic action of neonatal pituitary glands implanted in the rat brain. Am J Physiol 202:387–391

Krieger DT, Perlow MJ, Gibson MJ et al 1982 Brain grafts reverse hypogonadism of gonadotrophin releasing hormone deficiency. Nature (Lond) 298:468–471
Marciano FF, Wiegand SJ, Sladek JR, Gash DM 1989 Fetal hypothalamic transplants promote survival and functional regeneration of axotomised adult supraoptic magnocellular neurons. Brain Res 483:135–142
Mason AJ, Hayflick JS, Zoeller RT et al 1986 A deletion truncating the gonadotropin-releasing hormone gene is responsible for hypogonadism in the hpg mouse. Science (Wash DC) 234:1366–1371
Meijer JH, Rietveld WJ 1989 Neurophysiology of the suprachiasmatic circadian pacemaker in rodents. Physiol Rev 69:671–707
Ralph MR, Lehman MN 1991 Transplantation: a new tool in the analysis of the mammalian hypothalamic circadian pacemaker. Trends Neurosci 14:362–365
Ralph MR, Foster RG, Davis FC, Menaker M 1990 Transplanted suprachiasmatic nucleus determines circadian period. Science (Wash DC) 247:975–977
Schechter J, Gash D, Ahmad N 1987 Rathke's pouch grafts in adult brain sites. Am J Anat 178:55–64
Schmale H, Richter D 1984 Single base deletion in the vasopressin gene is the cause of diabetes insipidus in Brattleboro rats. Nature (Lond) 308:705–709
Setalo G, Horvath J, Schally AV, Arimura A, Flerko B 1977 Effect of the isolated removal of the median eminence (ME) and pituitary stalk (PS) on the immunohistology and hormone release of the anterior pituitary gland grafted into the hypophysiotrophic area (HTA) and/or of the in situ pituitary gland. Acta Biol Acad Sci Hung 28:333–349
Silverman AJ, Zimmerman EA, Gibson MJ et al 1985 Implantation of normal fetal preoptic area into hypogonadal mutant mice: temporal relationships of the growth of gonadotropin-releasing hormone neurons and the development of the pituitary/testicular axis. Neuroscience 16:69–84
Silverman R, Gibson M, Charlton HM, Silverman AJ 1990 Are neurones of the arcuate nucleus necessary for pathfinding by GnRH fibres arising from third ventricular grafts? Exp Neurol 109:204–213
Swanson LW 1987 The hypothalamus. In: Björklund A, Hökfelt T, Swanson LW (eds) Handbook of chemical neuroanatomy, vol 5: Integrated systems of the CNS. Part 1. Elsevier, Amsterdam, p 1–14
Valtin H 1976 Hereditary hypothalamic diabetes insipidus. J Pathol 83:633–636
Valtin H 1982 The discovery of the Brattleboro rat, recommended nomenclature and the question of proper controls. Ann NY Acad Sci 394:1–9
Young LS, Detta A, Clayton RN, Jones A, Charlton HM 1985 Pituitary and gonadal function in hypogonadotrophic hypogonadal (hpg) mice bearing hypothalamic implants. J Reprod Fertil 74:247–255
Wray S, Gähwiler BH, Gainer H 1988 Slice cultures of LHRH neurones in the presence and absence of brainstem and pituitary. Peptides (NY) 9:1151–1175

DISCUSSION

Dreifuss: It was shown several years ago (Gash et al 1980, Sladek et al 1982) that a transplant of hypothalamic tissue from a normal rat fetus into the third ventricle of an adult homozygous Brattleboro rat was able to reduce the polyuria of the host animal, which, being incapable of synthesizing vasopressin, suffers from hypothalamic diabetes insipidus. Could you please comment on these results obtained by grafting presumptive vasopressin-producing neurons?

Charlton: Nobody else has been able to repeat them. When you put a graft into the ventricle, by definition, you are putting in some cells that will die. Then you get inflammation in the graft. When we tried to repeat the experiments with Brattleboro rats given supraoptic nucleus grafts, the animals didn't drink very much for the first 3–4 days but nor did any of the animals with a control graft. Perhaps the inflammatory response affected the drinking behaviour for a few days.

Lechan: It is particularly fascinating in your transplantation models that sprouting axons containing GnRH are targeted to a specific locus in the median eminence. Could the endothelium contained within the portal capillary system itself be responsible for the specific targeting, perhaps by producing growth factors?

Charlton: From the experiments of Kadota et al (1990), it appears that several tissues containing fenestrated capillaries are capable of attracting neurosecretory axons. However, it is still difficult to say this with certainty. Interestingly, releasing factor axons project to the external zone of the median eminence and do not reach the posterior lobe, whereas oxytocinergic and vasopressinergic axons project to the posterior lobe. There appears to be something distinct about the median eminence, which may be blood vessels, compared to the posterior lobe.

Gainer: As I understand it, in the hypogonadal (hpg) mouse there is still the original complement of GnRH cells, although they don't make GnRH. Could the transplanted GnRH cells be following the axons of the endogenous mutant GnRH cells?

Charlton: They are not following any neurons in the arcuate nucleus, because in monosodium glutamate-treated animals, in which the vast majority of neurons in the arcuate nucleus are destroyed, GnRH neurons still grow to the median eminence.

Clarke: To come back to the issue of blood vessels—it is interesting that there are projections to the organum vasculosum of the lamina terminalis and to the median eminence, which are very highly vascular regions. There may be something special in the peripheral circulation that is not in or may be depleted in the circulation of the posterior lobe at the time those projections are being made.

Charlton: The posterior lobe lies outside the blood–brain barrier, so any factor will leak out into the surrounding tissue. Your suggestion implies that there is differential filtration by fenestrated vessels of the posterior lobe.

Clarke: To take that one step further, the reason the transplantation of preoptic area tissue and the pituitary grafts into the lateral ventricle didn't work may have been that there is a choroid plexus there and the cells were just confused because there were so many blood vessels around.

Charlton: They didn't grow to the choroid plexus.

Makara: In our pituitary stalk compression model, we compress the pituitary stalk at the specific site where it enters the gland (Makara, this volume). There is a denervated adult posterior lobe and the axons don't grow in. I don't know why. They sprout in the stump of the stalk which is just rostral to the site of compression. For up to 4–6 weeks we seldom see a fibre going down to the neural lobe, which is a few hundred microns away. In experiments done by others the axons grow into adrenal, posterior lobe tissue and pineal transplanted in the pituitary fassa.

When we look at brains six weeks after such compression using classical staining techniques, we see neurosecretory fibres growing along the ventral surface of the brain rather than into the posterior lobe. They radiate laterally and in a sheet-like manner; they outline the dura mater and the interface between the brain and the dura. So there may be something in the dura mater as well as in the blood vessels that attracts axons from the supraoptic and the paraventricular nuclei.

Theodosis: Could the axons be attracted to different types of glial cells in these structures? Glial cells in the posterior pituitary are not like the glial cells in the median eminence.

Barraclough: Daikoku-Ishido et al (1990) transplanted nasal pit tissue from 12.5 day rat embryos into normal cyclic adult female rats. When only primordial nasal pit tissue was implanted, very few cells were detected that contained immunoreactive gonadotropin-associated peptide (GAP) and no fibres were observed. However, when nasal pit tissue was transplanted with fragments of medial basal hypothalamic tissue, numerous immunoreactive GAP cells were observed with elongated fibres. These fibres contacted the blood vessels that had grown into the graft, which were fenestrated. Daikoku-Ishido et al suggest that specific factors exist in the medial basal hypothalamic tissue which dictate how GnRH neurons will migrate and develop in hypothalamic tissue *in vivo*.

Charlton: It would be very nice to take some of Dr Mellon's immortalized GnRH neurons (Mellon et al, this volume) and graft those into the CNS of hpg mice to see whether they grow any axons out to the median eminence.

Mellon: Since in the transgenic mice our cells don't grow axons to the median eminence, I would say that perhaps they are not developmentally mature enough to recognize those signals. They can migrate during development from the olfactory placode but they probably can't maintain or develop the axons to the median eminence.

Charlton: Yes, but if they arrived at the preoptic area too late and the pathway they were due to follow had been covered over by another pathway, then the cells would not know where to go. By the grafting technique you can put those cells right next to the median eminence and ask whether they can grow there.

Mellon: But in the grafting experiments, aren't the grafts put into adult animals, in which, if that pathway does disappear, it would already be gone, yet the axons still go to the median eminence? Doesn't that say that the pathway must be maintained?

Charlton: Our results say the attraction towards the median eminence is maintained, not the pathway from the preoptic area. GnRH axonal growth occurs only if grafts are placed in the third ventricle in close apposition to the median eminence. The grafts are taken up into a needle and during this process the tissue must lose its general cytoarchitecture. Yet within that environment, GnRH cell bodies develop long straight axons towards the the median eminence, which branch when they reach this target. So there is also something in the median eminence which tells axons to branch there and increase their surface area. The axons don't waste time branching anywhere else in that graft.

Poulain: Can you graft median eminence into another region, somewhere in the cerebellum for instance?

Charlton: I could, I haven't done it. At the moment it's much easier to use the lateral ventricle as a control site because then the grafts are more easily retained. When you put large grafts into the brain parenchyma, some of the graft tends to extrude onto the brain surface.

Dreifuss: How much innervation is there of the graft by the host brain in the hpg mouse?

Charlton: Not enough to transduce whatever it is that causes cycles. I don't see how there could be a direct pathway from the vagina to the cell bodies within the grafts. Ann-Judith Silverman has put DiI into the parenchyma either side of the graft and none of the dye tracked along axons into the donor tissue. Joe Hodgkiss did some electrophysiological slice studies on grafted hpg mice and found one hydroxylase-containing fibre growing across from the host. However, Ann-Judith Silverman has found that some tyrosine hydroxylase-positive axons enter the graft from the arcuate nucleus, so there is some input. With the grafts of suprachiasmatic nucleus tissue, nobody has been able to make the grafted animals entrain to a light–dark cycle when they put them back into a lights on–lights off regime.

Mellon: Is there any evidence for GnRH neurons connecting with each other within the graft?

Charlton: It is difficult to decide this. We take up a graft of the preoptic area in which the GnRH cells are arranged diffusely, and then we push it out into the third ventricle. Are we therefore disrupting contacts between GnRH cells? I think we probably are.

Plant: How do you think ovulation is being induced in these animals?

Charlton: In rodents the mating stimulus gives a surge of prolactin twice daily. That is a reflex release. So the nerve terminals affecting prolactin release are being instructed to secrete something into the portal vasculature. I think whatever is happening there will fire axons from our grafts running across towards the portal vessels. In the literature people have co-cultured median eminence with corticotropin-releasing factor (CRF) in the culture medium and found less GnRH secreted from the explants. Local action at the median eminence is, to me, the simplest possibility, rather than specific wiring inside the graft.

Barraclough: Prolactin won't cause ovulation.

Charlton: I know that, but it is reflexly released by the mating stimulus.

Barraclough: Rats with spontaneous persistent oestrus resemble the animal model you described in that they have polyfollicular ovaries and no LH surges. When they are mated they have LH surges and ovulate. It is not just the vaginal cervix stimulation that elicits this response, it is the whole act of mating, including grasping and mounting, which causes LH release.

Charlton: This example could also be explained by the fact that any female rat that mates has the pathways to release prolactin reflexly and this mating also causes a release of LH sufficient to elicit ovulation by mechanisms purely at the level of the median eminence.

Makara: The redundancy of the hypothalamo-hypophysial system is remarkable. In your case very few GnRH cells seem to be sufficient to induce full function. 10 years ago we made hypothalamic cuts in rats. In a few, we probably eliminated 95% of CRF input to the median eminence, but there was still an almost complete response in terms of adrenocorticotropic hormone (ACTH) release. At that time we were unable to differentiate between an almost completely deafferented animal and a non-deafferented one. The difference was clear after cutting the final 5% of the fibres—the response disappeared. So there is tremendous redundancy in the CRF system. In my paper (this volume), I showed that the 10% of CRF detectable by radioimmunoassay left in the median eminence after paraventricular lesion causes more or less normal ACTH release under some conditions when the nucleus itself is not needed there.

Plant: The absence of a negative feedback action of testosterone at the pituitary of the hpg mouse is a little surprising, because in the rat the pituitary is an important site of negative feedback. Has anybody used other mouse models and come to the same conclusion that there is no testicular feedback at the pituitary? That is like the primate (Plant & Dubey 1984).

Charlton: I was only talking of the mouse and even then I wouldn't say that testosterone did nothing at the level of the pituitary. It may do something, but we couldn't see it in our experiments.

Plant: One could just pulse in GnRH for this study.

Charlton: In our hpg mouse model, we have given multiple injections of GnRH with testosterone and there is no repression of the LH and FSH synthesis stimulated by GnRH.

Plant: What about oestradiol?

Charlton: In the female, oestradiol turns off gonadotropic hormone synthesis and secretion and therefore it is difficult to determine where the site of feedback is.

Lightman: In view of the fascinating results you have, with remarkably few neurons effectively being able to organize LH secretion and sexual maturation, how do you react to the extremely complex patterns of neurotransmitters and neuromodulators that have been shown to be important in the regulation of LH secretion?

Charlton: When I lecture students on this I always write 'fail-safe mechanisms' at the top of the blackboard. I think most of physiology has fail-safe mechanisms. They are all important. It is important, for example, that there is oestrogen, not just to prime the brain and pituitary, but also to prime behaviour and the uterus and to act as a local mediator within the ovary. In our model, the hpg mouse can do without a full complement of GnRH and can cope with disruption of normal pathways impinging on the GnRH neuron. However, in the real world, normal mice would use all possible mechanisms. As a physician, in a sense, you bypass most of these things with some drug that works but which may bypass a whole host of physiological processes in so doing.

Lightman: Dr Barraclough, you gave us this lovely diagram of the possible interconnections that might organize the physiological regulation of LH (Barraclough, this volume, Fig. 5). Do you feel these are all vital for the LH surge or is one primary and the others are back-up mechanisms which are not usually very significant?

Barraclough: I don't know. I like to think that noradrenaline is the most important neurotransmitter and other neurotransmitters modulate noradrenaline release and/or GnRH neuronal responsiveness to noradrenaline.

Zoeller: When I look at the wiring diagram of neurons potentially affecting the GnRH neuron and consider changes in the GnRH mRNA and peptide levels, one possibility is that specific afferents regulate different parts of GnRH neuronal function. That suggestion might be rather complicated, but the number of synaptic contacts on GnRH neurons and the pharmacology of GnRH release seem to require a complicated explanation.

Plant: In the lesioned monkey or the human female with Kallman's syndrome, when the pituitary is clamped with an unchanging intermittent GnRH stimulation, there is a pretty normal menstrual cycle (Knobil et al 1980, Leyendecker et al 1990). So it would seem that as you go up the phylogenetic scale, animals become emancipated from certain environmental inputs. In the rat the circadian (neural) trigger is crucial to the timing of the LH surge but the fundamental mechanism in the primate may reside at the level of the gonadotroph.

Zoeller: If you consider the pituitary as an integrator of steroid signals and hypothalamic signals, could you clamp from one direction or the other? Could you clamp the pituitary with chronic oestrogen, then pulse GnRH to show that GnRH is the regulator of the pituitary? Or clamp GnRH pulses and change the oestrogen input to show that oestrogen is what is driving the pituitary? In the animal both changes occur at once and the pituitary integrates those changes to produce menstrual cycles.

Plant: But in the primate the ovary is the sole trigger; in the rat it's part of the trigger, the other part is the CNS circadian trigger. The question in the primate therefore becomes, does the ovary dictate gonadotropin secretion by an action on the hypothalamus to regulate the hypophysiotropic drive to the gonadotroph, or on the pituitary directly to control responsiveness to GnRH, or a combination of the two? You can answer these questions only by measuring GnRH secretion directly or by clamping the peptide input to the pituitary.

Lightman: We have had several papers on the levels of mRNA within cells and relating those to the activity of cells. Tony Plant described beautifully a situation in which there is GnRH mRNA at quite high levels but nothing much seems to be happening at the response end, although there seems to be peptide present (Perera & Plant, this volume). Are we all happy that within the economy of the cell itself we can actually interpret things on the basis of the levels of mRNA?

Charlton: Oestrogen treatment causes a reduction in the levels of mature pituitary LHβ mRNA, and castration causes a rise. In preliminary studies, we have extracted total RNA from oestrogen-treated female rats and have found that whilst there is a reduction in the amount of mature transcript, there is a massive amount of the unspliced precursor RNA. Thus the steroid has not turned off the gene but has affected the splicing out of introns. One would not see this in assays of cytoplasmic RNA or by *in situ* hybridization using cDNA probes.

Pfaff: One could use a probe that included an intron, which would detect the unspliced RNA. We have done this with preproenkephalin (PJ Brooks et al, unpublished 1991).

Gainer: In his paper, Dr Barraclough (this volume, see p 239–242) mentioned the increase in levels of mRNAs for neuropeptides that is induced by stimuli which activate the release of the neuropeptides themselves. Why does the cell need this burst of mRNA transcription for 15 minutes when there are terminals sitting there with an abundance of secretable hormone? Probably no more than 10% of peptide from the terminals is secreted during the surge anyway. So what is the biological meaning of this increase in mRNA for 15 minutes?

Barraclough: The cell bodies may not know that all this releasable hormone exists in the terminal. The stimulus which causes the release of the hormone may also activate intracellular mechanisms required to replace the hormone that is secreted. These events may occur independently in the cell in response to the same depolarizing stimulus.

Young: I think it's easier to look at the stimulus reaching a cell. When it gets to the cell surface it probably sends one signal to the nucleus and one down the axon to release hormone. The signals are not connected except that they are caused by the same stimulus. It is an efficient way of correcting for subsequent release.

Pfaff: An evolutionary approach to that kind of question is that the strategy might simply be not to let the amount of the precursor get too low.

Gainer: That's why they have the safety factor of the peptide in the median eminence. From Harry Charlton's work, four GnRH neurons, or let's say there were 10, had enough hormone in the terminal to sustain a whole pregnancy. The mouse doesn't need 800 of those cells, it really needs only four.

Barraclough: The rat also needs only about 10% of the LH that is released during pro-oestrus to cause complete ovulation. There seems to be a safety factor built in to this preovulatory surge mechanism that assures propagation of the species.

Gainer: That's my point. Is the mRNA change a red herring?

Zoeller: The idea of a red herring depends on how you interpret it. A change in the amount of mRNA means that there is a cellular response to some input. *In situ* hybridization is probably the only technique that enables you to look at cellular responses in individual GnRH cells under these physiological conditions or after manipulation of steroids. Hence, it is an important measurement. What relationship it has to other cellular processes needs to be evaluated the best way possible, which may differ from one neuronal system to another. The data for CRF and thyrotropin-releasing hormone clearly suggest a positive correlation between secretory activity and changes in message. With GnRH it seems a little different.

Gainer: My problem is this: we are now making a catechism. When we see a change in mRNA level, we equate it with secretory activity—that's the interpretation. But is it the correct interpretation? What is the meaning of that 15 minute increase in the RNA level? Does it really reflect stimulatory activity?

Poulain: There is evidence that in the neurohypophysis the stores of the hormone in the terminals are variable. Most of the hormone that is released is 'fresh' hormone (Nordmann 1985). One reason for an increase in mRNA synthesis in the cell body every time there is an increase in secretory activity may be to make sure that the store of 'fresh' hormone is constantly replenished.

Young: If you want to relate mRNA levels to secretory activity, you have to take into account the time period you are looking at. For example, if the neuron is stimulated just briefly, it could release its stores and simultaneously the signal to make more transcripts is reaching the nucleus. So the release may be complete by the time translation is really getting into gear. If you look at a more chronic stimulus, such as salt-loading, there can be continual release while continual synthesis is occurring.

Gainer: It will still take 2–4 hours for a protein synthesized in a CRF cell to get to the terminal in the median eminence. The transport period imposes a delay that's not accounted for by the immediate production of mRNA and prohormone. The transport rate for vasopressin in the magnocellular system was found to be about 40 mm/day (Fink et al 1981).

Robinson: I have a problem with the notion that activity always equals secretion, because we have forgotten about pre-terminal regulation of secretion. The idea that the cell body is bound to increase synthesis because it assumes that an action potential directed to the terminal is going to release a fixed amount of peptide is obviously wrong because you can have presynaptic inhibition of terminals. There must be an integration process going on as well, so that the cell body knows eventually how much peptide there is in the terminals and how much more to synthesize and transport. We know that in the neurohypophysis, for instance, the balance between synthesis and storage is maintained over a much longer period than the day-to-day fluctuations in the amount of hormone stored in the terminals.

Makara: For me it makes good sense that whenever a cell starts to fire, say for more than a minute, it should increase its rate of synthesis of peptide transmitter. The cell cannot know how long the firing will last. The delay due to transport and the relatively slow retrograde transmission of information about how much peptide is stored in the terminal may make it necessary to start replenishing very early, if the cell wants to maintain peptide release for a long time. If gene expression is turned off after a short firing episode, then the newly synthesized hormone may take the place of the old hormone and the old hormone can be eliminated. Because of the time lag and also because of the lack of precise information on the amount of peptide in storage in the terminal, it makes sense for the cell to start synthesizing as soon as it starts sustained firing.

Clarke: But only about 10% of the GnRH in a GnRH neuron is released during an LH surge. Across oestrous or menstrual cycles, the changes in the median eminence are small. There is a vast excess of GnRH stored in the terminal and the amount released is vanishingly small.

One thing that seems to happen at the terminal is that GnRH is held in releasable or non-releasable forms. If the pulse frequency of GnRH secretion increases, the amplitude of those pulses tends to be small. If there are low frequency pulses, their amplitude is big. So the time between pulses dictates the quantum that is released.

Another line of evidence comes from NMDA stimulation studies on animals that are anoestrous as opposed to breeding. In the anoestrous season there are long periods between secretory events. If you give those animals NMDA, you get a big response as opposed to the small one in breeding animals.

Coming back to this issue of whether or not the terminals are talking to the cell body and saying that there is a reduction in stores, I really don't think that's necessary to make the system work, at least for GnRH.

Nagy: Do we know that all the time the same 10% of GnRH immunopositive neurons release the hormone? For example, if you treat pituitary cells with GnRH then measure hormone secretion from LH cells, you find that at any time only about 40% of the cells are preferentially responding to GnRH. These cells change over a few hours and another subpopulation becomes responsive to GnRH. So there is a certain cycle of the cells and different subsets of the cells do the job at different times; that's why the animal needs more than the minimum required for a single response.

Barraclough: Locus coeruleus neurons fire at a very steady rate of 5–20 per second constantly. Anatomically, they have a recurrent collateral with soma-dendritic synapses. As noradrenaline secretion from the locus coeruleus increases, the noradrenaline can suppress locus coeruleus neuronal activity via α-adrenoreceptors. This arrangement can dictate the frequency of depolarization of locus coeruleus neurons. I don't know if any such anatomical arrangement exists for peptidergic neurons or whether collaterals exist on such neurons.

Young: Locus coeruleus neurons are peptidergic. Do you mean peptidergic synapses?

Barraclough: I was thinking of thyrotropin-releasing hormone, growth hormone-releasing hormone, GnRH and CRF types of peptidergic neurons.

Hatton: Zambrano & DeRobertis (1966) proposed that the redundancy is there so that different groups of cells can go through a cycle of activity and inactivity. Yukitake et al (1977) then showed that cells in the paraventricular nuclei had different morphological characteristics at different stages of the activity cycle. If you accept that, you don't have to postulate that the cell knows how much peptide it has in the terminal; it just has a finite time when it is active and when it runs out of peptide it stops and becomes refractory, and then other cells take over.

Epelbaum: Has this increase in mRNA synthesis been shown for other transmitters, such as glutamate?

Lightman: We have shown changes in hypothalamic CRF and proenkephalin mRNA levels within two hours; the response of early response genes such as c-*fos* is very much more rapid than this.

Makara: J.P. Hermann has shown using intronic probes that within 15 minutes of exposure to stress, the amount of detectable CRF mRNA increases.

Let us look at the problem in a different way. What is triggering the increase of the transcript in the nucleus? Is it Ca^{2+}? If it is cytosolic Ca^{2+}, as for example with prolactin cells in the pituitary, then whenever the cell is activated, the cytosolic Ca^{2+} concentration will increase and once it reaches a given level, it activates transcription in the nucleus. It may be an automatic thing that just happens without any teleological reason.

Zoeller: I don't think it's quite that simple. For example, Van Nguyen et al (1990) have shown a synergism between Ca^{2+} and cAMP-dependent protein kinase A, which interact on specific regulatory sequences of the proenkephalin promoter. It seems to be more than a natural consequence of changes in the Ca^{2+} concentration because of this synergy.

Makara: The same argument holds for cAMP. How does the cell know that one molecule of cAMP will have the purpose of activating cell secretion and another molecule of cAMP will activate the nucleus? I don't think there are separate sets of membrane receptors, CRF receptors for example, to activate secretion and synthesis. I was arguing that simply stimulating the cell will activate one pathway that has two end products—an increase in the rate of hormone synthesis and the extrusion of hormone-containing granules.

Hatton: This may be one function of co-localization: one of the products may activate one of these pathways and another a different pathway. For example, if the cell released only glutamate, there might be an action potential going down and causing release of hormone without turning on synthesis, but if the cell released glutamate plus a peptide, there might also be a second messenger effect.

Lightman: I don't think we are going to answer this question but I think we would all agree that it's an extremely important issue!

References

Barraclough CA 1992 Neural control of the synthesis and release of luteinizing hormone-releasing hormone. In: Functional anatomy of the neuroendocrine hypothalamus. Wiley, Chichester (Ciba Found Symp 168) p 233–251

Daikoku-Ishido H, Okamura Y, Yanahara N, Daikoku S 1990 Development of the hypothalamic luteinizing hormone-releasing hormone-containing neuron systems in the rat: in vivo and in transplantation studies. Dev Biol 140:374–387

Fink DJ, Russell JT, Gainer H, Brownstein MJ, Baumgold J 1981 Multiple-rate components of axonally transported proteins in the hypothalamo-neurohypophysial system of the rat. J Neurobiol 12:487–503

Gash D, Sladek JR, Sladek CD 1980 Functional development of grafted vasopressin neurons. Science (Wash DC) 210:1367–1369

Kadota Y, Pettigrew K, Brightman MW 1990 Regrowth of damaged neurosecretory axons to fenestrated vessels of implanted peripheral tissues. Synapse (NY) 5:175–189

Knobil E, Plant TM, Wildt L, Belchetz PE, Marshall G 1980 Neuroendocrine control of the rhesus monkey menstrual cycle: permissive role of the hypothalamic gonadotropin-releasing hormone (GnRH). Science (Wash DC) 207:1371–1372

Leyendecker G, Waibel-Treber S, Wildt L 1990 The central control of follicular maturation and ovulation in the human. Oxford Rev Reprod Biol 12:93–146

Makara GB 1992 The relative importance of hypothalamic neurons containing corticotropin-releasing factor or vasopressin in the regulation of adrenocorticotropic hormone secretion. In: Functional anatomy of the neuroendocrine hypothalamus. Wiley, Chichester (Ciba Found Symp 168) p 43–53

Mellon PL, Wetsel WC, Windle JJ et al 1992 Immortalized hypothalamic gonadotropin-releasing hormone neurons. In: Functional anatomy of the neuroendocrine hypothalamus. Wiley, Chichester (Ciba Found Symp 168) p 104–126

Nordmann JJ 1985 Hormone content and movement of neurosecretory granules in the rat neural lobe during and after dehydration. Neuroendocrinology 40:25–32

Perera AD, Plant TM 1992 The neurobiology of primate puberty. In: Functional anatomy of the neuroendocrine hypothalamus. Wiley, Chichester (Ciba Found Symp 168) p 252–267

Plant TM, Dubey AK 1984 Evidence from the rhesus monkey (*Macaca mulatta*) for the view that negative feedback control of luteinizing hormone secretion by the testis is mediated by deceleration of hypothalamic gonadotropin-releasing hormone pulse frequency. Endocrinology 115:2145–2153

Sladek JP, Schöler J, Notter MD, Gash DM 1982 Immunohistochemical analysis of vasopressin neurons transplanted into the Brattleboro rat. Ann NY Acad Sci 394:102–115

Van Nguyen T, Kobierski L, Comb M, Hyman SE 1990 The effect of depolarization on expression of the human proenkephalin gene is synergistic with cAMP and dependent on a cAMP-inducible enhancer. J Neurosci 10:2825–2833

Yukitake, Y, Taniguchi Y, Kurosumi K 1977 Ultrastructural studies on the secretory cycle of the neurosecretory cells and the formation of Herring bodies in the paraventricular nucleus of the rat. Cell Tiss Res 177:1–8

Zambrano D, DeRobertis E 1966 The secretory cycle of the supraoptic neurons in the rat: a structural functional correlation. Z Zellforsch Mikrosk Anat 73:414–431

Summary

Professor S. L. Lightman

Neuroendocrinology Unit, Charing Cross & Westminster Medical School, Charing Cross Hospital, Fulham Palace Road, London W6 8RF, UK

This conference has taken as its theme the functional anatomy of the endocrine hypothalamus. We have been able to discuss the most recent advances in knowledge from the level of histochemical studies of cellular anatomy to the regulation of neuropeptide mRNA.

Speakers and discussants have described considerable functional plasticity in the hypothalamus as demonstrated by glial–neuronal interrelationships, receptor distribution and even return of function following transplantation. There has also been the demonstration of a massive scope for altering information input and output, owing to the multitude of neuropeptides within hypothalamic systems, with many of these neuropeptides being co-localized in the same cells. We have heard evidence for the differential regulation of these peptides and even of the mRNA for different peptides within the same cell. Together with changes in receptor complements, this represents a remarkably adaptable system.

We have touched on some very interesting areas concerning the interrelationships of feedback inhibition and the trans-synaptic activation of different cell groups in the hypothalamus, again pointing to the complexities of the integration of multiple sources of information by hypothalamic cells. We have discussed in detail several specific areas of hypothalamic organization that still evade clear explanation. There is, for instance, limited understanding of the factors which determine the timing of the onset of puberty, and indeed the mechanism for the generation of gonadotropin-releasing hormone pulses.

The demonstration of a putative rhythm in transgenic gonadotropin-releasing hormone cells in culture certainly provides an important new model for further study. Similarly, the mechanism of cross-talk between oxytocin cells during milk let-down remains as elusive as it is fascinating, while the ability of transplanted suprachiasmatic cells to impose their own endogenous rhythm on the host animal reinforces the primacy of individual cell groups in the generation of endogenous pace-making mechanisms. The hypothalamus has clearly opened up many major areas of neuroscience and endocrinology, and shows every sign of continuing to do so well into the foreseeable future.

Finally, I should like to express my appreciation to everyone who has helped make this meeting so worthwhile scientifically. The speakers and discussants have been full of stimulating ideas and questions and have made the job of Chairman a considerably easier one—for which, many thanks! I should also like to express the gratitude of all the participants to the Ciba Foundation for bringing us together for this unique meeting and, of course, to Professor Halász and the Hungarian Academy for providing us with the facilities for the conference and for their incredible hospitality during our stay in Budapest.

Index of contributors

Non-participating co-authors are indicated by asterisks. Entries in bold type indicate papers; other entries refer to discussion contributions.

Indexes compiled by John Rivers

Subject index